Animal Welfare and Meat Production
2nd Edition

WITHDRAWN

Mixed Sources
Product group from well-managed
forests and other controlled sources
FSC www.fsc.org

D1339504

Animal Welfare and Meat Production
2nd Edition

_____

Neville G. Gregory

*BBSRC and Royal Veterinary College,
UK*

and a chapter by

Temple Grandin

Colorado State University, USA

www.cabi.org

The Sheffield College
Learning Resources - Hillsborough
☎ (0114) 260 2254

6 00068963

Coutts	25/6/09
636.0832	£37-50
	L

CABI is a trading name of CAB International

CABI Head Office
Nosworthy Way
Wallingford
Oxfordshire OX10 8DE
UK

Tel: +44 (0)1491 832111
Fax: +44 (0)1491 833508
E-mail: cabi@cabi.org
Website: www.cabi.org

CABI North American Office
875 Massachusetts Avenue
7th Floor
Cambridge, MA 02139
USA

Tel: +1 617 395 4056
Fax: +1 617 354 6875
E-mail: cabi-nao@cabi.org

© N.G. Gregory 2007. All rights reserved. No part of this publication may be reproduced in any form or by any means, electronically, mechanically, by photocopying, recording or otherwise, without the prior permission of the copyright owners.

A catalogue record for this book is available from the British Library, London, UK.

A catalogue record for this book is available from the Library of Congress, Washington, DC.

ISBN-13: 978 1 84593 215 2

Typeset by AMA DataSet Ltd, UK.
Printed and bound in the UK by Cromwell, Trowbridge.

The paper used for the text pages in this book is FSC certified. The FSC (Forest Stewardship Council) is an international network to promote responsible management of the world's forests.

Contents

_____

Preface

_____

About 20 years ago there was a debate on the radio about slaughter methods. During a pause in the proceedings, someone from the gallery shouted, 'Animals cannot vote.' At the time this comment seemed irrelevant, and it was ignored. But, no doubt, it was relevant in the heckler's mind. The point he was making was that animals' interests are poorly represented. Psychologists would call his cryptic remark a condensing symbol, because it captured his point in a slick and provocative way.

The heckler's concern was about justice, and it is shared in varying degrees by others. Some question whether it is fair to slaughter farm animals without stunning, others are more concerned about animals kept in confinement in cages or about farming in ways that require amputation procedures. Subsumed in all these 'Is it fair?' questions, are two other questions: 'Is it necessary?' and 'Are there alternatives?' Deciding what is fair hinges on a common sense of justice.

In legal circles, alleged injustice is resolved through lodging an injunction or appeal. Children, people with learning difficulties and animals are not in a position to lodge an appeal, but others can lodge appeals on their behalf. Assuming responsibility or guardianship when lodging an appeal assumes an understanding of what is in the individual's best interests, and it may even presume what is going on in their minds. In the case of children and people with learning difficulties, agreement can usually be reached about their best interests. In the case of animals, we are obliged to make greater assumptions in forming opinions. Fundamental disagreements stem from assumptions about dissimilarities between the animal and human mind and from differences in the status and value that we attach to animals.

This book discusses a number of practices that some feel are unjust. The book does not try to defend or criticize the practices. It simply provides information that will allow readers to form their own views.

This book is an extension of a previous book by the same author (_Animal Welfare and Meat Science_). The previous book was criticized for being repetitive and having too few references. These criticisms have been addressed in the present book, and it has less emphasis on meat science and more on animal genetics, animal welfare auditing, aquaculture and livestock production in developing countries. It tries not to repeat what was said in _Animal Welfare and Meat Science_ and focuses on recent research findings. There is considerably more referencing on meat science, as this was one of the main complaints.

Animal welfare science sometimes uses human analogy in assessing the plight of animals. Some would criticize this as being anthropomorphic. This is not always a valid point. Anthropomorphism is the projection of human personality or mannerisms to animals, and the criticism would be relevant, for example, if someone said, 'The camel is a haughty animal because it looks down its nose at you,' whereas it is not anthropomorphic to diagnose pain or suffering in an animal by bringing in one's own experience of pain in a comparable situation. One is using human analogy, not anthropomorphism. This book uses human analogy sparingly and in appropriate places. We hope you find the book helpful, interesting and informative.

Neville G. Gregory
Royal Veterinary College, United Kingdom

Abbreviations and Synonyms

_____

Abbreviations

a	Measure of surface redness in meat
AC	Alternating current
AFTA	ASEAN Free Trade Area
AI	Artificial insemination
AIP	Acute interstitial pneumonia
ASEAN	Association of Southeast Asian Nations
ATP	Adenosine triphosphate
BRD	Bovine respiratory disease
BSE	Bovine spongiform encephalopathy
CAR	Central African Republic
CARICOM	Caribbean Community
CBPP	Contagious bovine pleuropneumonia
CLA	Caseous lymphadenitis
COMESA	Common Market for Eastern and Southern Africa
CP	Creatine phosphate
CPET	Cloning plus embryo transfer
CV	Coefficient of variation
DC	Direct current
DCB	Dark-cutting beef
DFD	Dark, firm, dry
EEG	Electroencephalogram
EFTA	European Free Trade Association
ET	Embryo transfer
EU	European Union
FCR	Feed conversion ratio
GH	Growth hormone
GM	Genetically modified
GMMAS	Gene mapping and marker-assisted selection
GNP	Gross national product
h^2	Heritability
HABs	Harmful algal blooms
HOP	Health of the Planet (survey)
ILCA	International Livestock Centre for Africa
JIVET	Juvenile _in vitro_ embryo transfer

KKCF	Kidney knob and channel fat
L	Measure of surface reflectance and paleness in meat
L : D	Light : dark
LFA	Live-weight for age
MAP	Modified atmosphere packaging
MERCOSUR	Southern Cone Common Market
MGA	Melengestrol acetate
MT	17-α-methyltestosterone
NAFTA	North America Free Trade Agreement
nn, Nn and *NN*	Pig halothane genotypes: double recessive (*nn*), heterozygotes (*Nn*) and homozygous dominant (*NN*)
OIE	Office International des Épizooties
PETA	People for the Ethical Treatment of Animals
pH$_{45 \, min}$	pH of muscle at 45 minutes *post mortem*
pH$_{ult}$	Ultimate pH of meat
PSE	Pale, soft, exudative
PUFA	Polyunsaturated fatty acids
RH	Relative humidity
RN	Rendement Napole gene
RV	Right ventricle
SADC	Southern African Development Community
SDS	Sudden death syndrome
TBARS	Thiobarbituric acid reactive substances
TD	Tibial dyschondroplasia
TDS	Total dissolved solids
TLU	Tropical livestock units
TMR	Total mixed ration
TSE	Transmissible spongiform encephalopathy
WOF	Warmed-over flavour
WTO	World Trade Organization

Synonyms

Cut-and-carry	zero-grazing
Dry sow stall	gestation crate
Durbar	checker plate
Force-feeding	cramming
Hatcher	setter
Head-bail	neck yoke
Heel bar	step-over bar
Lairage	stockyard, holding pens
Lift	elevator
Pithing	fiddling
Pugging	poaching
Sale yard	livestock market
Skip	dumpster
Store	stocker
Straddle conveyor	centre-track conveyor, double-rail conveyor
Stunning pen	knocking box
Weasand	oesophagus

1

Animal Welfare and the Meat Market

This chapter covers five topics. It looks at:

- the contribution that each livestock species makes to overall world meat production
- views about livestock farming and the meat industry in today's society
- the aims and some of the problems in keeping livestock in developing countries
- environmental hazards in intensive livestock farming
- distinguishing between acceptable and unacceptable farming standards

These topics are discussed in different sections in this book, and the aim here is to introduce them and bring together some general points.

World Livestock and Meat Production

The meat and livestock industry has an important role in feeding the world's population. It is a major source of dietary protein, and on average it provides about 8% of our energy intake. However, there are immense differences between nations in the importance that meat has in the diet. For example, in Bangladesh, meat and offal contribute as little as 0.7% of dietary energy whereas in Mongolia they make up 27%. The countries that have a large per capita meat consumption (about 100 kg/capita/year) are Australia, Austria, the Bahamas, Cyprus, Denmark, France, Luxembourg, Mongolia, Spain and the USA, while the nations that consume the least meat that is produced on farms (about 5 kg/capita/year) include Bangladesh, Burundi, the Democratic Republic of Congo, the Gambia, Guinea, India, Malawi, Mozambique, Rwanda and Sri Lanka.

There are three countries that make a very large contribution to overall meat production. They are China, the USA and Brazil. These countries produce about 40% of the world's farm animals used for meat consumption and over half the world's meat. They also account for 49% of the world's meat consumption. The three countries with the highest levels of meat export per capita of population are Denmark, Ireland and New Zealand.

Census figures for farmed animals in the world during 2003 are given in Table 1.1. This shows the importance of each species in terms of the number of animals it contributes to meat consumption. It is only a rough guide, as it does not allow for offtake rate or slaughter interval. **Offtake** rate is the proportion of animals that are sold or used for meat consumption every year. **Slaughter interval** is the time between birth (or hatching) and slaughter, in animals not used for breeding. The table underestimates the role of poultry because they have a short slaughter interval and poultry farms produce more than one crop a year. It overestimates the role of cattle, which have a slaughter interval that is usually more than 1 year as well as a low offtake rate.

The offtake (cropping) of cattle is quite complex, as can be seen from Table 1.2. Countries such as the Confederation of Independent States (CIS), the USA and China have high offtake rates. Over 35% of cattle in each of these countries are slaughtered every year. On the other hand, in India, which has the largest cattle population, fewer than 7% of the cattle are slaughtered each year. The slaughter of female cattle for meat consumption is not widely practised in India for religious reasons.

©N. Gregory 2007. *Animal Welfare and Meat Production*
(N. Gregory)

In Sudan and Ethiopia, the offtake rate for cattle is also about 7% because they are valued for monetary and security reasons, as explained later in this chapter. None of the European countries feature as big cattle producers, but the European Union (EU) as a whole is the third largest beef-exporting collective. The EU country with the largest number of cattle is France (about 20 million head) and its

Table 1.1.   Approximate number of farmed animals in the world (from FAO Yearbook 2004, No. 177).

	Million animals
Red meat species	
Cattle	1,371
Sheep	1,024
Pigs	956
Goats	768
Buffaloes	171
Camels	19
Deer	5
Equines	
Horses	55
Asses	40
Mules	13
White meat species	
Chickens	16,605
Ducks	1,086
Turkeys	274
Rabbits	Not known

offtake rate is about 30%, some of the cattle being slaughtered as veal.

The figures given in Table 1.1 do not include farmed fish and hunted animals. Aquaculture is now a substantial contributor to world meat supply. Between 1996 and 2003, production of farmed fish rose from 27 to 42 million tonnes, and it now accounts for 32% of total fish meat production. About 84% of aquaculture production is in seven Asian countries, with China and India making the biggest contributions. Not all fish are destined for human consumption. About 21% of fisheries plus aquaculture production goes to other outlets.

There are few reliable statistics on the number of hunted animals that are eaten, but the overall consumption of bushmeat in Central Africa is significant and has become a concern in wildlife conservation (Table 1.3). In the case of the slow-reproducing species, the hunting is no longer sustainable and their meat is becoming a delicacy or a treat and fetches high prices (Wilkie and Carpenter, 1999). The Democratic Republic of Congo (DRC) and Equatorial Guinea have a particularly high per capita consumption of bushmeat.

Livestock production systems and animal welfare

From an animal welfare perspective it can be helpful to think of livestock farming as having three production systems:

Table 1.2.   Cattle numbers in those countries with the largest cattle populations. (from FAO Yearbook 2004, No. 177)

Country	Number of cattle (1000)	Number of cattle slaughtered per year (1000)
World	1,371,117	291,891
India	226,100	14,470
Brazil	189,513	35,500
China	103,470	44,901
USA	96,100	36,686
Argentina	50,869	12,600
Sudan	38,325	2,680
Ethiopia	35,500	2,805
Mexico	30,800	6,800
Australia	27,215	9,229
CIS	26,254	12,300

Table 1.3. Estimated bushmeat consumption in the Congo Basin in 1998.

	Bushmeat eaten	
Country	kg/year	kg/km²/year
Cameroon	78,077,000	503
CAR	12,976,500	248
DRC	1,067,873,500	897
Equatorial Guinea	9,763,000	574
Gabon	11,380,500	50
Congo	16,325,500	77
Total	1,196,396,000	645

- Intensive livestock farming.
- Industrialized livestock farming.
- Subsistence livestock farming.

The causes of animal welfare issues in these sectors are different.

Intensive farming methods involve high levels of input whilst aiming for high levels of output. They are found mainly where there is a limiting resource, such as land or labour, but there is adequate capital to manage the limitation in other ways (e.g. investment in buildings, fencing or mechanization). Intensive livestock farming is the main supplier of the world's meat. It is a substantial source of pork, beef, lamb and farmed fish. The animal welfare issues in intensive livestock farming are the ones that are discussed most in this book, and they are usually connected with trying to make animals conform to particular management systems. They include sow aggression in the pig industry, castration and tail docking in sheep and handling stress in cattle.

In industrialized livestock farming, the farm is treated as a business whilst applying modern and often innovative manufacturing principles and approaches. Industrialized livestock farming operates on a large scale. It occurs when a company focuses on a particular commodity and there is usually a high level of control of either the quality or the supply of animals producing that commodity. Often it manages to economize by operating as a vertically integrated business. In the view of the chief executive of one of the largest animal advocacy groups, industrialized pig farming is

'an abomination'. This comment had more to do with feelings about mass production and anti-corporatism than animal welfare, but it could indicate future attitudes towards this agribusiness sector. The animal welfare issues in industrialized livestock farming are similar to those in intensive farming, but in addition there are problems connected with managing unusually large numbers of animals, such as inspecting grower pigs kept in large groups, handling methods for chicks in hatcheries and individual animal health care in feedlots. In fairness to the intensive livestock sector, the issues that are specific to industrialized production methods can be considered separately.

In subsistence livestock farming, animals are usually kept as a means for survival rather than profit. Very little of the produce is sold, and inputs other than labour are limited by lack of money. Of the three livestock farming sectors, subsistence farming occupies the most people, but its productivity is often the lowest because it occurs on marginal land. In some countries, its future is being threatened by population growth. When smallholdings are passed from generation to generation they are often subdivided between the offspring. The size of a subdivided holding reaches a stage where it can no longer sustain livestock, and instead the parcels of land are used for crops and vegetables or for housing. In other countries, however, the shift away from small-scale livestock production is being checked by urbanization of the population, as this is helping to reduce the rate of smallholding subdivision. Elsewhere, a growing number of livestock on small subdivided properties

are confined in pens or hutches, or they are tethered, and this is changing some of the priorities in
animal welfare. Nevertheless, underfeeding is the
predominant animal welfare and production concern in subsistence livestock farming.

Modern Views about Livestock Farming and the Meat Industry

Intensive livestock farming has the difficult task of
producing food cheaply and profitably whilst satisfying the newer expectations on environment, conservation, amenity, food safety and animal welfare.
In developed countries, it is facing the following
challenges and opportunities:

- moves towards industrialized livestock farming
- food safety issues
- demand for organic produce
- pressure from animal rights groups
- public concern about advances in animal
  genetics
- stricter control over environmental hazards
- decreasing profit margins per unit of product
  leaving the farm

This section briefly discusses the first five topics,
and the impacts that attitudes amongst the public
are having on the meat and livestock industry.

Industrialized livestock farming

Some animal advocates claim that industrialized
livestock farming leads to a loss of compassion
and empathy for animals. It is said that compassion is at a low ebb where large numbers of animals are managed and where there is limited
individual contact or care. In a survey of farmers
in the west of France, this was confirmed to be
the case. The farm staff said they had too many
animals to look after. Farming had become a victim of its own efficiency, and people within the
industry were unhappy with the situation, as were
those who viewed it from outside (Porcher *et al.*,
2004).

This situation has similarities to human
health care. Health care workers are complaining
that they have to look after too many patients, and
as a result they are becoming indifferent to the
patients. Indifference emerges when large numbers

of people or animals have to be served or cared
for. Women have more positive views about the
situation, and compassion plus empathy persist
better than in men. Regardless of gender, educated staff are more indifferent about their animals
than the less educated.

When stock people have a busy work schedule and have less time to watch and interact with
their animals on a one-to-one basis, the animals
are prone to being less manageable during routine
handling procedures. The animals are more nervous and in extreme cases this can lead to escape
behaviour and self-inflicted injuries. In general the
facilities used at farms, markets and lairages are
designed to control escape attempts and minimize
injury risks, but accidents happen, especially where
facilities are not well maintained and a batch of
unruly animals comes through. From the producer's
point of view, the incentives to spend time training
or familiarizing the animals with being handled
are not strong. The commercial meat quality benefits are not usually great enough to warrant the
additional expenditure in labour (Lensink *et al.*,
2000). In addition, from the producer's perspective, handling difficulties are usually someone else's
problem. They are worst when the animals are
confronted with a novel environment, such as a
market or abattoir.

Another concern about industrialized farming is that it focuses on treating animals as convert-
ers of inexpensive feeds into more valuable meat.
The criticism here is that animals are viewed as
items instead of sentient beings. It is claimed that
this outlook is amoral rather than immoral because
of indifference towards animals as individuals
(Singer, 1990).

Food safety

Indifference was also one of the criticisms raised
against the meat and fast-food industries in their
attitude to food safety during the early 1990s. In
the USA and Europe there were several serious
food scares at this time. In Europe, the scares
started with algal toxins in Crustacea and shellfish,
carcinogens in packaging films, *Salmonella* in
fresh chicken, *Listeria* in precooked chilled chicken
and *Botulinum* in pork. They got worse as bovine
spongiform encephalopathy (BSE) and *Escherichia
coli* O157 emerged in beef and veal, and the

decade ended with worries about the hazards of genetically modified foods.

On top of this, there was public criticism in Europe about farm animal welfare, and in particular about keeping dry sows in stalls, long-distance transport in sheep, lameness in broilers and confining layer hens in cages. The meat industry became a focus for attention in general protest movements, and the public became disenchanted with farmers and meat processors. At times, the image and prospects of the meat and livestock industry seemed depressing.

The general protest movement is now directing its attention at other targets, but it caused some lasting effects on the way things are managed in the meat and livestock industry. The food safety issues of the 1990s led to **traceability** systems. Traceability was originally intended as a way of tracking the origins of a food from the consumer back to its farm of origin. This could allow transfer of culpability for a food scare, whilst implementing food safety control measures. An important consequence of this development has been reduced market access to high-value markets for subsistence livestock farming, because this sector is not in a strong position to introduce the support needed for traceability systems, especially in poorer countries. In industrialized countries, traceability has resulted in an abundance of animal ear tagging and paperwork aimed at monitoring livestock movement and changes in ownership.

The introduction of livestock traceability was supported by governments as a means of controlling outbreaks of exotic diseases. In the EU traceability is achieved through animal passports and movement records. Animal passports and movement records are potentially useful for:

- tracing and identifying contact or suspect animals in disease control programmes
- providing evidence of the vaccination status of animals
- tracing products that pose a risk to human health back to their farm of origin
- identifying animals that have been treated with drugs that preclude the carcass from human consumption
- predicting the spread of a disease
- limiting the number of animals that need to be confined or slaughtered during an epidemic

Animal passports were seen as a way of strengthening national biosecurity in a free trade economy.

There have, however, been animal welfare disadvantages that have offset the benefits. The additional handling required in checking that an ear tag corresponds to a passport is stressful for the animal. Cattle are held in crushes to read the ear tag, and as with most crush work this can upset the animal and lead to bruising. Inconsistencies between ear tag and passport numbers at busy livestock markets can cause delays in unloading other stock, and tempers can get taxed. Questions get asked whether risk management has to impose such interference with otherwise simple procedures. Risk management is seen by some as too intrusive and disruptive. Others accept it and get on with it, but need reassuring that there are benefits.

The food scares of the 1990s had other long-term effects. Consumer confidence in food safety has shifted away from government control towards trusted branded products. Some companies have benefited, whilst others have had their reputation ruined by a food scare and have gone out of business. The threat of closure added impetus to adopting **auditing** systems, which checked on standards used further up the supply chain. Animal welfare concerns are being managed in a similar way, using the auditing approach adopted in food safety. In-house animal welfare accreditation schemes are now common and they are often requested by bulk-buying customers. Some major meat retailers and fast-food companies have recognized their role in maintaining standards in animal welfare as well as food safety, and they have exerted pressure in adopting recognized welfare standards through market access entitlements. In this way, the meat industry is becoming more self-regulated on animal welfare as well as food safety, and government authorities in some countries are now less concerned with defending or inspecting the commercial sector. Concerns about food safety have also been a strong impetus in promoting healthy eating and organic foods, and those markets are now flourishing in many countries, particularly for fruit and vegetables.

Healthy eating

Meat is often considered by meat eaters as 'good for health', but meat does not enjoy any status as a health food. It is not stocked on the shelves of health food shops and there are no health food counters that promote 'healthy meat'. Amongst the different meats, pork has the least reputation for being good for health, because of its associated layer of fat, whereas poultry has a higher health image (Verbeke *et al.*, 1999).

Meat consumers are moving away from buying fresh meat towards buying processed products and eating out. They are relinquishing control over the ingredients they are eating and this is not entirely consistent with healthy eating. Processed foods are not synonymous with healthy foods, in terms of the amounts of fat, salt and synthetic ingredients they contain. With the exception of fish, it seems unlikely that meat will gain a strong position as a health food whilst this trend towards processed products continues.

Organic and free-range livestock farming

In the EU, about 3% of farmland is used in organic farming. It is not a large part of the agricultural industry, but the biggest contributors are in Germany, Italy, France and the UK.

The original aims in the organic movement were developed during the early and mid-20th century. The concepts included:

- biodynamic farming, in which there is an emphasis on rotational cropping, balanced nutrition for both plants and animals and rejection of inorganic fertilizers
- composting organic wastes and non-chemical control of plant pests and diseases
- autonomy of the farm, including self-reliance, recycling of nutrients, minimum cultivation methods and the role of local farms serving local communities

The present focus in organic farming is in producing crops, milk and meat that are pure and free from contamination with agrochemicals or veterinary residues, and in a manner that is ecologically sustainable. In some organic associations the marketing aims have broadened to encompass animal welfare. For example, in EU countries, pigmeat that is labelled organic should come from pigs that have had access to an outdoor area and are provided with more space and litter than in conventional intensive systems. The boundaries and distinctions between organic and welfare-friendly farming methods could become blurred in the future, especially if branded products acquire archetypal labels and images.

The images that organic foods have amongst the public are one step removed from the organic farming methods promoted by the organic food associations. The present images amongst the public have more to do with purity of the food and health benefits than with the production method, but the two are not always synonymous. To the consumer, beliefs about health, safety and meat flavour are more important than farming methods, the environmental benefits and animal welfare (McCarthy *et al.*, 2003).

There are also discrepancies between organic farming standards and animal welfare standards. The following animal welfare concerns have been raised about organic farming methods, but not all of them have been substantiated:

- poorer control of endo- and ectoparasites
- greater risk of injuries because organic producers are less likely to dehorn cattle
- greater distress when cows and calves are separated in dairy herds, because this usually occurs at least 2 days after calving in organic farming systems, by which time bonding has developed
- increased risk of uncontrolled mastitis, because of reduced use of antibiotics
- greater risk of injuries, lameness, hypothermia and sunburn in organic (outdoor) pig units, but this is not exclusive to organic production systems
- greater risk of high mortality in layer hen flocks through disease and no beak trimming on organic farms

These potential disadvantages are offset by providing for behavioural needs in organic farming systems. There can also be some less obvious benefits, such as reduced aggression in outdoor organic pig finishing systems (Cox and Cooper, 2001).

Organic aquaculture has just started, and the standards are based on natural breeding, natural pond walls or banks, feed sources that grow within the pond, using organic feedstuffs as supplementary feed, and an emphasis on homeopathic disease prevention rather than relying on chemical treatment. It is too early to guess how this will evolve.

Should the meat and livestock industry be worried about animal rights?

When the EU was established in 1957, animals were referred to as goods or agricultural products.

This was because the original treaty was concerned with setting up a trading partnership between countries. It was not a document that considered the rights of animals. In 1995 the EU Parliament modified its position on the status of animals in society. Animals became recognized as sentient beings. This may at first sight seem academic, but in future it could have legal ramifications. It will broaden the dimension of legal arguments on animal rights and expectations for animals.

Many farmers in Europe have a deep-seated concern about these developments and about the animal rights movement. First, they are worried by the militant methods that are sometimes used by animal rights activists and the harm this does to property and even to some animals. Secondly, there is concern about the concept of giving rights to animals. It is seen as an attack on present values, and it is difficult to comprehend. Thirdly, supporters of animal rights are considered by some as having an oversimplified view about animals and animal care. The moral reasoning of the animal rights supporter is said to be distorted and lacking in practical experience. However, the evidence on the last point gives a different perspective. When a sample of animal rights believers were questioned, they showed comparable and in some respects higher levels of moral reasoning when presented with a broad range of moral dilemmas, not just those connected with animals (Block, 2003).

A minority of animal rights protagonists adopt threatening and violent strategies in pursuing their aims. To them, their use of violence is similar to using violence in a political uprising. Whether violence is justified in correcting an injustice depends on whether it is a last or reasonable resort. In the case of animal rights, society in general considers that the stage of last resort has not been reached, and that reasoned and reasonable discussion is more appropriate than violence or terrorism.

Nevertheless, society has already adopted the concept of animal rights. The five freedoms are a form of rights, although some argue that they are responsibilities for those in charge of animals rather than rights for the animals (Gregory, 1998). In legal terms the distinction between responsibilities and entitlements is known as a **jural relation**. The jural relations are listed in Table 1.4.

Examples of how jural relations work are as follows. In most countries a child has a right to education, and society has a duty to provide education.

Table 1.4.   Jural relations in human rights.

Right	Duty
Liberty	No right
Power	Liability
Immunity	Not allowed

Similarly, a child should be immune from unwarranted brutal punishment, and a parent should not be allowed to administer that type of punishment. The distinction between jural relations (e.g. between rights and duties) is not semantic when it comes to interpreting and implementing the law.

Similar jural relations already exist for animals, and they are usually empowered through codes rather than primary legislation. Like human rights, they are often framed in terms of people's responsibilities, and some examples are shown in Table 1.5. In these terms, giving animals rights is not an attack on present values. Instead, it is an extension of existing recognized responsibilities.

This does not mean that animals have the same rights as humans. In fact, equal rights is irrelevant in a social and legal sense. Virtually everyone agrees that animals have a right to consideration, but some find this too minimalistic. What Singer (1991) and many others argue for is equality in terms of intolerance of suffering: 'If a being suffers there can be no moral justification for refusing to take that suffering into consideration. No matter what the nature of the being, the principle of equality requires that its suffering be counted equally.' Nibert (2002) tried to explain why there is inequality in tolerance for suffering. He claimed that oppression and devaluing humans and animals are fundamental causes: 'Oppression of humans causes much of the mistreatment of other animals, and the awful treatment of other animals fuels human exploitation.' In addition, cruelty has the same moral basis for humans and animals, but it may not share the same level of importance or penal standards.

The right to consideration does not always apply to welfare compromises connected with livestock farming. This is because there is incomplete agreement as to when suffering occurs. For example, opinions about the acceptability of amputations without anaesthetic and about confinement systems vary considerably even amongst experts. Some people have completely different views about what constitutes suffering and in agreeing on which

Table 1.5. Examples of jural relations in animal rights.

Paradigm	Example
Duty – right	The stockperson must provide feed and water for pigs at least once a day
No right – liberty	A farmer has no right to deny an animal's freedom to find shelter during a storm
Liability – power	The dairy farmer has assumed liability for regularly milking the cows, in the absence of a suckling calf
Not allowed – immunity	The person in charge of a draught animal is not allowed to unreasonably beat or whip the animal

practices cause unacceptable suffering. For example, to some, performing a rectal examination on a cow for pregnancy diagnosis would be a welfare insult, whereas to others it is a normal management procedure that does not warrant concern.

The main conclusions from this discussion on animal rights are as follows:

- There is little need for the meat and livestock industries to worry about the concept of animal rights.
- Animals already have legal rights through the legal requirements made on their owners. The requirements are framed in a similar way to those for human rights, and animal rights are implicit rather than explicit. Those legal requirements are already established, and are generally accepted.
- Violent animal rights activism is inappropriate in present society. It is operating outside the law and there are non-violent alternatives.
- People and animals have comparable consideration in terms of unnecessary forms of suffering (cruelty). There are differences in recognizing the causes and signs of suffering, and this leads to disagreements about the rights or protection that should be extended to animals.

Understanding the animal rights approach

Knowing how fundamentalists within a philosophy think and behave allows others to understand their position and this in turn enables discussion. Without open discussion, there is a danger that attitudes become polarized, and in the case of the animal rights issues this can spill over into frustration and violence.

Animal rights fundamentalists hold that individuals do not have a right to use animals for their own pleasure or interests, regardless of the benefits to society, whereas less extreme animal rights supporters accept that animals are used by society, but that people in charge of animals have a moral responsibility to eliminate suffering in those animals. The pragmatist who does not subscribe to animal rights would be more inclined to accept that suffering is an everyday event, and that society should strive to control suffering, but it may be unrealistic to eliminate it. These divisions are somewhat narrow as they focus on the moral outlook towards suffering. There are other features, such as making a living from animals, recreation and companionship, which should be included when making ethical decisions.

When there is a controversial debate on animal rights, it is often easy to recognize the approaches used by the separate sides. The animal welfare and animal rights advocates have an almost predictable modus operandi (Table 1.6). Similarly, the people they are focusing on, whether they are livestock trade organizations or people supporting animal research, have particular stereotypes that they use in their defence.

Animal rights arguments usually have an 'absolutist' moral orientation, which supports a high level of idealism and low level of relativism (Galvin and Herzog, 1992). Animal rights activists are sometimes criticized for being unduly emotional about animals and for not listening to reasoning. This is not always received as a criticism as, in the words of one activist: 'I don't think there should be a division between emotion and philosophical issues . . . You are an emotional, thinking person there's nothing wrong with being emotional about something you believe in.' Animal rights advocates make a distinction between 'learned emotion' and 'rational emotion'. Learned emotion is

Table 1.6.   Stereotypes in debates over livestock welfare issues.

Approaches used by animal advocates	Approaches used by the livestock trade
Appeal to public opinion	Consultation with experts
Emphasis on intuition	Emphasis on knowledge
Risks portrayed in broad terms incorporating political and social factors	Narrow framing of the risk analysis, considering only measurable uncertainties
Broad symbolic and political goals	Narrow business goals, and task of remaining competitive
Emotion	Rationality
Openness	Secrecy
Suspicion of standards	Trust in standards

Table 1.7.   Prevalence of attitudes to animal treatment amongst animal activists and non-activists (after Plous, 1991, 1998).

	Question: What should the animal rights movement focus on?		
	Response during		
	1990		1996
	Activists %	Non-activists %	Activists %
Treatment of			
Animals used in research	54	26	38
Animals used for food	24	8	48
Animals used for clothing or fashion	12	22	5
Animals in the wild	5	30	3
Animals used in sport or entertainment	4	14	5
Animals used in education	1	0	2

synonymous with understanding what is involved in particular forms of suffering, and using that knowledge to heighten one's own emotional response. Rational emotion stems from consolidating one's attitudes by reading philosophical works on animal rights, such as Tom Regan's book, *The Case for Animal Rights*.

The messages that animal activists promulgate are presented in a variety of ways. The activists often use moral shock for attracting attention and support for the cause. The most effective forms of moral shock are those that create a sense of outrage and those that are embodied in a powerful condensing symbol. Condensing symbols are pivotal in providing cognitive plus emotional

impact. They can be verbal or visual images, and neatly capture a range of meanings. They are often far removed from the original meaning of the symbol. Some examples of verbal condensing symbols are 'Animals are my friends, and I don't eat my friends' and 'Warning: This Package Contains Dead Animals'.

The animal rights movement is beginning to lose some of the popularity it held during the 1990s. The present situation is similar to that during the 1910s. Antivivisectionism was strong at the beginning of the 20th century, but by 1920 it had declined, and it took two generations before it re-emerged. This pattern conforms with the 'declining interest hypothesis', and a useful indicator of contemporary interest is the turnout at public rallies, marches and demonstrations (Herzog, 1996). Recent trends have shown that attendance has been lower than protagonists had planned. Along with this, there seems to be a realignment in the focus of animal rights activists (Plous, 1991, 1998). There has been a softening in attitude towards animal research but a growing concern about livestock welfare (Table 1.7). This change in emphasis was, however, linked to a higher proportion of vegans being represented in the more recent survey.

Value orientations in the green movement

There is a close association between concern for animals and concern for the planet. This link exists at practical and the more radical levels. For example, Peek *et al.* (1996) found that men and women who take a pro-environmental stance were more likely to endorse animal rights.

Genuine attitudes towards the environment are deep-rooted in society, and a brief examination of some of the extremes in attitude can be helpful in understanding the fundamental outlooks. Some of the most extreme views about the environment are held by neopagans. They:

- have a romantic attachment to nature
- range from people who practise nature religion to those who place a personal spiritual slant on the green movement
- tend to believe that nature is in some way alive or sacred, and their values for nature are closely connected with archetypal images of ecology and the environment

Modern environmentalists often share the outlook that environmental degradation has stemmed from a society that has faith in science and technology, believes in progress and abundance and adopts a laissez-faire economy. In some individuals this may be blended with a mild form of nature reverence. The exact outlook depends to some extent on the individual's **value orientation**.

Gunter and Furnham (1992) described people's value orientation on environmental issues as being of two types: 'internal' or 'external'. Internally oriented people generally consider that their own destiny is up to themselves. They regard events that happen to them as due to their own

efforts and abilities, and they tend to want to control their own lives as much as possible. Externals are more likely to attribute events to chance or to decisions made by other people who are in control, and they are prone to letting fate control their lives. People with an internal value orientation are more likely to play an active part in environmental issues. Homer and Kahle (1988) found that internally oriented people were also more likely to be natural food shoppers. People who rarely purchased natural foods were externally oriented.

The distinction between internals and externals is useful. It helps us recognize whether information that is put forward about animal welfare and environmental issues is likely to be accepted. The externals are more likely to take notice of conciliatory messages, whereas internals are more likely to respond to information that portrays personal or social benefit. Clearly, a single message is not going to appeal to everyone.

It is often said by people in developing countries that worries about environmental quality and animal welfare are a luxury that largely concerns the wealthy nations. In the case of environmental issues that claim was challenged by the Health of the Planet (HOP) Survey (Dunlap and Mertig, 1995). This survey was conducted in 24 nations and it is one of the most comprehensive studies of its kind. It showed that people living in low gross national product per capita (GNP) countries consider other problems besides environmental issues as pressing. Those issues include hunger, homelessness, crime, violence, poor health care, high cost of living and racial/ethnic/religious prejudice or discrimination. Environmental issues are taken more seriously by people in wealthy nations when they are compared with socio-economic issues; the perceived seriousness of environmental problems relative to other problems was positively correlated with GNP ($r = 0.70$; Dunlap, 1997). However, the concern for environmental quality in low GNP nations was broad-ranging. The HOP survey showed that poorer nations were more likely to see environmental problems as health threats ($r = -0.70$), but they believed that environmental problems had not affected their health in the past. The old assumption that non-industrialized nations do not worry about environmental protection is incorrect.

Many people claim that they are environmentally and animal welfare aware. However, those concerns do not always translate into buying habits.

Marketers have found that consumers, despite their professed beliefs, are still extremely price-sensitive when it comes to buying green and welfare-friendly goods. People with strong pro-environmental beliefs are very focused. They buy particular items that fit their particular beliefs. In comparison, welfare-friendly purchases may depend more on what is available.

Genetic manipulation

The consumer concern that emerged during the late 1990s was food from genetically modified (GM) organisms. There is in fact a range of concerns, and these were identified by Macer (1992) before the issue attracted the media attention that distorted people's outlooks. Macer conducted two surveys, one in New Zealand and the other in Japan. He found that about half of the New Zealanders (48%) and Japanese (55%) were concerned about the prospect of eating meat from genetically modified animals. There was slightly less concern about consuming GM medicines and vegetables than GM meat. The reasons for the concern about GM meat are summarized in Table 1.8. There was uncertainty about health risks, the 'unnaturalness' of the meat and uncertainty about the adequacy of testing for risks. In the case of 'unnaturalness', there was the feeling that GM foods are against the law of nature, that the original taste of the food will

disappear, and that everything will be the same, which in Japanese culture is called 'shimofuri'. Meat is perceived as a commodity that has variety, and this provides interest in itself. Under quality and purity, some people said that 'artificial meat is frightening'. There was also a perception that potentially harmful foreign genes might be present. Reference was made to thalidomide and cancer as examples of unforeseen disorders. In general, there was close similarity in the reasons for concern amongst Japanese and New Zealanders. One difference was that New Zealanders were concerned about knowing what they were consuming, unlike the Japanese. This could either reflect differences between the cultures in awareness about the origins of meat, or it could be due to the greater tendency in Japanese culture to focus on benefits, rather than looking for hidden dangers.

Comments about economic, ethical and political concerns included 'I don't trust the safety standard which is decided by the government or industry', 'Misuse', 'Cannot trust the results of research looking at the effects', 'Can we morally accept artificial animals?' Animal welfare was not an important issue; only 1% of the Japanese and 5% of the New Zealanders who were worried about GM foods thought that it was relevant. This means that the main driver that could prevent or delay the arrival of the hypothetical GM chicken is its image as a less natural product. The public is not likely to buy it if they have a free choice

Table 1.8. Reasons for concern about eating GM meat products within Japan and New Zealand.

	Japan	New Zealand
Number of people expressing concern	278	981
Reason for concern (% of those expressing concern)		
Unknown health effects or risks	18	21
Unnatural feeling (including taste)	16	27
Doubts about safety	11	9
Do not know what we are eating	1	15
Information is being hidden	7	9
Side effects	4	5
Quality and purity cannot be guaranteed	2	4
Potential for new diseases	4	3
Environmental or ecological effects	1	–
Economic, ethical or political concerns	2	–

(Frewer *et al.*, 1996). Lack of acceptance is closely linked to perceived unnaturalness.

Nevertheless, GM meats are getting closer to the marketplace. The first European patent for a genetically modified animal intended for meat production was granted in 2001. It was for an Atlantic salmon that had an additional gene that imparted faster growth. At the same time, the parent company owning the patent applied for approval in the USA to market GM salmon. Tilapia could be next. Transgenic growth hormone (GH) tilapia are almost three times larger than non-transgenic tilapia, partly because of their superior feed conversion efficiency. This provides a strong commercial advantage because non-transgenic tilapia tend to be undersized through competition for feed.

Livestock Farming in Developing Countries

This section introduces some of the issues in subsistence livestock farming that impact on animal welfare. They are dealt with in greater detail in Chapter 2.

Reasons for keeping livestock

In industrialized countries the main reason for keeping livestock is to make money from producing meat, milk, wool or eggs. The situation in poorer countries is quite different. In Africa and Asia, rearing ruminants, particularly cattle, is one of the best ways for rural people to accumulate wealth. Many rural communities in developing countries do not use banking facilities. Instead, they invest their savings in livestock, and the aim is to increase wealth by breeding more animals. Livestock also provide some security against crop failure and currency fluctuations, if the need arises. Producing cattle for meat is often a subsidiary aim. However, cattle are used for meat consumption when:

- they have reached the end of their working life and need to be replaced
- the owner needs to sell them to raise money

In many of the Sahelian countries of Africa, the annual offtake for cattle is less than 10%. This increases the risk of overstocking. When hard times hit, such as a drought or political unrest, the accumulated wealth in the form of livestock is protected by the owner for as long as possible. Selling or slaughtering the stock during such times is a last resort. However, failure to sell early during a severe drought has catastrophic consequences for animal as well as human welfare.

Cattle, sheep and goats serve quite different functions, as can be seen for rural people in Lesotho from Tables 1.9 and 1.10. Cattle have an important role in trading and as gifts. They are used for local consumption mainly on important ceremonial occasions, whereas sheep, goats and poultry are more regular sources of meat in village life (Swallow *et al.*, 1987). Mortality is the main

Table 1.9.   Uses of livestock in Lesotho rural households during 1985.

	Cattle	Sheep	Goats
Number of households	462	250	235
Number of animals	3,447	13,654	8,088
Number not retained	515	2,367	992
Reasons for disposal (%)			
Died	34	14	27
Sold or traded	19	23	19
Gift/bride price	18	6	16
Stolen	11	21	10
Lost	–	–	3
Used as payment of herdsman	1	11	–
Slaughtered	17	26	26

Table 1.10.   Reasons for livestock slaughter in rural Lesotho in 1985.

	Number of animals		
	Cattle	Sheep	Goats
Reason for slaughter			
Home consumption	21	369	173
Funerals	39	56	11
Welcome baby celebrations	5	41	8
Initiation ceremonies	4	15	14
Weddings	9	13	3
Sale of products	1	–	1
Balimo ceremonies	13	35	23
Other	4	40	24
Total	96	569	257

reason livestock are not retained for the next year. Over one-third of non-retained animals are either traded, given away or used in debt settlement (Table 1.9).

During favourable times, animals are continuously bought, sold and exchanged, and the market is very active. Where livestock have investment value and the owner is looking to profit from bartering, it is in his interests to look after the stock. However, stature is more important than body condition in determining price, and so emaciation is not so heavily penalized as it would be in other farming communities. In addition, the incentive to care for stock is not so strong amongst livestock dealers or stock people who do not own the animals.

In countries such as Morocco, sheep are a way of banking capital, but they are not looked upon as an investment. In other words, they are a financial reserve and there is less emphasis on trying to increase their value. This is reflected in the way they are fed. During the dry season they subsist on crop by-products such as stubble and weeds growing on fallow.

Investing in livestock is particularly important during periods of monetary inflation. Suppose a young man moves temporarily from his village to earn a wage in a town or city. He is likely to save his earnings in the form of livestock, which are kept at his home village, rather than holding cash that is losing value. Later on, he will move back to the village and live off the wealth he has accumulated as livestock. In this situation, livestock are a good form of investment, provided they are sold or exchanged before they die. Their value can be more easily realized than a house or mortgage, which would be the equivalent investment in developed countries. In addition, there are good opportunities to gain from the market by buying low and selling high.

Traditionally, livestock in Africa have been readily exchanged when paying debts or purchasing goods such as maize, millet and beans. They are also used for settling secondary education fees or for the loan of a milch cow. In the past, trading caravans specialized in servicing some of the trading needs in cashless remote communities by exchanging commodities between the pastoral and agricultural sectors, and livestock were an important trading commodity. Now there are regional markets, which have replaced the nomadic livestock traders.

During the last two centuries there have been two important influences that have modified the role of livestock as a form of investment. First, there has been greater need for monetary currency, especially for paying taxes. Money has also become a substitute for settling debts. Secondly, acquisition of better land for crop production has left pastoralists with the poorer land and there is less opportunity to range widely and move stock to other regions when the need arises. As a result the impact of localized droughts is potentially more serious now than it was about 150 years ago, and the risks associated with keeping livestock are greater.

In regions such as the highly populated East African highlands, land subdivision is placing pressure on feed resources, and zero-grazing napier fodder (*Pennisetum purpureum*) and crop residues are becoming more common. The animal welfare benefits of changing from free-grazing to cut-and-carry systems are less tick infestation and fewer tick-borne disease problems, but there is greater risk of underfeeding and reduced cow fertility.

Productivity in **cut-and-carry** systems is limited by the availability of labour and the quality of the feed. Some sheep farmers spend as much as 2 h a day collecting enough forage for one sheep. Most tropical grass species have a high fibre content, and their feeding value is low in protein and minerals. Goats or lambs fed napier fodder alone can grow at 20 to 25 g per day and this can be increased to about 50 g per day if they are offered tree legume foliage or wilted cassava leaves as a protein supplement (Johnson and Djajanegara, 1989).

Cut-and-carry systems are becoming very common in Indonesia and India. Here, the animals may be penned throughout the year, or during periods when there is a risk of livestock damage to crops. In humid regions this risk period extends for most of the year. In Indonesia small stock are generally well cared for, even though they have limited freedom to exercise. They are kept in groups of five to eight in raised pens under thatched roofs, and they are separated from their dung and urine by a slatted floor (Fig. 1.1). They are usually given more feed than they need because the farmers want excess feed to fall into the pit below the pen where it will soak up urine. Farmers value the manure as much as the animals (Tanner *et al.*, 2001).

In time, livestock could be replaced by other forms of saving and investment, and when this happens there will be more emphasis on making money from farming livestock instead of keeping them as security.

Overstocking

Overstocking is the main environmental hazard in keeping livestock in developing countries. It leads to loss of ground cover, and this impacts on the welfare of animals when it reaches the stage of chronic underfeeding. Loss of vegetation occurs in the following ways:

• Insufficient opportunity to allow the plants to recover between grazings.
• Physical damage such as trampling and pugging.
• Tearing the growing points out of the plant.
• Urine scald.

The grazing habit of sheep is particularly damaging because they are bottom grazers. In other words, they eat pasture close to the ground, and in pasture species that have their growing point above ground level this can be very destructive. Sheep grazing encourages the survival of less-productive pasture species, or, in the absence of these, the land can become bare or infested with inedible species, and the topsoil is prone to erosion, especially from wind or landslip. Erosion of the topsoil leaves behind a less fertile subsoil or bare rock.

In rangeland farming, four approaches can be used when managing stocking density and trying to avoid overstocking:

• The land is stocked at a fixed rate according to its perceived long-term carrying capacity.
• Livestock are cropped according to current feed availability or imminent changes in feed availability (e.g. seasonal slaughter).
• Stock are moved within the rangeland according to water and feed availability (e.g. transhumance herding, nomadic and semi-nomadic pastoral systems).
• Stock are thinned on a needs-must basis such as raising cash or paying debts.

The chosen method depends on the economic position of the pastoralist, his/her attitude towards

Fig. 1.1. Type of raised shed used for sheep and goats in East Asia.

risk, the resilience of the vegetation to short-term overgrazing, and variability of the weather in that region.

Labour requirements can be demanding in subsistence pastoralism. Labour enables better distribution of grazing pressure, which helps limit localized overgrazing. Nomadic pastoralism is an extreme form of this type of droving management, but it has been declining in recent decades because of greater enforcement of country borders and animal health control boundaries. More often, children are responsible for stock movement when grazing near home, and adolescent or adult men supervise stock that are trekked and grazed over a wider area.

Transhumance, agistment and semi-nomadic farming

About one billion people live in the world's arid and semi-arid regions (Squires and Sidahmed, 1997). These regions provide a livelihood for about 40 million pastoralists plus an unknown number of people involved in seasonal transhumance systems. Being semi-arid, these regions do not support productive cropping, but instead they are home for large numbers of rangeland cattle, sheep, goats, camels and wild ungulates. These animals subsist for about 6 months of the year during the dry season. This means they are half-starved, they mature slowly, start producing young later in life and produce limited milk.

Seasonal feed shortages, droughts and disease are the three main reasons for transhumance. Transhumance involves moving stock to another area where there is more feed or the disease risk is lower. It is a form of supervised migration where the herdsman directs the stock between areas of pasture and water sources. A transhumance period may last for weeks or months, and the herdsman stays out at night with his livestock or at a cattle post. It is a temporary semi-nomadic existence. Agistment is the transfer of stock to distant grazing land and paying a fee for the care and feeding of the animals.

In recent times both nomadic and semi-nomadic pastoralists have been able to adopt a more sedentary existence because of the reduced incidence of trypanosomiasis. The downturn in this tsetse fly-borne disease is partly due to active control programmes in the southern half of Africa and because of declining wildlife, which acts as a natural food source for the fly, plus reduced rainfall in some parts leading to less favourable conditions for the fly. Instead, the fly is cycling within the cattle population, and so it has been easier to control, especially where tolerant zebu cattle are kept.

The change to a sedentary lifestyle has had social repercussions. Some nomadic people have in the past had a poor relationship with agriculturalists. Since they have been nomadic, there has been little need for them to get on socially with other people. Now they are living closer to other people, who, over the years, have accumulated grudges against the former nomads. Social grievances such as these sometimes come to a head when straying stock damage crops, and animal maltreatment is one outlet for anger or frustration.

In other cases, nomads have in the past provided valued services for agricultural communities. Some nomads have been traders, and their arrival at a village provided an opportunity to buy the goods they dealt in. Nomads also had valued skills such as castrating cattle, and provided remedies or health care for the agriculturalists' sick animals. The interdependence of nomads and sedentary agriculturalists and communities is now declining.

Absence of proper veterinary care is a welfare issue in transhumance systems. Veterinary care is also improvised and rudimentary because of the cost of registered veterinary medicines. Sometimes livestock owners claim that they do not have any animal health problems, but this may be due to lack of awareness of the signs. In addition, in some districts farmers are not aware that there are community veterinary services and that animal diseases can be treated with modern drugs.

Enclosure of common land

Common land is used by landless poor people for grazing cattle, sheep, goats and, to a lesser extent, pigs. This form of pastoralism has virtually disappeared in Europe, but it is still important within Africa, the Indian subcontinent and Central Asia, especially for goats. When common land has been enclosed, it has often been distributed to farmers and people in the lower income group. Some of their holdings have been purchased and amalgamated into larger farms. In this way, large amounts of land that were used for grazing and for collecting fuel have been privatized and converted into cropping land, particularly in regions that can support irrigation. Hens and goats are assuming greater importance on those smallholdings that continue to keep livestock, especially in areas with high populations.

Four systems are used for goats:

- confined throughout the year
- free-range
- seasonally confined
- tethered

The last three apply to goats kept on common land. Seasonal confinement is becoming the most common system, and confinement coincides with the critical period when crops need protecting. This often extends from when the crops are sown until the harvest is taken.

Environmental Hazards Created by Livestock Farming

The environmental hazards that need to be managed in livestock farming include:

- pollution of open waterways with dung and urine
- pollution of groundwater with dung and urine
- changes in the composition of water leading to poisoning of aquatic life
- soil erosion
- aerial pollution including unpleasant odours
- dust emission
- inappropriate disposal of dead stock
- noise
- encouragement of flies and other unwanted animals
- release of agro- and veterinary chemicals
- destruction of particular plants

Ammonia accumulation is the predominant environmental hazard in intensive livestock farming. Release of ammonia is an environmental hazard because of:

- smell
- indirect effects of ammonium cation deposition from the atmosphere on:
  - conversion of moorland into low-value grassland

- tree damage
- leaching of potassium, magnesium and calcium cations from soil and leaves

Ammonia is implicated in three animal health and welfare features. They are:

- skin lesions such as hock burn
- chemical degradation of hoof keratin and subsequent lameness
- inhalation of ammonia, and respiratory tract damage and disease

Industrialized and intensive livestock farming systems need to reduce emissions of atmospheric ammonia in order to comply with international environmental standards. Most of the ammonia comes from feeding high-protein feeds. For example, about 60 to 70% of the nitrogen in pig rations is excreted in faeces and urine. The nitrogen in faeces is mainly present as protein, and nitrogen in urine is mainly in the form of urea. One way of reducing ammonia emission is to reduce dietary protein levels. This may involve some sacrifice in growth performance of the animals, but there can be a cost saving as protein is one of the more expensive feed components.

Ammonia emissions from urine occur when urea is hydrolysed by the enzyme urease. This enzyme is abundant in topsoil, in livestock bedding and on concrete floors. Ammonia release can be particularly pronounced with straw bedding systems (Jeppsson, 1999), but emission rates as high as 1 g $NH_3/m^2/h$ have been recorded from concrete floors in cattle barns (Braam et al., 1997).

Ammonia can be toxic and it has a remarkably rapid uptake rate into tissues. It is fat-soluble when in its non-ionized form, and water-soluble when present as the ammonium cation. Its fat solubility allows very rapid uptake by cell membranes and subsequent cell poisoning, and its water solubility as NH_4^+ allows prompt irritation of moist tissue surfaces, such as the cornea and upper respiratory tract. Calves and cattle avoid putting their muzzles on bedding in the normal sleeping position if the ammonia concentration in bedding is high (Fig. 1.2). As a result they may spend less time sleeping.

Ways of reducing ammonia emissions include:

- reducing nitrogen intake in livestock feed
- lowering the temperature of stored effluent
- covering stored effluent

Fig. 1.2.   Normal posture in cattle during deep sleep.

Table 1.11.   Examples of controversial farming practices.

- Failing to give animals access to water every day
- Confining sheep and goats in slatted floor pens for their entire lives
- Killing male Holstein calves at birth because they are unwanted
- Not using local anaesthesia plus a non-steroidal anti-inflammatory drug when disbudding calves
- Not providing assistance in cases of dystocia

- reducing the pH of stored effluent or manure within animal houses
- drying manure promptly
- prompt dilution of effluent with water
- direct incorporation of effluent into soil, instead of broadcasting

In aquatic species, high ammonia concentrations in water reduce the excretion rate of ammonia from gills and increase the uptake of ammonia. This can lead to systemic elevations in ammonia concentration in the haemolymph, reduced concentrations of the respiratory pigment haemocyanin and suppressed oxygen transport. These effects are compromised further by the increased oxygen consumption rate associated with a switch from anaerobic carbohydrate metabolism to lipid metabolism during ammonia intoxication. In spite of this, fish do not always react negatively to ammonia in water. Sometimes they enter water containing appreciable concentrations of ammonia, even at levels which can lead to poisoning and death (Erichsen Jones, 1952).

Distinguishing between Acceptable and Unacceptable Farming Standards

There are many uncertainties about what is acceptable practice in livestock farming and what is unacceptable, and in some cases it will be difficult to get agreement on where common standards should be set. Take the examples listed in Table 1.11. These practices are justified in some circumstances, whereas in others they are considered inappropriate. Research can help in decision making, and one area where it has been particularly useful is in determining which routine husbandry amputation procedures are least painful or distressing for

the animal. The section on amputations summarizes the key findings from that research. Knowledge founded on good science such as this helps to harmonize agreement about what is acceptable and what is unacceptable. The section on easy-care systems examines the conflicts and paradoxes between 'natural' extensive farming systems and welfare compromise. Decision-making is more complex with this issue and depends on the practical context as well as welfare considerations. The final section considers some accepted standards on lifespan and mortality from disease in the intensive livestock industry. Here again, harmonization of expectations may be difficult.

Amputations

There are six types of potentially painful husbandry procedure (Table 1.12). Whether they are painful depends on how they are done, and whether an analgesic is given. The age at which they are done can also influence pain severity.

Where alternative methods serve the same purpose, the more painful procedures are being discouraged. For example, research using plasma cortisol as a pain and distress indicator has shown that it is more painful to castrate lambs with a knife compared with a rubber ring. The differences between some of the alternative methods are summarized in Table 1.13.

Three types of pain and discomfort are present with amputation injuries. They are acute pain, inflammatory pain and neuropathic pain. Each requires a different approach when applying

Table 1.12.   Potentially painful husbandry procedures.

Category	Purpose
Cautery	Branding, disbudding, tail docking, debeaking, cauterizing tissues for haemostasis or pain control (e.g. horn amputation wounds, tail stumps)
Cryocautery	Freeze-branding, disbudding
Cutting or piercing	Horn/antler amputation, castration (knife), tail docking (knife), mulesing, pizzle dropping, ear notching or tagging, dubbing, desnooding, spur removal, toe removal, teeth clipping, beak trimming, nose ringing
Constriction with a ring	Castration, tail docking, spiker desensitization
Crushing	Castration (clamp)
Corrosion with chemicals	Disbudding, castration

Table 1.13.   Relative levels of acute pain and distress following different methods used for the same amputation purpose.

Tail-docking lambs
    Ring = cauterizing iron <<< cutting
Tail-docking calves
    Ring = cauterizing iron
Castrating lambs
    Ring = crushing <<< cutting
Castrating calves
    Ring = crushing = cutting (with cautery)
Disbudding and dehorning
    Cautery <<< scoop disbudding = guillotine shears = saw = embryotomy wire

pain control. Acute pain and discomfort can be alleviated with a local anaesthetic or nerve block, and inflammatory pain can be treated with anti-inflammatory drugs (Mellor and Stafford, 1999). Inflammatory pain is less likely when using cauterizing methods if the heat deactivates the nerves in the wound. Neuropathic pain is more difficult to treat and can occur when there is sensitization – for example, if neuromas develop in the severed or crushed nerves. The amputation procedures that cause chronic neuropathic pain are not fully understood.

Figure 1.3 gives a schematic presentation of the plasma cortisol responses during acute and inflammatory pain with a cutting procedure. Acute pain occurs at the time the cut is made and when free nerve endings in the wound are stimulated – for example, by rubbing – whereas, inflammatory pain takes several hours to develop and is provoked by chemicals, such as bradykinin and serotonin, which accumulate at the site of the wound.

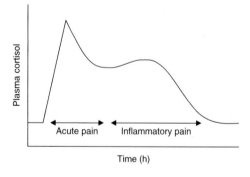

Fig. 1.3.   Schematic curve for plasma cortisol showing the two pain components following an amputation procedure.

If the intention is to control both types of pain, a local anaesthetic plus an anti-inflammatory drug will need to be given before the cut is made. If only a local anaesthetic is given, acute pain will be diminished, but inflammatory pain will still be present. If a long-acting local anaesthetic is given, the inflammatory pain is not eliminated, but instead it is delayed (McMeekan *et al.*, 1998). In some farming situations the cost of giving both a local anaesthetic and an anti-inflammatory drug would be considered prohibitive.

Easy-care systems

The following situations emphasize the need to consider the context of a farming practice when making a judgement about what is acceptable and what is unacceptable.

Many farming systems make use of compensatory growth. **Compensatory growth** is the acceleration in growth that occurs when a period of growth inhibition ends and favourable conditions are restored (Fig. 1.4). Growth inhibition is often due to reduced feed intake. Suppose a litter of piglets is weaned abruptly at a very early age. There is a growth check during weaning, and that check can be greater than if the litter had been weaned later on when they had already learnt to eat solid feed. However, if all goes well, the early-weaned pigs will catch up, and during the period of compensation their growth rate is higher than normal. The final live-weight and overall growth rate of the early- and late-weaned litters can be the same, even though the early-weaned litter had a more difficult weaning period.

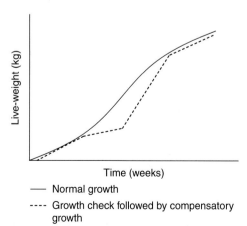

— Normal growth

---- Growth check followed by compensatory growth

Fig. 1.4. Schematic growth curves for normal animals and animals that experienced a check in growth followed by compensatory growth.

In other cases, making use of compensatory growth is an appropriate thing to do, and it has good outcomes. An example where it is an acceptable practice is in broiler chickens where growth is constrained during the second week of life to improve subsequent walking ability. The reduced growth during that period is made up when the birds are returned to full feeding.

Compensatory growth is very common in subsistence livestock farming. There is seasonal underfeeding, when pasture is not growing during the dry season or during winter. The animals lose or fail to gain weight, but growth accelerates when feed becomes available again. This situation is a fact of life in subsistence farming, and discussion about its acceptability is immaterial.

Reliance on compensatory growth can, however, encourage stock keepers to be less considerate and to tolerate lack of care. Examples of this, besides early weaning, are:

● cost-saving by underfeeding during wintertime, in store and breeding animals
● failing to provide protection against inclement weather
● optimizing feed conversion efficiency in growing animals by allowing occasional periods of feed shortage or deprivation
● overstocking in order to achieve high output, leading to exaggerated seasonal underfeeding
● cost-saving on veterinary treatment and allowing non-lethal disease to take its course whilst untreated

Periods of harsh weather, poor feed quality and inadequate feed availability are inevitable, but one of the skills in livestock farming is adjusting livestock numbers according to these conditions, and knowing which type of stock needs greater feeding or care. For example, should the farmer be favouring young growing animals, replacement females or breeding adults during a temporary feed shortage?

Underfeeding may also be tolerated in the knowledge that animals make use of feed more efficiently if they are underfed. When feed intake is restricted, the efficiency of digestion increases provided feed intake is not below the animal's maintenance requirement (Doreau *et al.*, 2004).

Easy-care management also has a welfare dimension at the time of parturition. Failure to assist an animal that is unable to expel a fetus because of an abnormal presentation is considered by some to be an act of negligence. Others view it as part of natural selection that is inevitable in easy-care systems, and it is for the greater good in the long term that the animal does not survive. It will not be easy to harmonize these different outlooks.

Longevity, sickness and mortality from disease

The health of a nation is often assessed from its life expectancy and infant mortality rate. Similar figures are helpful when assessing the standard of health care and management in farm animals. When mortality rates exceed the levels shown in Table 1.14, there would often be cause for concern amongst stock owners.

Stock owners will usually take action when mortality rates reach these levels. Their primary aim might be to reduce losses and improve profitability, but there will inevitably be other considerations, depending on circumstances and the individual's motives. The ethical question is where the interface is between an acceptable and an unacceptable level of mortality, and the answer is 'It depends on . . .'. In applied ethics a judgement is reached for a given situation, and the decision has to be based on a range of considerations connected with individual circumstances. For example, acceptable mortality from disease on one farm

may be unacceptable on a second farm, simply because of differences in ease of control of the underlying cause. Ethics brings together all the prevailing interests in reaching the 'right' decision. In the present context the ethical decision would include feasibility of disease control, financial resources of the stock owner to introduce treatment, likely effectiveness of the control measure, impact on profit margin, effect on morale from achieving better control, expectations amongst the rest of the community and legal obligations, as well as concern about suffering experienced by affected animals. Ethics encompasses everything, not just the moral issue.

The acceptability of early death in breeding stock has a similar dimension. One of the simplest signs that something is going wrong at a farm is reduced longevity in the breeding animals. In pig herds in the USA, sows last on average for 1.8 parities, producing 17 pigs (Guo et al., 2001). In Europe survival is higher, at 3.3 parities and 37 pigs per sow. The primary causes of early culling are breeding failure or poor reproductive performance and leg disorders. It has been suggested that modern genetic selection is to blame, but this may be an oversimplification. The genetic basis for these problems is complex because of the range of conditions that contribute to early culling, and the heritability (h^2) for longevity in dairy cows, sows, beef cows and dairy buffaloes is generally low ($h^2 = 0.02$ to 0.27). This implies that it may not

Table 1.14. Mortality rates that might be tolerated in industrialized nations before concerns are raised about the welfare of the animals or profitability of the enterprise.

Type of stock or situation	Mortality rate (%)
Lambs – birth to weaning[a]	10–25
Calves reared on milk replacer	10
Piglets – birth to weaning[a]	18
Breeding sheep	5–10
Sheep transported for slaughter	0.02–0.04
Feedlot cattle	4
Fat cattle transported for slaughter	0.005
Breeding cattle – ranching system	4
Breeding pigs	4–8
Pigs transported for slaughter	0.5–1.0
Broilers – placement to 40 days	3–4
Broilers transported for slaughter	0.5

[a]Excluding stillborn animals.

be possible to reduce early culling genetically whilst maintaining selection pressure for productivity. Instead, the main hope is with managing the individual causes of the early losses. In general, there is no clear genetic strategy for improving longevity in breeding stock. Reduced longevity may or may not be a welfare compromise, depending on the condition that causes early rejection. If it is lameness, then it clearly has a welfare dimension, but, if it is breeding failure, there may be no suffering, simply a shorter lifespan.

Good health care is more than just calling in a vet when a problem develops. It is also a state of mind that incorporates disease prevention into the daily routine. It has been reported that veal farmers who have a positive attitude towards animal care have lower disease levels in their stock and greater productivity, and this was presumably linked to better prevention of disease (Lensink *et al.*, 2001b). The signs of ill health in animals can be subtle and need experience and time in recognizing them. For example, cattle and pigs with nematode infestations show very few obvious signs except they are less active. They often spend less time feeding and more time resting, neither of which seems abnormal in itself (Forbes *et al.*, 2004). When reduced appetite can be recognized, it is a useful indicator in a wide range of disorders (Bareille *et al.*, 2003). This calls for skill and experience in those who look after stock, and in the longer term a willingness to pass on that knowledge in training the next generation of stock people.

2

Animal Welfare in Developing Countries

This chapter considers some of the animal welfare problems that occur in developing countries, and it discusses possible solutions. Some people argue that concerns about animal welfare in developing countries are misguided. Human suffering is more important and more relevant to the needs of poor people. A typical comment is: 'Animal welfare is a concern that can only be afforded by wealthy nations.' A counter-argument is: 'What is good for animals kept in poorer countries is also good for their owners.'

There is a strong incentive to keep livestock alive and well. Livestock sustain the community, and without them, life is difficult, especially in rural communities where there may be no alternative employment other than keeping livestock. Take the situation in poorer parts of Africa. Livestock are an important means of accumulating wealth and asserting social status. They provide fertilizer, make cultivation and transport less onerous, and are used as bride price. In other words, there is a dependence on livestock in everyday life, and if the animals are not fit for purpose, the benefits diminish. If livestock welfare is compromised, for example through starvation or disease, people lose out as well.

In wealthy nations, only a minority in the community depend on livestock. Most people have investment savings accounts at their banks, and so they do not need livestock to store their wealth in case of hard times. Inorganic fertilizer can be bought from agrochemical companies, and there is less dependency on livestock for draught work. Instead, the main reason for worrying about animal welfare is moral concern. There is a sense of obligation in looking after domesticated animals because of their dependency on people. This moral value is similar to the one afforded to children. The strength of the concern, however, varies between individuals.

In developing countries, concern for animal welfare is based on self-interest rather than moral responsibility. For example, in communities where livestock are valued with pride because of the social status that goes with ownership, it might be undignified to mistreat an animal rather than immoral.

The welfare problems vary with production system, and a good starting point is to identify the main production systems in developing countries. Most of the systems shown in Table 2.1 apply to cattle, sheep and goats. The boundaries between the systems are flexible. For example, many cattle owners split their stock between a transhumant herd and a milking herd kept near the main settlement. Animals are periodically swapped between the two herds. Other farmers may move the whole herd between systems, on a needs-must basis. For example, transhumant pastoralists may become nomadic during a drought or disease epidemic.

In Africa, transhumance is usually on foot, but in West Asia and Mediterranean regions it is often by truck or train. Transhumance by truck could become more common where right of passage becomes difficult because of population growth. In many heavily populated regions, grazing systems are being displaced by cropping, but grazing is increasing in Central Asia and the far north following de-collectivization, and in South America where forests have been converted to savannah.

©N. Gregory 2007. *Animal Welfare and Meat Production*
(N. Gregory)

Table 2.1.   Types of livestock production in developing countries.

Category	
Exclusive pastoralists	Grow no crops. Depend on sales of dairy products and animals for buying grain for human consumption. May move the stock long distances in search of feed and establish settlements in different grazing regions
Transhumant pastoralists	Have a permanent base at which older members of the community remain. Part or all of the milking herd may be kept at this base (compound dairying). The main herd is moved in response to seasonal changes in feed availability and disease challenge (e.g. diseases transmitted by tsetse flies). Stock at the base are brought in at night (kraaled). Surplus male animals are sold to buy grain and other basic necessities
Agro-pastoralists	Semi-settled pastoralists who hold land rights. Grow grain, and keep small herds of livestock. Cattle are kept for traction, manure, dairying and fattening
Traction animals	Often stall-fed and kept separately, but sometimes managed with fattening or dairy cattle
Seasonal tethering	Stock tethered to a stake during the dry season to manure cropping land. May be sold at the end of this period
Permanent tethering or stalling	Small stock kept in a hutch or pen, or permanently tethered, often in an urban setting
Fattening	E.g. stall-feeding under shelter. Fattening period can be from 6 to 24 months. Fed conserved fodder (e.g. stover) or cut-and-carry fodder. Sold to meet needs rather than to exploit seasonal price differences
Peri-urban and 'modern' cattle production	Cattle owned by wealthy businessmen. Animals used as an investment or for prestige
Scavenging	Cattle considered as common property and scavenge feed whilst roaming freely about the village. Their management depends on committee decisions

The main animal welfare problems in the systems listed in Table 2.1 are:

- seasonal underfeeding
- underfeeding because of droughts
- overstocking leading to underfeeding from loss of productive plant species
- infrequent access to water
- insufficient milk for calves because of offtake for domestic consumption
- inappropriate methods used for managing animals
- inadequate or inappropriate treatment of diseases and disorders

Overstocking and underfeeding underpin many of the problems, and they are considered under the following headings:

- managing seasonal underfeeding and over-stocking

- choice of species and breed
- ranching as a more intensive alternative to rangeland farming
- stock management during droughts
- urban livestock farming

What Are the Problems?

Seasonal underfeeding

About 20% of the world's land area is rangeland. Most of the rangeland is in semi-arid regions, and, where rainfall is seasonal, livestock subsist for 6 months of the year at about their maintenance level. Stock deteriorate if the dry season is severe and if it is followed by a disappointing wet season. Stock numbers may be too high for the amount of feed available, and, if there is a disease outbreak, they will be weakened further. It is combinations such as these that make 'a bad year' for farming and precipitate catastrophic welfare problems.

The suffering associated with underfeeding and overstocking is obvious. In people, the forms of suffering associated with chronic underfeeding include:

- frustration and irritability
- a feeling of being 'sick in the stomach'
- weakness and loss of motivation
- difficulty in getting to sleep and disturbed sleep
- sensitivity to cold

These signs can progress to frustration, irritability and a sense of hopelessness (Gregory, 2004). Presumably, comparable effects occur in underfed animals, although the balance between the various states may be different.

Physical weakness in underfed animals has the following adverse effects. It:

- compromises their ability to forage
- makes them more dependent on regular watering
- may contribute to susceptibility to disease
- makes them more prone to accidents and injury

Maintaining motivation to search for feed is important for survival, and ensuring an adequate supply of drinking water helps maintain that motivation. Ultimately the weakened animals die from emaciation. Normally these dead animals would be removed for human consumption, but dead cattle left in rangeland is a sure sign that a famine has been particularly severe. The excess carcasses are left where the animals died.

The changes in **grazing behaviour** as feed becomes scarce are important. When feed is plentiful, stock can afford to be selective in what they eat. Cattle choose higher-quality forage and are less likely to select browse, dried leaves or stems. When the dry season sets in and feed is in short supply, they spend more time grazing and searching for better-quality feed. Typically, time spent grazing in both cattle and sheep increases from about 8 h per day to 13 h. Grazing at night increases, but darkness does not influence what they eat. They select feed of the same nutritional quality at night as during the day (Ayantunde *et al.*, 2001). But, at the upper limit (13 h grazing per day), they cease to put additional effort into searching for better feed, and they lose weight.

Under poor feeding conditions, up to 20% of total grazing time can occur at night (Smith, 1959), whereas, under more favourable conditions, night-time grazing may occupy only 8% of daily grazing time. In communal farming systems stock are usually kraaled at night, and so night grazing is not allowed. If night-time grazing were allowed, there would be longer overall grazing time and greater feed intake (Nicholson, 1987). It would allow cattle to graze during cooler periods, and take advantage of shade during hotter hours. In addition, early morning grazing is likely to carry dew, and this makes grazing easier with less need for trekking to watering points. Kraaling, however, is essential for avoiding stock losses from predation and theft.

Cattle have an inefficient feeding style when feed is sparse. Compare Maasai zebu cattle with the wildebeest (*Connochaetes taurinus*). They both occupy similar rangeland in parts of East Africa. They have a similar mature size (180 kg) and their normal daily walking distance whilst grazing is about the same (16 km) (Homewood *et al.*, 1987). Wildebeest, however, are better adapted to these conditions. They range further whilst looking for feed, whereas cattle lose weight when they walk more than about 16 km a day. The difference lies in their walking styles. Wildebeest trot between grazing patches, and this is more efficient than the steady pace used by cattle. The Australian Merino also trots between grazing patches during the dry season.

Walking capacity varies between types of cattle, and it is determined partly by their size. This is well recognized by rangeland pastoralists, and, where labour permits, the cattle are split into different mobs according to walking capacity as well as regional conditions. Matching the animals to the terrain is an acquired skill, and some local considerations that influence decision-making are as follows:

- The availability of water relative to feed. In the wet season, there are pools of surface water, and walking distance between water and feed is relatively short, whereas, in the dry season, water is often limited to occasional boreholes where the water has to be raised by hand or draught power.
- Lactating zebu cows that are allowed water once every 3 days are at their tolerance limit under most hot arid conditions. The lactating

animals would normally be kept nearest the water source.

- Young calves and sick animals cannot walk far, and may need to be kept separately, either in kraals or tethered near the settlement.
- Distances that need to be covered to reach feed may increase when pasture has been burnt at the end of the dry season.

The risk of underfeeding is also influenced by longer-term management strategies. For example, in Afghanistan, nomadic sheep flocks that move within lowland regions and sedentary flocks that remain in the lowlands are more prone to winter live-weight loss if the lowland regions are heavily grazed in summer, whereas nomadic flocks that spend the summer in the hills and return to less heavily stocked lowland for the winter show smaller wintertime weight loss (McArthur, 1980).

Droughts

During severe droughts cattle have to be slaughtered or moved to regions where there is a better chance of survival. As the drought sets in, livestock prices plummet as there is an oversupply of stock sold through 'distress sales'. A farmer has to sell more animals to buy the same amount of maize and millet to supply the household. The problem is exacerbated if the price of grain is also inflated because of simultaneous poor crop yields. This combination is particularly difficult for poorer farmers, and it is their herds that often experience high mortality rates during a drought (Roderick, 1995).

The impacts of a drought are progressive. Old animals and young males are usually moved out or slaughtered first, and, as the herd gets smaller, non-lactating females have to be used for draught work in place of males. This creates problems with body condition and overall fertility of the nucleus breeding herd (Chimonyo *et al.*, 2000). Animals die slowly, with the weaker ones succumbing first, and **catastrophic mortalities** develop if stock owners are unable to secure sufficient feed. In the case of the 1968 to 1973 Sahel droughts, 85% of cattle in the region are thought to have died. The welfare and production impacts were not solely connected with the deaths. The remaining stock had depressed growth, small mature weights and poor reproductive performance, all of which contributed

to reduced long-term output. Weakened stock are also prone to accidents because they take less care in looking after themselves (Roderick, 1995).

The impacts of the 1984/85 and 1998 droughts in southern Ethiopia were similar, and they were monitored by Donaldson (1986) and Desta and Oba (2004). In the 1984/85 drought overall mortality rate in cattle was about 30%. Many non-lactating animals were trekked hundreds of kilometres northwards into the highlands and south into Kenya to escape the drought. The milking herds that stayed behind experienced heavy losses. The average calving interval rose by over 20%, and calving rate decreased from 75% to less than 10%. Milk offtake fell by only 35% because of strong demand for human consumption, and cow condition fell to the lowest measurable score in all the mature cows. The consumption of 'bush foods' and cattle blood by rural people rose. The cattle were bled monthly, with about 2.5 l being removed at each bleeding, and the blood was either eaten fresh or cooked with milk. Cattle are the preferred source of blood, but small stock are also bled, often from a cut above the eye. Offtake from small ruminants ranges from about 1 l per month to less than 0.25 l per month.

The lessons from the 1998 drought were much the same. Cattle mortality rate rose to 26%, and 83% of the deaths were from feed scarcity. Cattle offtake rate during the drought was low (8%), and so the recent experience of the 1980s droughts did not seem to encourage early destocking. Cows were worst affected and experienced the greatest mortality. There was a dependency on lactating cows for human survival, and some cows died as a result.

When a drought sets in, newborn animals are often killed at birth to allow a greater offtake of milk for the household. If instead, the young are retained and weaned early, or if they are weaned into the feed shortage, their chances of survival may not be good.

Emaciated animals that survive the drought can experience heavy losses if the following winter is harsh. Thin cows require 6 to 10% more energy to maintain their live-weight through the winter in a cold environment, as compared with cows with a moderate to high body condition score.

Seasonal and drought **transhumance** usually takes one of two forms. Stock are moved to higher ground where it is cooler, or to wetter regions where there are more feed and water. This could be part of a regular altitudinal transhumance. Alternatively, it may be a semi-nomadic search for feed and water. Transhumance is becoming more difficult in some countries where disease control and landownership boundaries are being introduced.

When there is a succession of droughts, livestock farming can dwindle to almost nothing. Some pastoralists have moved away and become hired herders in other regions. There have been reports of poor work commitment amongst some hired herders, and the standard of care in looking after the animals has been criticized. Some situations have been rescued by villagers, who removed stock from the hired pastoralists' care and set themselves up as herders instead (Blench and Marriage, 1999). In other situations the dispossessed farmers have avoided the indignity of the loss of stock by committing suicide.

The problem facing many government agricultural departments is knowing how to maintain or increase productivity on arid rangeland without degrading the land and water resources, and whilst minimizing animal losses when a drought occurs. The approach used in West Asia and North Africa is to rely on supplemented feed. This is costly and it can degrade land if it delays destocking. The alternative is to invest in infrastructure that enables the offtake and sale of animals for slaughter. Some of the money could go towards restocking when conditions improve.

Water

During droughts, livestock generally die of starvation rather than dehydration. However, water quality inevitably deteriorates. Surface water becomes brackish as its total dissolved solids (TDS) content rises, and bore water that is drawn from deeper aquifers is often more saline. Poor-quality water can depress water intake. People prefer not to drink water with more than 2500 p.p.m. TDS. When the TDS exceeds 3000 p.p.m., growing stock may start to scour and growth can be checked, but this depends on the composition of the salts, as in some cases stock have withstood drinking 15,000 p.p.m. TDS (Whitmore, 2000). When animals become progressively weaker as a drought worsens, their tolerance of saline drinking water declines.

The length of time that an animal can forgo water determines its suitability for living in a semi-arid environment. In parts of Africa where there have been repeated droughts, some farmers have been changing from cattle to keeping sheep and goats. This is because sheep and goats cope with a lower **drinking frequency**. Goats also show better recovery following a drought.

During the dry season, cattle require frequent visits to a water supply (Table 2.2). They need to drink at least once every 3 days. Goats can withstand longer periods without water, but they require more frequent visits than camels, even though they both store water in their forestomachs. Typically, a herd of goats will walk 8 to 15 km per day during the dry season whilst browsing and grazing, and will lose up to 25 g live-weight per animal per day. The herdsman aims to take his goats to a well or waterhole sufficiently frequently to avoid suppression of feed intake. If the interval between watering days is too long, the goats lose their drive to feed during the non-watering days, and weight loss accelerates. In this way he learns how hard he can push his animals.

Limiting drinking to once every 3 days may seem harsh by standards accepted in industrialized countries. However, it is an integral part of the success of pastoral rangeland systems in some semi-arid countries. On a once-in-3-day drinking frequency, cattle can graze up to 21 km away from the water supply, assuming a maximum daily walking distance of 14 km. This gives overall access to 1385 km^2 of rangeland, which is nine times the area available to cattle that go to a water supply every day. Reduced drinking frequency is associated with lower overall water consumption, by as much as 30%, and this can allow greater sharing of water with a larger number of stock and wildlife.

The once-in-3-day drinking frequency requirement should not be interpreted too rigidly. A study with Boran cattle in East Africa showed that the

Table 2.2. Maximum interval between visits to a water supply in a semi-arid tropical environment during the dry season.

Species	Maximum interval for water restriction (days)
Cattle	3
Sheep	5
Goats	5
Camels	10

lower water intake in a 3-day system suppressed feed intake, resulting in greater weight loss in the cows (Table 2.3; Nicholson, 1987). Calf growth was also lower, and 210-day weaning weight was reduced by 9 kg when cattle drank water every 2 days and by 14 kg when they drank every 3 days.

In some regions access to water is becoming difficult where property owners have been assuming ownership of water rights either through enclosure awards or from constructing wells or water tanks. Until recently, every household had the right to draw water for animals and domestic use. Now they have to pay for it either with cash or in exchange for livestock. Water debts are often accumulated in the dry season, and then paid off when cash becomes available from the sale of milk or stock.

Overstocking

There is growing concern that overstocking is resulting in desertification of large areas of rangeland. It is said that some of the deserts of West Asia and the northern coastline of Africa arose from over-exploitation, but it is not clear whether livestock were the main contributor to the erosion. The strongest evidence that lays blame on livestock comes from South Africa and Australia, but it is circumstantial. The livestock **carrying capacity** in parts of South Africa and Australia is now lower than it was in the latter part of the 19th and the early part of the 20th centuries (Dean and Macdonald, 1994). The rangeland has lost some of its diversity and the abundance of vegetation has declined. Loss of plant cover has increased the risk of erosion. The risk to animal welfare is increased

through poor feeding, depending on how stocking density is being managed. The implication is that this situation arose from overstocking.

Graziers are reminded about the link between grazing pressure and desertification when they see loss of vegetation around ponds and wells or near settlements. The usual outcome is soil loss from wind erosion, leaving bare stony ground. Localized land damage also occurs where supplementary feed is put out for stock (Mack, 1996).

One way of assessing whether an area is overstocked is to estimate the consumable feed that can be produced in a region, and from that to derive the sustainable stocking rate. The sustainable stocking rate is then compared with the observed stocking rate. However, it can be difficult to arrive at an accurate estimate of carrying capacity, and, when the coefficient of variation for annual rainfall exceeds about 30%, it is unrealistic to even attempt an estimate. Nevertheless, the exercise has been helpful in pointing out high-risk situations where the observed stocking rate is well above carrying capacity (de Leeuw and Rey, 1995).

The estimated carrying capacities for different agro-ecological zones in Africa are shown in Table 2.4. Regions at greatest risk of unsustainable stocking rates are the heavily populated parts of the semi-arid zone of northern Nigeria and the highlands of Ethiopia and Kenya. Stocking rates in these regions are about 100 tropical livestock units (TLU) per km^2 and far exceed the optimum stocking densities. The threat of over-exploitation may go beyond these three regions, as about 49% of the world's arid plus semi-arid zones are said to be overpopulated.

Several other approaches have been used when determining imbalances between sustainable

Table 2.3. Effect of drinking frequency on water consumption, feed intake and live-weight loss in Boran cattle herded in pastoral rangeland of southern Ethiopia and northern Kenya.

Drinking water frequency (once every)	Water consumption (ml/kg/day ± SD)	Dry matter intake (g/kg$^{0.75}$/day)	Live-weight loss during the dry season (kg)
1 day – Lactating cows	79.4 ± 17.6	68.0	83.4
– Dry cows	57.3 ± 11.3		
2 days – Lactating cows	75.5 ± 18.1	65.6	68.4
– Dry cows	54.1 ± 9.2		
3 days – Lactating cows	52.2 ± 14.7	61.8	111.4
– Dry cows	44.5 ± 12.5		
		± 1.58 SE	

Table 2.4. Consumable feed production and sustainable stocking rates in different agro-ecological zones of Africa.

Zone	Consumable feed (t dry matter per ha)	Estimated sustainable stocking rate (TLU per km²)
Arid	0.19	8
Semi-arid	0.51	22
Subhumid	0.72	31
Highland	0.76	33

and existing stocking rates, and the conclusions have often been the same (Kreuter and Workman, 1994). In principle, stocking pressure is a better measure than stocking rate. Stocking pressure is the measure of livestock density in an area during the period of peak use. It relates more closely to the risk of plant damage and erosion.

Some pasture species are sensitive to high stocking pressures whilst others withstand high pressures but rely on rest periods between successive grazings to regenerate. *Phalaris* persists well under dry conditions whilst subjected to periodic high pressures, but needs adequate rest periods to regenerate. Tussock-dominated rangeland is sensitive to high stocking pressure on localized patches. Patch grazing in tussock rangeland often occurs in the more fertile places and near water-points, and once they start to erode there can be rapid decline in carrying capacity, animal performance and animal condition (Ash *et al.*, 2004). In humid tropical regions, if communal grazing land is overgrazed because of demands for cropping, there may be a shift in sward composition towards less productive unpalatable species. Correcting such changes calls for a consensus and commitment from the whole community, and this has not always been easy because of the split in interest between keeping livestock and growing crops.

In Africa, the traditional approach to managing overstocking has been through consensus within the community. One way has been to limit the number of livestock that owners can put on to communal land. An alternative has been to set grazing aside for dry periods. The right to graze a region has usually been based on heritage, and setting pasture aside as standing hay has been determined either by the local leader or by a grazing committee.

The risk of welfare problems from overstocking increases during harsh winters and sometimes it is worse in larger flocks or herds. This was demonstrated in a study in Lesotho on the effects of stocking density on sheep production (Mpiti-Shakhane *et al.*, 2002). Lamb mortality during cold nights was high, particularly in heavily stocked rangeland (Table 2.5). In Senegal, the effects of overstocking were more obvious in larger herds (> 100 cattle) because of the way they were managed. Cattle in larger herds had to walk further to obtain feed and water, they were kept on poorer-quality range for longer periods, they were taken to arable land to feed on crop residues later in the year, and they had poorer overall body condition (Ezanno *et al.*, 2003).

There are several other factors arising from local conditions that contribute to excessive stocking pressure. In the Ethiopian highlands, high stock numbers are in part due to reliance on cattle and equines for tillage and transport. Overstocking is also becoming common amongst landless farmers (de Haan, 1995). Some herdsmen deliberately overstock pastures that are prone to ingress of woodland and thicket, and as a way of controlling potential tsetse habitat.

Milk offtake

In communal farming systems the lactating herd is usually grazed close to the settlement and the cows provide milk for both human and calf consumption. The calf is presented to the cow to stimulate milk let-down, but its intake is controlled by allowing it to suck on only one teat or by limiting its intake to what is left after hand-milking. For the rest of the day the calf and cow are kept separate to prevent suckling.

There is inevitable competition between the household and the calf for milk. In Kenya, about 40% of the first 90 days' lactation is used for

Table 2.5.   Effect of rangeland stocking density on growth and mortality in lambs and ewe live-weight.

TLU/ha/year	Weaning weight @ 12 weeks age (kg/ha ± SE)	Mortality (% birth to weaning)	Final ewe live-weight (kg ± SE)
3.5	20.2[a] ± 0.004*	40	40.0[a] ± 4.11
5.6	41.4[b] ± 0.002	58	39.0[a] ± 3.75
14.0	69.4[c] ± 0.01	64	36.0[b] ± 4.10

TLU, tropical livestock units. Means with a different superscript letter were significantly different at $p = 0.05$.

human consumption and the remainder is taken by the calf (Roderick *et al.*, 1999). At this time the cow may be producing 2.7 l/day. In Ethiopia colostrum is also taken for human consumption, and this might deprive the calf of immune protection. The calves are usually tethered to a ground stake inside the family hut, and taken to the cows for feeding. When the calves are 1 month old, they are allowed out of the hut to graze unattended for 4 to 5 h a day, and, at 3 months of age, water is offered for the first time (Donaldson, 1986). If milk intake is lower than usual, water may be offered at an earlier age. Calf growth would be greater if it were penned with the cow at night during the nursing period. Cows are more inclined to allow suckling at night, whereas during the day they are more likely to ignore the calf by continuing to walk and graze during suckling attempts.

Goats are often regarded as the poor man's cow. They are a more reliable source of milk than cattle under harsh conditions, and they are more likely to be over-milked during droughts. For example, during the 1984/85 drought in Ethiopia, average monthly milk offtake per goat held constant throughout the drought, whereas milk offtake per cow fell. In Mexico there are problems with extending lactation for too long. If does do not have time to restore body condition before the next mating season, they may abort during the subsequent pregnancy.

In cows that are also used for draught work, underfeeding has a greater impact on milk production than on capacity to perform the draught work. However, as draught work increases, body weight loss becomes greater and this can delay conception by lengthening the anoestrous period.

Managing animal behaviour

The inappropriate methods that are sometimes used in managing livestock behaviour can be divided into three categories. They are animal abuse, inappropriate restraint and confinement and inappropriate routine husbandry procedures.

Animal abuse has evolved as a normal part of life in some communities. Three factors contribute to this: aggression, assertion and intimidation. First, some cultures have developed an aggressive outlook to people, animals and society in general. For example, some nomadic and itinerant people have little need to cultivate harmonious relationships outside their own family groups. They have the option of moving on. Secondly, lifestyles that require an assertive and decisive personality can be linked to callousness. Some people living under continuous threats from weather, predators and animals have acquired robust and aggressive personalities. Thirdly, and perhaps most importantly, there is the need to control livestock, and abusive methods are sometimes the accepted approach. With wilful animals, control can be exerted through domination. For example, the threatening bull and the animal that persistently attempts to raid unfenced crops are controlled by intimidation.

Assertion of dominance and intimidation are reinforced by beating the culprit or victim. For example, Fulani people in Africa are encouraged from childhood to confront threats or wilful behaviour in livestock by hitting them with a stick (Lott and Hart, 1977). Failure to strike an animal is seen as cowardice and unwarranted beatings are generally ignored by colleagues in the community. Fulani culture favours aggressive domination of livestock, and the people believe that they are courageous in being aggressive.

In other cultures there are paradoxes between what people believe and what they practise. For example, in India, cattle are deified and abused by the same people. Lack of cooperation by the animal is a common cause of abuse – for example, when animals refuse to move in the required direction (Rahman, 2004). The sanctity of cattle is partly symbolic, in that it does not apply in all circumstances, or it may only apply to breeding cows.

Some forms of mistreatment have developed out of perceived necessity. This is the case with animals trucked to abattoirs on the backs of open-top vehicles. To prevent them escaping, they are tied down on their sides to the floor. Unloading is an awkward and hazardous procedure (see Fig. 10.3).

In densely populated regions, small ruminants and poultry are often kept in slatted-floor hutches or pens. This occurs in countries between Indonesia in the east and Senegal in the west. **Confinement** has the following advantages for the stock owner:

- making efficient use of feed in cut-and-carry feeding systems in suburban districts
- avoiding damage to unprotected crops
- manure collection in a pen or hutch, or deposition on an unfenced field
- utilizing unfenced common land
- protection against road traffic accidents
- protection of stock from night-time predators
- protection against theft
- reduces the need for constant supervision of stock
- confinement can allow protection from fly attack during the wet season, especially when combined with wood smoke

However, tethering and confinement in pens or hutches introduces inevitable welfare hazards. The following are some examples:

- raised wooden hutches used for goats and sheep are prone to develop damaged floors, which can lead to leg injuries
- foot problems develop in stock made to stand in their own manure for long periods
- tethering young or naive stock requires close supervision to begin with, until the animal learns the limitations of its own movement. Adequate supervision depends on time availability of the owner's family and this will vary
- risks of rubbing and pressure sores
- infrequent watering where tethered stock have to be taken to water
- inadequate protection from sun in stock tethered whilst feeding on crop residues in arable fields
- erratic feeding, e.g. with cut-and-carry during monsoon season; insufficient labour to move tethered animals regularly to a fresh patch of feed

- reduced grazing time during periods of poor feed availability, when confined in night-time kraals.

Small ruminants confined within compounds are sometimes tethered to control fighting and bullying, and to reduce the risk of escape. They often experience lower early mortality than transhumant flocks because in sedentary flocks newborn lambs and kids are not forced to walk long distances. Transhumant sheep flocks are often composed of the more prolific breeds to compensate for their higher mortality.

Cattle are more likely to be tethered than confined. Exceptions are the night-time kraals used in rangeland pastoralism, and Muturu cattle in parts of Côte d'Ivoire, which are kept permanently confined in sheds whilst stall-fed.

The routine **husbandry procedures** used in developing countries that have unfortunate welfare consequences and are described elsewhere in this book include:

- hot-iron branding as a disease treatment method
- tethering and confining methods
- castration methods
- fostering methods
- drying-off methods

In many developing countries there is little advantage in castrating male cattle. Castration is only performed if the animals become troublesome, and this may not emerge until they are 3 to 4 years old. When surgical castration methods are used, they are done without anaesthetic but in the cooler seasons when there are fewer flies.

Some animals are made to walk long distances to markets. In the past, trekking between markets was common in regions served by numerous small 'bush' markets, rather than a centralized market. Dealers and vendors walked the stock between markets searching for a better price, and in the process stock lost condition.

Disease and disease treatment

A simple way of assessing the importance of animal disease in a community is to examine mortality figures. Unfortunately there are very few estimates of livestock mortality rates from disease in developing countries, but two reports are worth noting.

In a study of traditional cattle herds in Ethiopia, mortality from disease to 24 months of age was 19%, and, in Cameroon, goat mortality from disease to 5 months of age was 22% (Donaldson, 1986; Manjeli et al., 1996).

Many livestock diseases in developing countries have a seasonal incidence, and a common policy is to cull the worst-affected animals and to wait for the change in season to allow less severe cases to improve. This is often the way with dermatophilosis. In tropical regions of Nigeria, Bunaji cattle are particularly prone to dermatophilosis. It flares up during the wet season and is one of the main reasons for culling to improve productivity (Table 2.6). Conjunctivitis, pneumonia, gastro-enteritis and diarrhoea are also encountered mainly in the rainy reason when conditions favour transmission and survival of the primary pathogens.

Some diseases and disorders emerge during the dry season when stock are likely to be weakest. For example, phosphorus deficiency expresses itself towards the end of the dry season when the combination of lactation and low phosphorus intake take effect. Leptospirosis can erupt when waterholes dry up and livestock and wildlife concentrate at the remaining water sources (Horn et al., 2003). Internal parasites often become a problem when a drought breaks, whereas ectoparasites can be a problem whilst the drought lasts. Understanding and evading these risks are a more important part of stock management than in intensive livestock farming, because there is less reliance on veterinary remedies once a disease or infestation gets a grip.

It is said that 80% of the world's cattle population is at risk from **ticks**, and in Senegal

Table 2.6. Culling, death and morbidity rates in 16 traditionally managed cattle herds in northern Nigeria, 1993–95 (after Kudi et al., 1998).

Causes of culling	Number culled	%
Dermatophilosis	120	65
Old age	30	16
Other diseases	22	12
Malnutrition	10	5
Reproductive failure	3	2

Causes of death	Number of deaths	Number diagnosed
Pneumonia	10	526
Poisoning	9	48
CBPP	6	54
Dermatophilosis	5	8
Trauma	4	112
Gastro-enteritis	3	104
Bloat	3	90
Foot-and-mouth disease	3	21
Diarrhoea	2	416
Septicaemia	2	15
Foot rot	1	74
Coccidiosis	1	8
Conjunctivitis	0	323
Mastitis	0	53
Retained placenta	0	30
Snakebite	0	19
Metritis	0	9

CBPP, contagious bovine pleuropneumonia.

and Tanzania ticks are recognized as the main constraint in cattle distribution during the rainy season. The level of control used against ticks varies considerably. Cattle can be treated with insecticide but more commonly the ticks are removed by hand on a daily basis or when they are noticed. The harvested ticks may be fed to poultry. Tick management is often neglected in goats, and, in parts of Nigeria, some herdsmen believe that ticks are harmless and so they are not removed even from cattle. Besides transmitting disease, ticks cause ill thrift or 'tick worry'. This is due to loss of blood, pain from the bite and subsequent inflammation, secondary bacterial infection and fly strike. Animals lick the affected part if they can reach it, and this can lead to raw areas of skin (Hazari et al., 1991). The development of skin papules and re-infestation with fresh larvae are particularly irritating and the cattle show restlessness and reduced feeding as part of their general discomfort. Ticks are the single most important cause of hide damage in Ethiopia and they are often concentrated at the sensitive areas in the groin and scrotum (Mersie and Bekele, 1994). If the scrotal skin becomes thickened in response to the ticks, thermoregulation can be compromised and bull fertility reduced (Mokantla et al., 2004). Production losses from tick worry depend on the animals' feeding level (Pegram et al., 1993). In well-fed animals, ticks inhibit feed intake and suppress growth, whereas in underfed stock the suppression of appetite is less pronounced and further growth suppression may not be obvious. Once the cattle learn where high numbers of larvae exist within an area, they avoid those regions if given a choice.

In West Africa, calf deaths in large herds are more likely to be due to disease, whereas underfeeding is a more important cause in smaller herds (Ezanno et al., 2005). In the large herds, most of the milk goes to the calf, and so underfeeding is often the lesser risk. In small herds, milk has to support household consumption as well as the calf, and the condition of the cow plus her feeding has a greater influence on calf health and survival.

Regular stock inspection is an important part of disease management. Many of the traditional ways of herding stock encourage inspection and allow early recognition of a disorder, whereas this may not be the case in some of the easy-care systems used in developed countries. For example, in Texas, 32% of cattle graziers inspect their stock fewer than four times a week, and half of these saw the stock only once a week (Godinez et al., 1993). The opportunity they have for recognizing and reacting to an animal disease and disorder is inevitably less.

When a particular disease is prevalent and it is decided to introduce a vaccination programme, experience has shown that it is helpful to maintain the programme well beyond the main risk period of the disease. Short-term **vaccination programmes** only bring short-term benefits. For example, during the rinderpest vaccination campaign in sub-Saharan Africa during the 1960s, large numbers of young stock were vaccinated, but the aid was not maintained, and so during 1984 and 1985 a rinderpest outbreak killed about one-third of the stock in West Africa (FAO, 2001). Another lesson from successful vaccination programmes has been that they allowed pastoralists and agro-pastoralists to increase their herd sizes. In some cases this resulted in increased stocking pressure, and it encouraged larger herds of poorly fed animals harbouring subclinical pathogens.

Most animal diseases that occur in African livestock are treatable, but not all animals are treated. Pastoralists still rely on traditional methods for controlling and evading livestock diseases. This includes ethno-veterinary treatment methods against ticks and worms, transhumance in the face of a rinderpest epizootic, and avoiding pastures where there is a seasonal risk of trypanosomiasis. Rinderpest and contagious bovine pleuropneumonia (CBPP) strike fear amongst pastoralists. When an outbreak occurs, they sometimes move their stock out, and this increases the risk of spread of the disease. Both diseases can take on plague proportions, and they cause serious morbidity with fever and inappetence. These diseases are feared in much the same way as droughts, because of the losses they cause.

Traditional health-care methods are used where there is no veterinary service. In some cases there is a sound basis for the methods that are used (Schillhorn van Veen, 1997). For example, in West Africa, herders expose their cattle to trypanosomiasis by 'seasoning' them in regions where the disease is enzootic. The aim is to achieve natural immunization, which allows the cattle to survive under higher tsetse/trypanosome pressure on the fringes between savannah and tsetse-infested forest. This carries inevitable risks, but, since it is practised in stock that have some genetic resistance to the disease, the risk usually pays off. Similar seasoning is used in camels against surra (trypanosomiasis), and, in Sudan, surviving camels have a higher market price as *afiq* ('salted') animals.

In some communities treating an animal infected with worms or a disease does not enter the owner's thinking. Education and information transfer would change this. In other cases, traditional medicines are preferred because of their familiarity and cheapness. Some traditional health-care methods defy conventional logic. For example, in parts of Africa, diarrhoea in cattle is treated by hot-iron branding the perineal region. Similar 'therapies' are used for treating skin disorders and hernias in camels, as described in a later chapter.

Sheep and goat flocks do not usually receive regular veterinary care. Instead, most animals either die or recover unassisted. Owners frequently try to sell sick animals before death occurs.

Intensive forms of subsistence farming in tropical countries

This section considers how intensification in developing countries is creating some cow and calf welfare issues.

In some countries, traditional free-grazing systems are giving way to more intensive forms of management. Smallholdings near towns are being subdivided through inheritance. As the holdings get smaller, farming methods are becoming more intensive to maintain satisfactory levels of production and income. In addition, land around towns and cities is needed for cropping. Where fencing is rudimentary, cropping is incompatible with free-grazing. The animals have to be tethered on patches where they can graze, or held in pens where they are given fresh fodder that has been cut and carried.

An example of a system that has become common in Nigeria is as follows. During the wet season, cows and followers are tethered on fallow or uncultivated pasture with a rope around the horns, neck or leg. They are taken out in the morning and tethered around 7 a.m. until midday, when they are watered and rested under shade. They are tethered again from 3 p.m. until 5 or 6 p.m. Although unsupervised during the day, the cattle may need to be moved to fresh pasture at least once during the morning and afternoon, especially if they are lactating. At night, the herd is brought back to the owner's compound and tethered for security. A newly calved cow would be

allowed to graze more freely close to the house provided there was no risk of crop damage. Later, young calves would be tethered in the compound during the day while the cows graze, and the calves are given access to the cows at night. This is an intensive system in terms of labour demands, and it depends on the diligence of the people controlling the stock.

In **cut-and-carry** systems, the livestock are fed crop residues, napier fodder and other fodders purchased from neighbours. The cut feed is taken to the animals, which are either held in pens or tethered at a convenient site. Watering is by leading the cattle to a water source or carrying water to the stock. Again this is labour-intensive, and on some days during the wet season the stock may receive limited feed.

Cut-and-carry systems help to relieve some of the impacts of heavy stocking pressure. They are used in heavily populated regions and where forage is dense enough to be cut. For example, such a system is common in highland regions of Kenya and Indonesia. As population increases in the future there will be greater reliance on cut-and-carry systems. This will direct the animal welfare focus towards confinement and tethering systems, lack of shade, insufficient watering, erratic feeding and skin sores from the tethering equipment.

A welfare advantage of cut-and-carry is that in humid regions there is less risk of tick infestation compared with grazing. Linked to this, there is less risk of tick-borne diseases such as East Coast fever and anaplasmosis. Some other welfare disadvantages are:

- greater risk of lameness in penned animals standing continuously in slurry, and from claw overgrowth
- greater risk of inbreeding, which can be expressed as reduced viability, ill thrift and poorer fertility in the offspring

In traditional free-grazing systems, mating is often opportunistic. Where confinement has been introduced, there has been poorer management of breeding, including inbreeding.

Besides inbreeding, there is often insufficient focus on breeding management in small herds. Sometimes this is intentional, as the aim is for an extended lactation, and so service may be delayed for as long as possible. In other situations, there are few bulls in the neighbourhood. The loan of a bull depends on maintaining a harmonious relationship with the bull's owner, and, where neighbourhood disputes exist, future production may be jeopardized. For the future, the main problems that will face intensive small-scale tropical cattle farming will be greater subdivision of land and the supply of replacement stock.

With cut-and-carry and tethering systems there is a greater dependence on physically managing feed, water and the animals. In the past, this type of work has often been shared amongst the children in the family. Children are now expected to attend school, and this is putting pressure on adults, who have other responsibilities or distractions. As a result, the level of care in watering, feeding and avoiding heat stress is not always ideal.

Managing the Problems

Some solutions for the animal welfare problems in developing countries lie in the following approaches:

- encouraging alternative forms of security besides livestock
- predicting droughts reliably, and distributing early drought warnings throughout the affected region
- providing infrastructure for capitalizing on the value of livestock when destocking (e.g. transport, processing facilities, meat preservation and storage facilities, meat export when in excess)
- improving awareness on how to maintain wealth whilst destocking and restocking
- better integration of crop and livestock farming
- better integration of pastoralism with tourism
- improving understanding about disease management

Managing seasonal underfeeding and overstocking

In recent years there has been a change in thinking about how overstocking issues should be managed. There is a move away from concern about maintaining rangeland plant composition towards how to use the rangeland to support a community. It is argued that setting the stocking rate at a level which avoids periodic overstocking would make livestock production uneconomic (Behnke and Kerven, 1995). Livestock population crashes are almost inevitable where rainfall is erratic. Instead, the shortfalls created by environmental fluctuations have to be buffered whilst minimizing the animal welfare insult. The traditional ways of achieving this are:

- transhumance or agistment
- sale of non-breeding stock at the start of a feed shortfall
- providing alternative supplementary feeds during shortfall periods

At the strategic level, the focus needs to be on providing infrastructure that facilitates these three responses. Depending on individual circumstances, this may take the form of improved transport infrastructure, better access to large livestock markets and consumer markets for meat, and uptake of methods for preserving meat. This is a strongly reactive and facilitating management approach, where the decisions depend on circumstances, and the community is not forced into a particular decision.

Personal needs within the community must fit into the overall strategy. Many livestock owners are obliged to maintain stock numbers for as long as possible. Sometimes they are a person's lifetime savings, and, in the absence of a suitable alternative reserve or investment opportunity, the owner has to hold on to his stock whilst keeping them alive for as long as possible. A useful long-term goal might be to encourage a trustworthy way of transferring those savings into another form, as this would allow slaughtering of unsustainable livestock numbers during times of feed unavailability.

Deciding when to reduce stock numbers would be determined by the economic position of the farmer, his attitude towards risk, the resilience of the vegetation to short-term overgrazing, and variability of the weather in the region. Labour availability may also be a consideration. Labour in pastoral systems allows better distribution of grazing pressure, which helps limit localized overgrazing, and it can postpone the decision to send some of the animals for slaughter.

For many pastoralists, **destocking** is a denigrating experience, but this outlook may change in the future, albeit slowly. There are two influences that are beginning to operate. First, most of the demand for meat comes from towns and cities.

Increasing urbanization will raise that demand, and livestock buyers who supply this trade are looking for good value in individual animals. In other words, they want healthy animals that carry a good yield of meat, and they should be prepared to pay a premium for higher-quality animals. This favours the offtake of well-grown animals and it should modify the pastoralists' priority away from large numbers towards individual fitness. Secondly, pastoralists will probably become more integrated with a monetary market, but in more remote regions that change will take time. It will depend on respect for cash, which will only exist where there is low inflation and a stable economy. The hope, for the environment and for the impact it has on livestock welfare, is that money will replace livestock as the preferred means of investing and storing wealth. This could lead to better management of livestock numbers, but whether this happens will also depend on what can be done with the money through alternative investment opportunities.

There are several other traditional values besides self-image, traction, manure, bride price and security that promote overstocking. In some households cattle are revered because they are associated with family spirit mediums, and this can act as a barrier to selling. In addition, in some parts cattle owned by women are less likely to be sold because they are the only property of economic importance which women own. They are more likely to hold on to them.

Pasture improvement could help offset feed shortages in some situations, but the incentive may be poor if it is communal land. Pasture burning has been the traditional method for regenerating rangeland pasture. The aim in burning is to remove excess dead plant matter and stimulate fresh green grazing ahead of the rainy season. Sward improvement through introduced species has benefited many places, but a common problem has been in maintaining persistence of the preferred species.

Choice of species and breed

In some drought-stricken regions of Africa, farmers have been changing the type of animals they have been keeping. They have been switching to cattle breeds that can survive on browse or from cattle to goats and sheep. Cattle breeds that can tolerate browse (e.g. Sokoto Gudali) are more drought-tolerant than breeds that rely on grass (e.g. Bunaji). Goats and sheep are favoured for the following reasons:

- they are better suited than cattle to common land that has been degraded by overstocking
- greater tolerance of water deprivation (see Table 2.2)
- goats can survive on browse, which can persist for longer during a drought
- pastoralists are more inclined to milk goats harder than cattle during a drought, for household consumption
- greater flexibility in selling off sheep and goats during a drought
- following the drought, goat and sheep numbers can be built up faster than with cattle because they reproduce at an earlier age

Although cattle are usually preferred for prestige, there are other species that have similar importance in some communities. Keeping camels has comparable social value in parts of India, Pakistan and the Middle East, and in the Asian Pacific owning pigs has an equivalent status symbol.

Goats are often preferred where animals are raised for meat and to supplement family income, rather than as an investment. Goat meat is acceptable to all castes and religions. As a result the offtake rates are higher than for most other ruminants. For example in India, over 40% of the goat population is slaughtered every year.

In Islamic communities sheep are favoured for slaughter during religious festivals, and goats are sometimes favoured for family celebrations. Sheep generally attract higher prices than goats, and they are trekked long distances to religious centres and urban areas during Eid-ul-Azha and other festivals. Rams are preferred and this reduces the need to castrate male lambs. Cattle are the favoured sacrificial animal in some other religions and cultures. For example, in Côte d'Ivoire, cattle are slaughtered for funeral ceremonies. The body of the deceased person is wrapped in the hide, and the meat consumed.

Labour availability sometimes determines which species a farmer keeps. Traditionally, many of the routine handling procedures such as herding, tethering, watering and providing cut-and-carry feed have been done by children. Where schooling has taken that labour source away, some livestock keepers have switched to pigs because of their low labour requirement. Pigs, like cattle, can be destructive on crops and so they are usually confined.

Indigenous poultry breeds are preferred in non-intensive systems because they depend less on supplementary feed. They are also more broody and aggressive and show stronger parental care, which can be useful in rearing a clutch of replacements, and they can show some resistance to heat, protozoan diseases and ectoparasites (Khan, 1996). They are not always suitable for intensive conditions, as they have experienced higher mortality rates.

Ranching as a more intensive alternative to rangeland farming

This section considers the conflicting views about whether semi-arid pastoral land in poorer regions should be farmed as rangeland or ranched. The relative success of these two systems can influence survival of the cattle and so there is an animal welfare dimension.

A common objective in setting up ranches has been to improve commercialization and increase cattle offtake. When unimproved rangeland is converted to ranching, the land is subdivided by fencing and boreholes usually need to be sunk to supplement the water supplies and provide continuous access to water. Transhumance is not practised, and instead feed has to be set aside either as standing or conserved hay, or bought in to supplement what is available during the dry season. Grazing circuits may be established within the bounds of the ranch, for example to make use of crop residues as fodder, but generally the stock would be more free to roam than in rangeland managed by pastoralists. The overstorey is usually cleared or allowed to disappear without regeneration, and exotic forage species such as buffel grass (*Cenchrus ciliaris*) or stylo grass (*Stylosanthes guianensis*)

may be introduced. Pasture regeneration by burning is not generally practised, partly because of the risk to fencing. Breeds of cattle may be introduced to suit more lucrative markets, but they may need to be cross-bred with indigenous stock to ensure survival under the prevailing conditions and diseases. The stock are not usually kraaled at night. There is less need for supervised droving or herding, the calves have unrestricted access to the cows, little if any milk is taken from the cows for human consumption, and predatory wildlife is either excluded or controlled.

There is a fundamental difference in outlook in rangeland and ranch management, which impacts on animal welfare. Faced with a downturn in feed availability, the rancher would usually adjust stock numbers by selling off animals. The risk in terms of animal welfare depends on the way the rancher reacts, including rapidity of response and tolerance of underfeeding. The pastoralists' approach is different. They adjust their tenure system rather than herd size by laying claim to other areas that can sustain their stock. The animal welfare risks here depend on competition with other pastoralists and other circumstances that may be outside the pastoralists' direct control. In overcrowded regions, the pastoralists' approach is becoming less tenable.

The conversion from rangeland to intensive ranching has been successful in North and South America, but less so in Africa. During the 1960s a number of attempts were made in Africa. From that experience it was learnt that subsistence pastoralists often meet their production goals when operating as rangeland farmers, but their goals are quite different from those of commercial ranchers. They are less concerned about producing animals for meat consumption.

The overall levels of production in terms of offtake have often been higher in the ranching

Table 2.7. Comparison of production achieved under traditional communal pastoral system and ranching in Botswana.

	Communal production	Ranching
Calving (%)	46	74
Calf mortality (%)	10	9
Live-weight at 7 months (kg)	123	177
Post-weaning growth to 18 months (kg)	89	106
Weaner calf production (kg/cow/year)	51	120

systems (e.g. Table 2.7). This has been partly due to poorer feed during the dry season in communal rangeland farming systems, resulting in greater weight losses and longer post-partum anoestrus (Rennie *et al.*, 1977). However, ranching is prone to the effects of localized overstocking, and it is not appropriate for all semi-arid regions. For example, when the Ankole region in Uganda was converted to ranch land, obvious signs of overstocking developed. There was soil erosion, pasture composition deteriorated and the cattle failed to fatten when the stocking rate reached 0.6 ha per 300 kg liveweight (Harrington and Pratchett, 1974). Another important failing with attempts at ranching has been lapses in maintaining infrastructure support. Inadequate maintenance of fences, boreholes and wells has led to reversion to rangeland pastoralism.

Fencing is the major cost in establishing a ranch. The advantages and disadvantages of fencing in terms of its impacts are summarized in Table 2.8. Providing supplementary feed for ranched animals encourages them to congregate at the sites where feed is dispensed. This causes localized high stocking pressure and can lead to degradation of that area. In addition, if cattle calve under these conditions there is a higher risk of infectious disease in the calves (May *et al.*, 1999). Cows calving on green pasture are widely dispersed and so the risk is lower.

Clearly, there are advantages and disadvantages to rangeland pastoralism and to ranching, and there is no dogma that applies to all situations.

Table 2.8. Advantages and disadvantages of fencing rangeland.

Advantages
- Delineates landownership
- Controls access to land
- Contains livestock
- Excludes competing wildlife
- Conservation of feed for the dry season

Disadvantages
- Too expensive
- Limits transhumance for neighbours and truncates migratory routes
- Deprives other animals of important resources
- Conducive to localized high stocking pressure by resident stock
- Can entangle or injure stock

Experience has shown that privatized ranching does not fit well socially in Africa, and it has been suggested that instead there should be group ranching that is integrated with a form of stewardship (Behnke, 1995). This could take the form of collective or cooperative ranching, but it will need to fit comfortably with the social structure of the community. The land could be grazed under exclusive tenure and there would be a single decision-making body that would oversee transgressions, grazing crises and other problems. Whether this is successful depends on how fair the stewards are towards the graziers and how sensible they are in overseeing management of the various resources.

Whatever the system, finding the optimum stocking density will be a key issue in terms of sustainability, survival and animal welfare.

Stock management during droughts

When drought and overstocking coincide, ranch and rangeland degradation can set in very quickly. In the Chihuahuan desert, it set in within 2 years (Holechek *et al.*, 2003). The welfare problems that developed depended on how stock owners reacted, once it was recognized that a drought was setting in. They had five choices:

- do nothing and hope for rain
- agist some stock
- provide supplementary feed
- early weaning
- sell or slaughter some of the stock

It is important to establish a plan early on in a drought. The plan must aim at minimizing financial losses, and must try to balance forage demand with forage supply. A plan must be followed throughout the drought period. Otherwise, there is a risk that decision-making will be delayed and it will be too late to salvage the stock before they become emaciated. Most drought plans set dates or criteria at which decisions and actions will be made. For example, the decision to sell 20% of the stock will be taken when a particular area no longer supports any grazing, or 50% of the stock will be moved to another region if rain has not arrived by a certain date.

There has to be a systematic reduction in animal numbers when a catastrophic drought is looming. Non-breeding stock are usually sold first, to maintain the breeding nucleus, and calves may be slaughtered as soon as they are born, to maintain condition in the cows or allow milk to be used for domestic consumption. In Australia, Merino wethers are the first to go, and a drought is often a good time to cull out ewes with less than fine wool. The drought plan needs to define dates at which pasture assessment is made, and, if it falls below the required amount, a set proportion of the animals have to be culled or agisted. In developing countries this may be easier to plan than to execute, because the infrastructure for preserving and transporting a large number of carcasses to consumers is lacking, or because it is difficult to move stock to regions where feed is available. This is where aid could be directed. Local livestock markets are a necessity for destocking programmes in regions where there are a large number of small herds or flocks.

Not all famines are due to droughts. The 2005 crisis in Niger is a good example. This was due to a collapse in grain supply from the south because it was being channelled to other markets and uses, including pig and poultry farming. The shortage of grain passing northwards from Nigeria precipitated a human food shortage in Niger. Livestock prices collapsed because more stock had to be sold to raise sufficient money to buy grain at the inflated price. On this occasion, the animal welfare impacts were less severe than the repercussions on human welfare.

Urban livestock farming

In some countries where meat is considered expensive, low-income town dwellers produce it for themselves. This is done on a backyard basis, and according to local circumstances the animals may be allowed to roam the neighbourhood to search for feed during the day. They are not always slaughtered in official abattoirs, but instead in back-street sites or where they are kept at night. If, say, someone wishes to buy a pig for eating, he or she would book the meat with a trader, who would then source a suitable animal within the community. The trader would arrange the purchase, slaughter and dressing of the carcass.

The main species kept in urban situations for producing meat are:

- chickens
- pigs
- pigeons
- rabbits
- guinea pigs
- goats
- turkeys
- tortoises
- snails

Chickens are the most common urban livestock species. They are usually kept as free-ranging birds and their feed is supplemented with occasional supplies of grain. Goats and sheep are also kept as scavenging animals. They pick up some of their feed from waste ground, rubbish tips and the streets. In some countries, pigs also roam free in towns or villages, but as with goats they can be quite destructive feeders and in other places they have to be permanently penned.

Besides being a source of meat, these animals provide other benefits including:

- supplementary income
- milk, eggs and manure production
- guard animals
- act as pets
- serve as gifts
- use in celebrations and festivities

Some animal welfare issues in urban livestock keeping are:

- ignorance and inexperience in keeping and looking after animals
- unhygienic and inappropriate housing and holding conditions
- inadequate veterinary care
- confinement and tethering, resulting in inadequate space
- inadequate feed
- trauma

Goats adapt reasonably well to an urban environment. They are good at scavenging, but they have a fatal habit of eating plastic bags. Sheep cannot depend on scavenging feed and surviving by their own means in urban areas, and they require supplementary feed.

There are social hazards associated with keeping livestock in built-up areas. They include:

- perception that keeping animals is unhygienic and that animals are potential transmitters of disease, particularly to children
- danger to traffic, especially when grazing roadside verges

- damage to property, especially gardens and crop smallholdings
- theft of animals and disputed ownership
- odour and antisocial noise hazards

In the past the emphasis in livestock aid programmes has been on the three main grazing species (cattle, sheep and goats). Where there has been investment in large-scale livestock enterprises, the ventures have often failed because they relied too heavily on government support. In addition, the advantages of economies of scale in poorer countries are not the same as in industrialized nations, and large-scale livestock production is not always appropriate. The strategies used for supporting food production in poorer economies will change with the increasing urbanization of the world's population. In future, attention will probably be given to improving the standards used in raising backyard livestock as this is seen as an emerging growth sector.

3

Genetics and Animal Welfare

This chapter gives examples of how genetics can provide solutions for welfare and husbandry problems in modern livestock farming. It also describes some of the ways in which animal breeding has created problems for animal welfare, particularly through selection for production characteristics in meat-producing animals. It summarizes information on the heritability of disease susceptibility.

It is sometimes said that modern strains have been genetically selected beyond their sustainable form and physiology. For example, slaughter-generation broilers have crippling deformities that are semi-lethal. When broilers were allowed to grow beyond their normal slaughter age, the mortality rate suddenly got worse at about 70 days of age because of lameness (Goliomytis et al., 2003). The breeding lines are maintained by limiting their growth. Some of the criticisms directed at the animal breeding sector have, however, been speculative. For example, it has been said that selection for improved feed conversion efficiency has led to the birds spending less time and energy moving. In the case of the broiler chicken, this does not seem to be the case (Skinner-Noble et al., 2003). Birds with good feed conversion are actually less lethargic. By examining the evidence based on heritability and genetic correlation analysis, one can sort out fact from fiction and determine whether adverse changes in form and function are inherited through selection for performance.

In many genetic improvement programmes, selection and multiplication is achieved through a **breeding pyramid**. A small number of elite stock at the top of the pyramid are selected and interbred. Progeny suitable for commercial use are multiplied on a larger scale at a lower level in the pyramid, and their progeny or grand-progeny

form the final generation (Fig. 3.1). Distribution of the wanted traits can be amplified by artificial insemination. Incompatible or competing production traits can be maintained to mutual advantage by breeding separate sire and dam lines. For example, sire lines are often used for transmitting meat and growth traits, whilst fecundity and milking potential (and in some cases small size) are focused within dam lines. Where there is a genetic link between production and a trait that is undesirable from an animal welfare perspective, it is usually most pronounced in one of these lines.

Problems arise if an undesirable trait either is linked to a production trait or is undetected within the elite group of animals. The undesirable trait can be distributed along with the economically important traits, and this can impact on the welfare of the progeny. Some examples where this has happened include selection for:

- conformation and body size, in particular cattle breeds, and the effect this has had on dystocia
- muscularity in double-muscled cattle and exercise stress disorders
- growth rate in pigs and osteochondrosis
- muscularity in particular pig breeds and stress-induced deaths
- conformation and growth rate in poultry and leg disorders or lameness
- breast meat yield and growth rate in chickens and ascites

Domestication and reproduction

When an attempt is made at taming and domesticating wild animals, the emotional stresses they

©N. Gregory 2007. Animal Welfare and Meat Production (N. Gregory)

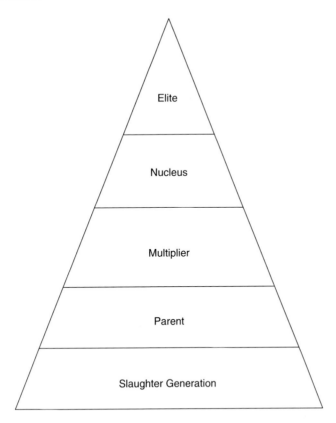

Fig. 3.1. Example of a five-tier breeding pyramid. Most of the genetic selection occurs in the great-great-grandparent (Elite) and great-grandparent (Nucleus) generations. Selection pressure may be maintained in the Multiplier (grandparent) generation, but will usually be relaxed in the Multiplier (grandparent) and Parent generations. The purpose of the Multiplier and Parent generations is to increase the number of breeding animals available for producing the slaughter generation. Lines from two breeding pyramids may be brought together at the parent generation, where there are separate sire and dam lines: for example, where a sire line has been selected for meat and growth traits and the dam line has been selected for prolificacy and for small size, which bestows low maintenance requirements.

experience can inhibit ovulation by suppressing luteinizing hormone (LH) release. Reductions in LH will delay puberty and disrupt the normal oestrous cycle. It is as though the heightened emotional reactivity of the wild type, which confers an advantage in avoiding predators and capture in nature, results in impaired reproductive success when it is chronically stimulated in captivity.

Delays in puberty have been seen when quail have been introduced to captivity. When wild-caught individuals were housed for the first time, the proportion of females that laid eggs at 20 weeks of age was 50%, but this increased to 61 and 67% in the second and third generations. In fully domesticated quail, it is normally 100%. Once the animals have been tamed and are no

longer stressed by captivity, reproduction can improve.

A similar outcome is evident in pigs, and it is probably linked to live-weight gain, which has an important influence on when puberty develops. Puberty occurs earlier in genetically improved lines, and the suitable age for first mating and hence age at first farrowing are linked to high growth rate ($r_g =$ −0.35 to −0.45; Serenius and Stalder, 2004).

There are cases where genetic selection for production performance has led to **mating failure**. When turkeys were genetically selected for larger breast muscles, they became top-heavy. Their leg structure had to be changed to accommodate the extra load and achieve better balance and gait. This was done by selecting for a wide span between the legs. This in turn made it difficult for the male to mount and balance on the female during mating, and most mating now has to be done by artificial insemination. The situation in broiler chickens is different. They too have been selected for increased breast muscle mass, but at the same time their legs have been mainly selected for straightness, to avoid angulated and rotated legs. In retrospect this has been less effective at correcting problems with lameness, and there is now a high prevalence of leg disorders in slaughter-generation broiler chickens. It has not, however, been associated with failure to mate in breeder generations. In pigs, selection for leanness could be causing poorer reproductive success following first farrowing, because the two are adversely correlated in some pig lines (Serenius *et al.*, 2004). Lame animals experience difficulty in mating and if the lameness is painful they are reluctant to mate.

Another example where self-sustained reproduction has been jeopardized by modern genetic selection is **broodiness**. When hens become broody they lay fewer eggs, and this defeats the purpose of keeping them for commercial egg production. Early on in the development of the modern layer hen, it was appreciated that the incidence of broodiness was relatively low in the White Leghorn breed, and that this trait was moderately heritable. Genetic selection by including the White Leghorn breed was a convenient solution to the broodiness problem.

Genetic Antagonisms

The **heritability** (h^2) of a trait is a measure of the extent to which that trait can be improved through genetic selection. It is calculated from the proportion of phenotypic variance in the trait that is explained by additive genetic variance. When the heritability is 0.4 or more, the trait is highly heritable. Between 0.2 and 0.4, the trait has moderate heritability, and, if it is less than 0.2, heritability is said to be low and genetic progress when selecting solely for that trait will be ineffective or slow.

Some production traits in livestock are genetically correlated with each other. For example, strains of pig that have long bodies often have reduced backfat thickness. The strength of the relationship is measured with the **genetic correlation** (r_g). This is calculated as the ratio of the genetic covariance of the two traits to the product of their genetic variation. An isolated estimate of a

genetic correlation should be treated with caution, as it is based on a ratio between values which are themselves estimated with uncertainty. However, genetic correlations are useful in identifying the nature of a genetic relationship (from the sign of the r_g), and where repeated studies arrive at similar estimates one can attach more confidence to it as a true measure of genetic strength.

There are two ways in which traits can be genetically related: linkage and pleiotropy. When there is **linkage**, different loci that control separate traits are situated close together on the same chromosome. The genes fail to segregate during meiosis, and so they are 'linked'. In **pleiotropy**, a single gene affects two or more different traits. In practice, individual production traits are usually controlled by more than one gene, and a genetic correlation between two traits is the net effect of pleiotropy and linkage.

Genetic interrelationships

Livestock breeders have a particular responsibility to animal welfare because their actions can lead to **genetic antagonisms**, which can affect a sizeable proportion of future populations. Genetic antagonisms occur when genetic selection for particular traits results in unwanted traits emerging in the progeny. This can occur either because the unwanted trait is genetically correlated to the wanted trait, or because insufficient attention is given to removing unwanted traits that inadvertently start increasing – for example, through line breeding.

The risk of a genetic antagonism arising from an unwanted trait being closely related to a wanted trait can be assessed from the heritability of the wanted trait plus the genetic correlation between the wanted and unwanted traits. See the example in Box 3.1.

Box 3.1

The heritability for milk yield in dairy cattle is usually in the range 0.22 to 0.50. In other words, it is moderately to strongly heritable. It is the single most important trait included in dairy breeding programmes. The genetic correlation between milk yield and number of services per conception is 0.41, and between milk yield and calving interval it is 0.54. From this, it can be concluded that when genetic progress is made in milk yield, it is likely that reproductive performance will decline.

Nevertheless, not all production disorders are due to genetics. Some may be due to intensification through high-plane feeding or high stocking density, and others can be due to particular housing conditions. Quantitative genetics is used to sort out the principal causes of welfare-related production disorders in a population of animals.

Applying quantitative genetics in examining genetic antagonisms

This section shows how to determine whether a welfare-related production disorder could be due to genetic selection for production performance. The data in Table 3.1 give the following conclusions.

Table 3.1. Genetic antagonisms and benefits for animal welfare that can occur when genetically selecting for production (or quality) traits.

1. Pigs

	h^2
Lean meat yield	0.29 to 0.57
Growth rate	0.25 to 0.37
Lameness	0.16 to 0.33
Osteochondrosis (femur)	0.31 to 0.32
Tail biting	0.27
	r_g
Growth rate and osteochondrosis	0.24 to 0.34
Lean meat yield and lameness	0.20 to 0.29
Backfat thickness and tail biting	−0.28

2. Beef cattle

	h^2
Birthweight	0.11 to > 0.8
Dystocia (linked to the calf)	0.12 to 0.72
Dystocia (linked to the cow)	0.05 to 0.40
	r_g
Birthweight and dystocia (linked to the calf)	0.81
Birthweight and dystocia (linked to the cow)	0.23

3. Dairy cattle

	h^2
Milk yield	0.22 to 0.50
Clinical mastitis	0.02 to 0.20
Somatic cell score	0.06 to 0.47
Milk fever	0.08 to 0.13
Ketosis	0.07 to 0.11
Cystic ovarian disease	0.10
Retained placenta	0.05
Displaced abomasum	0.00 to 0.30
Dystocia	0.01 to 0.18
Lameness	0.03 to 0.19
Udder injury	0.12
Udder oedema	0.05
Angular body shape	0.26
	r_g
Milk yield and clinical mastitis	−0.40 to 0.57
Milk yield and ketosis	0.65

Continued

Table 3.1. *Continued*

Milk yield and cystic ovarian disease	0.34
Milk yield and displaced abomasum	−0.15
Milk yield and dystocia	−0.56
Milk yield and lameness	0.24 to 0.32
Milk yield and udder injury	0.30
Milk yield and udder oedema	0.44
Angular body shape and body condition score	−0.47 to −0.77
Angular body shape and foot and leg disorders	−0.38 to −0.50

4. Sheep

	h^2
Growth rate	0.36 to 0.43
Wool yield	0.15 to 0.18
Faecal egg count	0.14 to 0.37
Dagginess	0.06 to 0.36

	r_g
Growth rate and faecal egg count	−0.39 to −0.68
Growth rate and dagginess	−0.55
Wool yield and faecal egg count (NZ Romney)	−0.53 to −0.66
Wool yield and dagginess (NZ Romney)	−0.48
Wool yield and faecal egg count (Australian Merino)	0.72 to 0.80
Fibre diameter and faecal egg count (Australian Merino)	−0.39 to −0.41

5. Broiler chickens

	h^2
Growth rate	0.4 to 0.8
Breast weight as % of live-weight	0.40
Ascites	0.12 to 0.41
Right ventricular (RV) hypertrophy	0.20 to 0.61

	r_g
Growth rate and ascites	0.22 to 0.35
Breast weight as % of live-weight and RV hypertrophy	0.75

6. Layer hens

	h^2
Growth rate in layer pullets	0.71
Feather pecking in layer hens	0.05 to 0.56

	r_g
Growth rate in layer pullets and feather pecking	−0.28 to −0.90

7. Farmed salmon

	h^2
Growth rate	> 0.5
Aggression	0.25

	r_g
Growth rate and prevalence of aggression	−0.47

Pigs

Osteochondrosis and other forms of lameness are moderately heritable, but more importantly they are linked to genetic selection for leanness and growth rate. Improved growth rate and leanness lead to more osteochondrosis, lameness and tail biting. These conditions are painful and when severe they can be debilitating. The fact that the genetic correlations are not high means that these features are not inevitably linked, and that it should be possible to achieve genetic improvement in growth and lean meat yield without the associated risks, provided the appropriate selection criteria or selection pressure is used.

Beef cattle

The heritability of birthweight varies considerably, and the exact value depends on breed. Genetic selection for improved growth performance is commonly linked with higher birthweights. Resulting high birthweights can be associated with dystocia, which causes suffering for the cow at the time of calving, and afterwards if there is permanent damage. Large-birthweight heifer calves are not strongly disposed to dystocia when they breed and produce calves.

Dairy cattle

Many of the health disorders experienced by dairy cattle have a low heritability, but when dairy cattle are genetically selected for milk yield some disorders can become more common. There can be an associated increase in genetic predisposition to mastitis, but the genetic association between milk yield and mastitis is erratic and often unpredictable. The two welfare-related issues that are most closely linked to genetic selection for milk yield are dystocia and udder oedema. The h^2 of the maternal component of dystocia is higher than that of the paternal component, but both are low (0.01 to 0.18). Dystocia is, however, linked to low milk production, and genetic improvement in milk yield can be associated with fewer calving problems. In comparison, udder oedema is more common in high-yielding lines. Udder oedema is a painful condition, and in severe cases the discomfort is obvious from the cow's behaviour.

Genetic improvement in milk yield is also likely to be associated with more cases of udder injury. This is a feature of the cows' larger udders. Higher genetic merit for milk yield can also be associated with more lameness and poorer body condition score ($r_g = -0.46$). Dairy cattle breeders prefer a cow that is angular in body shape. Angularity does not have a high heritability, but, where genetic selection for this trait has been successful, it has resulted in less lameness and a more emaciated appearance (low body condition score). Whether the low condition score is real (i.e. reduced fat and muscle mass) or perceived (e.g. redistribution of subcutaneous body fat to the internal depots) is not known.

Sheep

Growth rate is heritable in sheep, but wool yield less so. When sheep are genetically selected for growth rate there is generally less risk of heavy worm burdens and dags. Reduced dagginess is a benefit in terms of wool yield and it has welfare benefits because the sheep do not have to be mustered and handled so frequently for dag removal. Genetic selection for high-yielding fine wool production has increased the risk of high worm burdens. This can develop into a welfare problem if it is not treated adequately. This association would not necessarily apply in carpet wool lines of the New Zealand Romney.

Broiler chickens

Selection for growth rate and muscular breasts is likely to increase the risk of ascites in the birds. Ascites presumably develops from the increased metabolic demand from the additional muscle, relative to the competence of the cardiopulmonary system. There are a number of musculoskeletal and reproductive changes that are linked to production performance in broilers but the genetic correlations have not been reported. Genetic selection for rapid growth and breast meat yield has resulted in reduced propensity for multiple ovulation, reduced fertility in males and poorer walking ability in broilers (Kestin et al., 1992; Hocking and Robertson, 2005). Several types of deformity, physical imbalance and disorder contribute to poor gait, and so the genetics is not simple. Some of the simpler morphological defects such as curled toes, the bow-legged condition and leg angulation have low heritability ($h^2 = 0.00$ to 0.11; Singh and Trehan, 1994). Broilers selected for breast meat yield have not formed any mutations at the myostatin allele and so muscularity is not being achieved through the same system that is common in double-muscled livestock (Mott and Ivarie, 2002).

Layer hens

Layer hens have been indirectly genetically selected for small body size and low growth rate. This has been associated with selection for improvements in the efficiency of egg production. By selecting for low growth rate, the risk of feather pecking has increased.

Farmed salmon

Farmed salmon have been genetically selected for rapid growth rate. This has been beneficial both commercially and in terms of reducing the prevalence of aggression during rearing.

Genetics and Behaviour

Aggression

Aggression is often an unwanted trait in domestic animals. It is linked to social unrest and injuries in herd or flock mates, and it can contribute to uneven production or performance within a group, especially when the aggression is linked to domination of a feed source. Some aggressive behaviours are highly heritable. For example, the h^2 of cannibalism in layer hens is 0.65, and when layer breeders have been selecting for smaller mature body size there has been an increase in feather pecking ($r_g = -0.28$ to -0.90). In pigs, the h^2 for aggression by a sow towards its piglets varies between 0.25 to 0.50. It has been suggested that there is a risk that pig breeding companies are inadvertently selecting for pigs that are aggressive towards each other. These companies often use electronic feeder systems in their herds, which can promote competition and rivalry. Maternal aggression in cattle is not strongly heritable ($h^2 = 0.06$).

In the human, aggressive behaviour that is associated with depression and tendencies towards suicide has been linked to a low concentration of 5-hydroxyindoleacetic acid (5-HIAA) in cerebrospinal fluid (CSF). 5-HIAA is a metabolite of serotonin (5-hydroxytryptamine), and low levels of 5-HIAA indicate suppressed production and release of serotonin. Serotonin is synthesized from the amino acid tryptophan, which is a normal constituent in most diets, and one of its effects is to inhibit aggression. Tryptophan supplementation

of food and drinks has been used successfully in reducing aggression in humans and in broiler breeder hens.

Excitable behaviour and fear

Cattle that are easily excited during handling can have a disruptive effect on other animals as well as making tasks more difficult for cattlemen. Mustering, droving, trucking and handling the animals in yards at markets and meatworks can also be hazardous if the animals are unduly nervous or excitable. The best way of preventing this type of behaviour is familiarizing the animals with handling early in life, whilst they are calves. However, there is also a genetic component. The h^2 for temperament score, which is used as a measure of excitability, varies between 0.13 and 0.61 (Gauly et al., 2001). This wide range indicates that in some genotypes there may be opportunity for improving temperament through genetic selection, but in others genetic selection may be ineffective.

In pigs, fear can affect productivity by increasing piglet crushing if the sows are flighty, by making it more difficult for the stockperson to recognize oestrus and by increasing the risk of injuries. It is thought that fear has a low heritability in pigs. The h^2 for fearfulness in sows to the presence of a stockperson is only 0.08, but this response is genetically correlated to pre-weaning mortality rate in live-born piglets ($r_g = 0.37$; Grandinson et al., 2003).

Stereotypies

Stereotypies can be a sign of frustration. It is thought that the repetitive behaviour provides some relief from frustration by promoting a surge of endogenous opiates within the brain. The animal becomes dependent on inducing opiate release for its own pleasure or mental relief, and so the stereotypy becomes habitual. Propensity to display stereotypic behaviours can be inherited. In laboratory animals these behaviours can be genetically transmitted and they have been reduced by selective breeding.

Maternal care and neonatal survival

In general maternal care is not a highly heritable characteristic. For example, in beef cows the h^2 of maternal behaviour is about 0.06. In the sow, the h^2 of her reaction to a piglet's distress call is also 0.06, and the h^2 of the sow's alerting response to her piglets being handled by a person is almost zero (0.01). However, the expression of a genetic component can develop as the dam gets older. For example, in Scottish Blackface ewes, the h^2 of maternal behaviour score during early parities was about 0.13, but at the fourth parity it had increased to 0.40. In this example, maternal behaviour score was assessed by the ewe's reaction to a stockperson examining and handling the lamb. Low-scoring ewes retreated from the lamb and stockperson, whereas high-scoring ewes made physical contact with the lamb while it was being held. Not surprisingly, there was a positive genetic correlation between high maternal score and early lamb growth rate ($r_g = 0.40$; Lambe et al., 2001).

Lamb mortality rate is one of the most variable production traits in sheep production. It would be helpful if this feature could be improved genetically, and one way of approaching this is first to decide whether the focus should be on behaviour in the lamb or the ewe. Lamb survival is considerably more heritable as a trait of the lamb itself ($h^2 = 0.30$) than as a trait of the dam ($h^2 = 0.04$), which indicates that genetic progress would be quickest by concentrating on the lamb.

Carcass leanness could be associated with better chances of survival in newborn lambs. When Scottish Blackface sheep were genetically selected as fat and lean lines, the lean line had a number of survival advantages in terms of behaviour of the ewe and lamb at lambing time (Table 3.2). This was offset slightly by a lower birthweight in the lean-line lambs. No overall difference in lamb mortality between the lines has so far been demonstrated (Dwyer et al., 2001).

Table 3.2. Advantages of genetically lean sheep in terms of behaviour at lambing.

Fewer incorrect presentations

Shorter time to deliver second lamb when there were twins

Lower plasma β-hydroxybutyrate concentrations towards the end of pregnancy

Shorter time to recover from lambing before the start of grooming the lamb

Shorter righting time by the lamb after birth

Shorter time to sucking by the lamb

Pig sire lines are selected mainly for growth rate and low backfat thickness. These traits are not compatible with ideal maternal characteristics (Grandinson et al., 2005). Sows in these boar lines can experience the following problems. They:

- are more likely to experience lameness and be culled early
- have higher mortality rates
- have limited fat reserves at farrowing and so they can have poor milk yield and if this gets out of control some of the piglets can die from starvation
- are more restless, stand up and lie down more frequently, and are more prone to crushing their piglets

When in good condition, sows with a genetic predisposition to use fat reserves during lactation have higher piglet growth rates during lactation and lower piglet mortality rates (Table 3.3).

Genetic selection for heavier piglets at birth should lead to fewer crushed piglets. It would not, however, lead to fewer stillbirths and so it may not improve overall survival rate. The emphasis in Danish breeding programmes has been changing from selecting for litter size at birth towards the number of live piglets at 5 days after farrowing. This is placing more emphasis on early survival.

Genetics as a Solution to Husbandry and Welfare Problems

Genetic selection could solve some of the problems that develop from intensive farming. Presently, there is interest in using genetic approaches for overcoming aggression in layer hens and turkeys and leg weakness in broilers. However, making these changes may not be as easy as might be thought. There are two dilemmas. First, it is not always clear what selection criteria should be used.

Table 3.3. Genetic correlations between change in body weight in the sow during lactation and performance of the litter.

Litter performance trait	r_g
Piglet mortality between birth and weaning	0.45
Growth rate between birth and weaning	−0.85

For example, what trait should be used for opportunistic traits such as cannibalism in layer hens? Secondly, will introducing an additional selection trait dissipate some of the genetic gain already made in production traits? If so, the breeding company may not apply strong selection pressure for the additional criterion.

The example in Box 3.2 indicates whether genetic selection could be used as an alternative to mulesing. This study indicates that:

1.   Selecting sheep for a high susceptibility to fly strike (C × C) is likely to lead to fly strike in subsequent generations.
2.   Breeding for low susceptibility is not likely to eliminate all fly strike since 43% (51) of the A × A lambs showed intermediate levels of fly strike. This means that some other forms of control would be needed even if genetic selection were adopted as the main alternative to mulesing.
3.   The frequency distribution of susceptibility to fly strike in the progeny from an intermediate population (B × B) appeared to be normal. This implies that, for the type of sheep used in this study, the susceptibility to fly strike would neither improve nor deteriorate if no genetic selection pressure were introduced. It is not possible to state from this experiment whether an A × B cross would reduce susceptibility to fly strike because this was not tested. However, in view of **1** and **2** above, it would seem likely that the trend would be towards reduced susceptibility because the genes

from A would be replacing the genes from C or B in the parents of each generation.

Genetic selection for resistance to diseases and disorders

The effectiveness of genetically selecting animals for disease resistance depends on the type of disease. In some diseases, such as vibriosis in farmed salmon, the heritability of disease resistance has been moderately high (0.38), but in many other infectious diseases in farmed species it has been disappointingly low. Nevertheless, this area of science has attracted interest because of the potential benefits.

It should be possible to select for disease resistance by selecting for immune responses. Although immune responses interact with each other, the inheritance of the different immune mechanisms seems to be independent. For example, when chickens were selected for six generations for three different *in vivo* immune responses (high antibody response, high cell-mediated immune response and high phagocytic activity), the immune responses did not respond simultaneously. Each of the chicken lines was selected for one immune trait, and in all three cases there was a significant increase in immune capacity for that trait. However, improvement in one immune response was not associated with coincident improvement in one of the other immune responses.

Box 3.2 Breeding for reduced susceptibility to fly strike in Australian Merinos

It has been suggested that breeding for reduced susceptibility to fly strike would be a more humane approach to controlling fly strike than the conventional method of mulesing. Mulesing is the surgical removal, without anaesthetic, of the breech skinfolds in wrinkly Merinos where blowflies lay their eggs. In the early 1930s (before mulesing was adopted as a general husbandry method) there was a study called the Nyngan Breeding Experiment, which graded sheep for susceptibility to breech strike as follows:

Very susceptible – repeatedly struck each season	C
Relatively unsusceptible – rarely struck	A
Intermediate	B

When A, B and C ewes were put to rams of the same class as the ewes, and the lambs were graded for fly strike at 3 months, the findings were:

	Number of lambs in each strike grade		
	A	B	C
Lambs from A × A	66	51	2
Lambs from B × B	19	84	14
Lambs from C × C	2	21	21

Some features of the immune system are highly heritable. For example, the h^2 of heterophil–lymphocyte ratio in chickens is 0.59 (Campo and Davila, 2002). There are strain differences in susceptibility to malabsorption syndrome in broilers, which appear to be related to differences in total number and proportion of leucocyte subpopulations in the blood and interstitial mucosa and this could present opportunities for improving resistance to some specific diseases (Zekarias *et al.*, 2002). Genetic selection according to immune system indicators is already well developed in dairy cattle breeding. However, there are concerns that it is being used inappropriately. Genetic selection against somatic cell count, as a way of reducing the expression of mastitis, might compromise the animals' cellular immune response system.

Although the heritability of most infectious and metabolic diseases is low, disease susceptibility can vary between breeds. In other words, it seems to be inherited in particular breeds or lines but less so in the general population. This is discussed in the section on Genetic Diversity. Sometimes there has been false hope in improving disease resistance through breed introduction or substitution, as can be seen in the example in Box 3.3. This study

---

Box 3.3 Osteochondrosis in pigs

Osteochondrosis is a painful heritable leg disorder in pigs. It is usually managed by culling boars and gilts that show signs of abnormal stance or gait just before they reach puberty. From experience it is known that a high selection pressure is needed to bring about good control of the condition. It would be helpful to farmers if cross-breeding with osteochondrosis-free stock could be used instead of culling within existing herds. This would mean that genes from animals with outstanding performance could be retained, and osteochondrosis would be controlled by dilution through introduction of new stock.

The following study, which was conducted in Sweden, tested whether this approach would, theoretically, be a suitable alternative. A boar genotype (Wild boar) that was known to be free from osteochondrosis was crossed with a female line (Yorkshire) that historically had experienced osteochondrosis. The cross-bred progeny were assessed for osteochondrosis, and compared with pure-bred Yorkshires.

Osteochondrosis was evaluated *post mortem* on a 0 (best) to 5 (worst) point scale.

Table 1.   Mean live-weight for age (LFA) at slaughter in Wild boar × Yorkshire (WY) and pure-bred Yorkshire (YY) pigs.

	LFA (g/day)
YY	0.58
WY	0.39

Table 2.   Severity of osteochondrosis, presented as a frequency distribution of osteochondrosis score within breed, in two long bones.

	Humerus		Femur	
	YY	WY	YY	WY
$n =$	942	190	942	192
Score	Percentage of pigs			
0	51	56	35	24
1	29	30	30	23
2	15	10	18	28
3	4	3	17	23
4	1	1	0	2
5	0	0	0	0
Mean score	0.76	0.65	1.17	1.59

indicated that cross-bred pigs had a markedly decreased growth rate (about 30%), but the distribution, frequency and extent of the osteochondrosis were similar to the pure-bred Yorkshire. Pig farmers would be well advised to cull affected stock rather than relying on controlling this disease by cross-breeding with unaffected animals (Uhlhorn *et al.*, 1995).

Breeding cattle for the right environment

One of the maxims in animal breeding science is that:

> A character measured in two different environments should be regarded not as one character but as two.

In other words, developing a superior line for a particular trait might depend on selecting for that trait under the relevant production conditions. An example of a genetic selection experiment in beef that demonstrated this effect was conducted in northern Australia. It showed that:

1. Genetic selection for post-weaning growth rate in Hereford × Shorthorn cattle in a climate that has hot arid summers was ineffective in producing progeny with higher growth rate when protected from adverse heat and treated regularly for endo- and ectoparasites. However, it was effective in producing progeny with less suppression of growth rate under hot conditions. Under hot conditions, the progeny from the selected line ate more, panted less, sweated more and maintained lower rectal temperatures. They were more heat-tolerant and evaporated more of the water they drank, instead of eliminating it as urine. The implication is that they had better thermal comfort. They were also less severely affected by keratoconjunctivitis.
2. The higher growth rate in the selected line under hot conditions was due to their more efficient thermoregulation, and it led to a higher feed intake. In addition, they had a lower maintenance requirement.
3. The selected line appeared to have superior resistance to infestation by intestinal worms and to the effect of the worms.
4. They also had lower birthweights, and this might have helped reduce the prevalence of dystocia.

The overall conclusion was that selection under adverse (hot) conditions would improve the tolerance and performance of Hereford × Shorthorns under

those conditions. It would not necessarily benefit performance if the cattle were transferred to more temperate conditions (Frisch, 1981). Genetic improvement in growth rate can be achieved by different routes.

New technologies

New molecular technologies will allow livestock breeders to select on the basis of genotype rather than phenotype, and this will increase the efficiency of selection in farm animals and the rate at which genetic gains are achieved. **Gene mapping and marker-assisted selection** (GMMAS) will enable selection early in life, without any interference from environmental effects. This means that there will need to be a stronger focus on understanding the mechanisms responsible for particular traits in those traits that are controlled by more than one gene (polygenic traits). Selection for gender-specific traits could be applied equally to both sexes at the gene level, and it would not require costly trait evaluation, provided the genes have been reliably mapped. GMMAS will increase selection intensity and the accuracy of selection, and it will decrease the generation interval. However, it will also increase the speed of expression of detrimental genetic antagonisms if there is inadequate compensation in the selection strategy.

Genetics and Stress Responsiveness

Inherited overreaction to stress has been recognized in a number of species. The following are two examples:

1. In rainbow **trout**, there are two distinct types of physiological response to confinement stress. Some fish develop a high plasma cortisol response while others have a low cortisol response. The h^2 of the size of the cortisol response is 0.4 to 0.6. The high responders are more likely to show flight behaviour instead of freezing when confronted by an unfamiliar smaller intruder, and the low responders show less stress-induced suppression of appetite.
2. Stress sensitivity has been a common problem in certain breeds and strains of pig. Stress-sensitive types are prone to developing a severe fatal metabolic acidosis plus hyperthermia when aroused, and they experience a similar metabolic response to inhalation of the volatile anaesthetic halothane. They also have accelerated acid

production in their muscle *post mortem* and this results in an abnormally pale, soft, exudative (PSE) meat. These pigs are also highly muscular and produce exceptionally lean carcasses.

In the past, pig breeding companies have inadvertently increased the prevalence of the stress-sensitive gene in the pig population through an inappropriate breeding policy. They selected pigs for leanness and superior ham conformation and sold them as meat sire lines for crossing with sows to produce a lean slaughter generation. The meat sire lines were developed by including stress-sensitive breeds such as the Pietrain and Belgian Landrace, which have improved leanness and meaty hams. The aim was to produce lean slaughter pigs as offspring from the sires. The disadvantage with this approach was that it relied on all the offspring being slaughtered. If any offspring were retained for breeding, the semi-lethal stress-sensitive gene would be distributed in the breeding herds as *Nn* genotypes instead of being limited to the sire line and slaughter stock. Stress sensitivity is inherited as a double recessive trait (*nn*). It was believed that the heterozygote (*Nn*) is less prone to stress-induced deaths and to developing PSE meat after slaughter. However, experience showed that the heterozygote (*Nn*) was more prone to developing PSE meat and to death during transport than had originally been suspected. Once this was realized, the selection for *nn* was reversed. Lines that were *nn* were removed from the sire lines, and leanness was selected by other means. In retrospect, selection for *nn* as a quick way of improving leanness was an irresponsible decision, as it was known beforehand that *nn* is a semi-lethal gene with adverse effects on meat quality.

Genetic Diversity

Genetic progress in livestock production inevitably reduces genetic diversity. There are benefits from maintaining genetic diversity (Table 3.4). As livestock breeding becomes concentrated in specialized multinational companies, there is a greater risk of reduced genetic diversity.

Genetic diversity and maintaining minority breeds

One way of maintaining genetic diversity is by supporting the survival of minority breeds. There are many examples where minority breeds are claimed to have unique value because of a rare genetic make-up. Their survival depends on preserving nucleus stock. Examples of benefits from maintaining unusual breeds include the following:

- Dairy farmers who use natural mating for their heifers often prefer to use a beef breed of bull, in comparison with a dairy breed, to avoid dystocia and injury during mating. However, with the growing emphasis in beef breeding towards fast growth and large mature body size, the advantages in terms of freedom from dystocia are being lost. As a result, a

Table 3.4. Benefits from maintaining genetic diversity.

Benefit	Example
Prevention or reversal of inbreeding	AI mate-pair allocation programmes in dairy herds
Dilution or elimination of semi-lethal genes	Control of ascites in broiler chickens through breed substitution and selection against the condition
Combining specific genes important in production	Heat tolerance and tick resistance in cattle with beef characteristics of temperate breeds
Introducing specific genes because of changes in market demand	Growth in market for black and mixed-colour wool; reducing hen aggression with expansion in free-range egg production
Improving the efficiency of production in novel ways	Dwarf female lines in broiler breeders; bantam-cross chicken layer hen strains
Rapid introduction of changes in anatomy and breeding performance	Poll gene from Danish Red cattle; improvement in litter size with Meishan pig breed
Social value	Curiosity and archetypal values

AI, artificial insemination.

market has developed for small-sized, easy-calving beef bulls for use on dairy heifers. Such strains can be found in a range of the traditional beef breeds, as well as the more specialized Dexter breed.

- Poll strains of a number of dairy and beef breeds have been developed by introducing a poll breed, such as the Red Poll or Danish Red, and then backcrossing to the preferred dairy breed.
- A wide range of improved tropical and subtropical beef breeds have been developed from crosses between indigenous cattle breeds and faster-growing European breeds. Pure forms of the indigenous breeds have sometimes been difficult to access, but, once the new breeds have been stabilized, the need for sourcing nucleus herds of the indigenous breed has passed.
- The Duroc pig breed experienced a revival in the 1990s when it was found that genetic selection against fatness in pigs had caused a deterioration in meat succulence. The Duroc has unusually high levels of marbling fat, and this helped to correct the decline in meat quality, as well as imparting a dark skin and coat colour, which helped protect outdoor pigs from sunburn.
- The Meishan was introduced by European pig breeding companies as a quick way of improving prolificacy in maternal lines. In pig breeding, breed introduction or substitution is often a quicker way of improving a trait than selection within a breed.
- In some countries which have a Mediterranean or subtropical climate, the Wiltshire breed and Wiltshire crosses have attracted interest where it has been financially unrewarding to shear sheep. The Wiltshire has hair instead of wool. In addition, the Garole breed, which grazes in swampland of West Bengal, is attracting attention as a way of introducing resistance to foot rot. In Merinos the h^2 of foot rot is about 0.20 (Woolaston, 1993), and introduction of a low-prevalence breed followed by backcrossing might improve resistance.
- In most countries there is a distinct preference for broiler carcasses with white instead of yellow skin, as, traditionally, consumers have associated a yellow carcass with an old layer hen. Most of the breeds from which broiler strains have been developed were yellow-skinned. Skin whiteness was introduced by intercrossing with the Light Sussex, and retaining the white skin gene in subsequent selection.
- In the 1990s there was a revival of interest in bronze in place of white turkeys, as this represented a more traditional product.

An opposite argument is that many of the rare breeds are insufficiently diverse to warrant concern about their extinction. Some breeds are simply combinations of other breeds, with selection according to a specific coat or feather colour. They have added little in the way of useful diversity, and in some cases were only intended for niche or novelty markets. Nevertheless, some of those breeds have archetypal, social or cultural attachments for a community.

Inbreeding

Inbreeding is the mating together of individuals that are closely related to each other by ancestry. Animals are closely related if they have one or more ancestors in common in the first four generations of their pedigrees. Inbreeding develops through elimination of alleles rather than genes. In other words, it occurs as a herd or flock becomes more homozygous. This happens when a breed or species becomes rare, and the opportunity for mating with an individual of dissimilar ancestry is reduced. It could also occur when a population is replaced by clones, which are produced by asexual reproduction, and the clones are allowed to interbreed.

The level of inbreeding is estimated from the inbreeding coefficient, which is defined as the probability that two genes at any locus in an individual are identical by descent. It can be assessed from a comparison of relationship between individuals within a line and individuals taken at random from the whole population.

One of the earliest controlled experiments that examined inbreeding depression was conducted in guinea pigs. There were full-sib matings for more than 30 generations. During the first 12 generations a number of morphological peculiarities and malformations emerged, but more importantly there were more stillborn pups and depressions in fertility and viability. When divergent inbred lines were interbred at the 20th generation, there was pronounced hybrid vigour in the F_1 offspring, with increased resistance to tuberculosis, increased body size, improved litter size and growth rate pre-weaning. In addition, there were fewer stillbirths.

In young inbred animals, fitness is reduced and susceptibility to infections increased. Together, this raises mortality rate pre-weaning. In practice it has been difficult to identify inbreeding as a cause of raised mortality on commercial farms, but there are many experimental studies where inbreeding has produced this effect.

Examples of inbreeding depression that have been seen in experiments on livestock and poultry are given in Table 3.5. It is suspected that in the past inbreeding has occurred in the following situations:

- stud breeders trying to produce a line that is true to type
- closed herds or flocks – for example, no new breeding animals introduced for animal health reasons
- inappropriate use of a single sire, because of its genetic merit
- small herds or flocks of a rare breed
- limited availability of bulls

There are three ways in which controlled inbreeding is a useful and beneficial breeding technique:

1. Inbreeding can be used to identify and eliminate recessive hereditary defects.
2. Outbreeding inbred lines allows the expression of heterosis (hybrid vigour), and this has been one of the main reasons for line breeding livestock in the past.
3. Inbreeding allows stud producers to sell breeding stock that are true to type (consistent). The buyer knows what to expect, through experience.

New technologies

Concerns have been raised about two new breeding technologies which might lead to reduced diversity if used inappropriately. They are juvenile *in vitro* embryo transfer and cloning.

Juvenile *in vitro* embryo transfer (JIVET) is an emerging breeding technology in the dairy industry. It is a relatively simple way of increasing the number of animals in superior female lines. Multiple ovulation is induced in a heifer with anticipated high genetic merit, the unfertilized eggs are harvested by flushing them from the reproductive tract, the eggs are fertilized *in vitro* with semen from a high-genetic-merit dairy bull, and the subsequent embryos implanted into ordinary recipient cows. The offspring will not be identical, but they will be full siblings.

Cloning involves the replacement of maternal chromosomes in an ovum with those of a differentiated somatic cell. Because the somatic cell is diploid, fertilization is not necessary, and under appropriate conditions the reconstructed ovum can develop into an embryo, which is then implanted into a recipient by embryo transfer (ET). When used on a large scale, this approach has the potential to allow rapid replication of a single genotype. If taken to an extreme, it could jeopardize genetic diversity.

Asexual reproduction through cloning is widely used in the horticulture industry. For example, most fruit trees are propagated on to rootstocks using budding or grafting. One way it could be used in the dairy industry is as follows. The genetic merit of elite cows could be assessed in as few as 30 cloned daughters. Thousands of cloned embryos will then be produced from proven cows with high genetic merit, and sold in straws for implantation into commercial cows that have been synchronized.

Table 3.5. Examples of effects of inbreeding in production animals.

Species	Effects
Dairy cattle	Reduced fertility and birthweight; higher calf mortality, lower milk yield
Sheep	Reduced growth rate, lower fleece weight, shorter staple length
Pigs	Lower ovulation rate and reduced litter size at birth, increased mortality among piglets, decreased growth rate, delayed puberty, hernias, skeletal deformities
Poultry	Reduced egg hatchability, higher chick mortality, delayed onset of lay, reduced egg production; deformities including crooked breast bone, misshapen beaks and cleft palate

The resulting offspring would replace the existing herds, and frozen embryos from exceptional cows would be stored in cryobanks for future use. In this way, the prospect of loss of genetic diversity would be substantially greater than that experienced so far through AI.

An important feature of cloning is that it captures genetic improvement in an individual or a small number of individuals and it produces identical replicas of that source, which are distributed on a large scale. It has two risks. First, if the clone is widely distributed, inbreeding depression could occur when an attempt is made at outbreeding. Secondly, it captures genetic improvement in a limited number of animals and at a given point in time. It rapidly upgrades the national herd to the standard of those animals, but it limits further genetic changes. It diminishes the breed's ability to change through evolution because genetic diversity has been narrowed and because new genetic recombinations (for example, by epistasis) are avoided. The second risk could be serious if the genotype is required to react to an emerging disease or perform in a changing environment.

Moral considerations on new technologies

Historically, moral outlooks have often been based on religious and cultural traditions. In societies where these traditions have become less important, the trend has been towards basing moral arguments on technical issues or features and from canvassing the general public to learn about their immediate thoughts or feelings on an issue. When a survey was conducted in the Netherlands, 57% of the interviewees had an intuitively negative attitude concerning cloning of dairy cattle embryos. Common concerns can be divided into the following categories. Such cloning:

- could lead to misuse in humans
- might have negative effects on animal health and well-being
- might interfere with the integrity of the animal
- is likely to diminish genetic variation
- could lead to over-intensification of milk production, which in turn may impact on society and the environment

If cloning leads to irreversible loss of genetic variation, it has a moral dimension. Something that can be beneficial to the bovine species and to society is knowingly lost, and seemingly it cannot be replaced or reintroduced. However, the gravity of this moral issue depends on the answer to three questions:

1. How extensively will the cloning be used?
2. Will a non-inbred elite population be maintained which will be used for genetically selecting future clones, and will it not itself be subject to replication by cloning?
3. Which genotypes will be lost?

It is not inevitable that genetic diversity will decline as it depends on how cloning is used. However, the companies providing the cloning service would not be controlling the loss of genetic diversity. That would be determined by which animals were not allowed to breed sexually and instead were submitted as recipients for cloned embryos. In a regulatory sense, this would be difficult to control.

Inbreeding and future breeding technologies

In the short-term future, JIVET will probably be used to multiply superior sire–dam combinations. In the longer term, rapid multiplication of identical genotypes might be achieved with cloning plus embryo transfer (CPET). The risk of inbreeding depression developing from either JIVET or CPET would depend on how they are used commercially. Clearly, if the genetics of dairy cows internationally becomes concentrated in a limited number of highly uniform lines, which are separated only by competing studs, the risk would increase as those studs merge or get taken over. Where there is little difference between commercial lines, crossing one maternal line with another would have little impact on reducing genetic diversity.

Several industry-watch groups have called for control over cloning to ensure that genetic diversity is maintained. It is worth considering how this 'control' might operate. One approach is to set up a government-sanctioned authority with a legislative duty to review cloning project applications on a case-by-case basis. The authority would be expected to consult with appropriate bodies on the technical feasibility and implications of the project and to reach the 'right' conclusion as to whether the project should proceed. This process would not be entirely satisfactory as it transforms a moral and political decision into a management decision, and it is likely to evade the ethical issues. In addition, if the primary aim is to ensure that genetic diversity is maintained, there is a stronger case for monitoring the application of cloning commercially rather than the initiation of a commercial cloning venture. Presently, when AI is used by the dairy industry, the degree of inbreeding can be estimated, using nationally available databases, for any proposed mating. This helps farmers select semen from bulls that fit their production needs, while controlling the level of inbreeding. However, the use of AI operates in a free market, and in some situations inbreeding is not being thought about. Similarly, managing inbreeding during the application of cloning may be a secondary and sometimes forgotten feature.

Genetic diversity and resistance to diseases or production disorders

One of the strongest arguments put forward for conserving rare lines is that some lines have evolved, through natural selection, into being resistant or more tolerant to specific pathogens. Introduced improved lines sometimes show less resistance to indigenous diseases than the locally bred animals. One example is the N'Dama breed of cattle in western Africa, which is tolerant of trypanosomiasis. This breed is native to Senegal, and in its pure form is becoming scarce because of cross-breeding with introduced breeds, particularly zebu cattle.

This genetic dilution is making the cattle population more susceptible to trypanosomiasis, and there is a need to maintain a stock of pure-breds to preserve the gene in an easily accessible form. Other examples of genetic resistance are given in Table 3.6, and these genotypes are highly eligible for conservation.

Biodiversity and the ethics of gene ownership

In 1988 the US Patent and Trademark Office issued the first patent on a living animal. It was a transgenic mouse that contained a variety of genes derived from other species, including chickens and humans, and the purpose was to use the mouse as a model in cancer research. The patent was owned by Du Pont, and it gave the company ownership of a range of cancer-causing genes, along with the animals into which the genes were inserted. Since then, the patenting of genes plus recipients has expanded considerably and now includes some farmed species. Under US law, no species is exempt from patenting, except for the human, who cannot be owned because of infringement of antislavery legislation. However, a human gene inserted into an animal would be eligible for patenting.

This raises a number of ethical issues, but the focus here is on genetic diversity and ownership of that diversity. Experience in the horticulture

Table 3.6. Examples of breeds or strains that have enhanced resistance or tolerance to specific pathogens or conditions.

Type	Pathogen/condition
Cattle	
Droughtmaster, Belmont Red	Hot conditions
Zebu	*Mycobacterium bovis*
Specific allotypes	*Brucella abortus*
Nguni	Ticks (including *Amblyomma hebraeum*, *Boophilus decoloratus*, *Hyalomma* spp.)
Sheep	
Various breeds	Foot rot (*Dicelobacter nodosus*)
Red Maasai	Nematodes (including *Haemonchus contortus*)
Chickens	
Lines with a specific dominant autosomal gene	*Salmonella* spp.
Line 6_1	Marek's disease
Various lines	Gumboro disease

industries is showing that there is growing concern about the way in which ownership of unusual genes is being acquired and exploited. Plant biogeneticists are searching for genes that could impart commercial advantages when inserted into crop plants. Tropical plants have been a focus for the searches, because of their wide diversity. Discovery of a useful gene has enabled patenting, and use of that gene in traditional ways within the countries of origin could be subject to patent control. Presumably, this could also happen in future with livestock species, as can be seen from the hypothetical case in Box 3.4. This is just one example of the way in which ethical conflicts in bioengineering could arise in the future. As with any ethical debate, decisions will need to be reached by examining responsibilities, plus social and personal values. The social values attached to biodiversity and genetic diversity include appreciation of:

- the richness and variety of life
- the interdependency of species and co-evolution
- adaptation to particular environments
- adaptation to particular ecosystems
- the development of specialized forms and abilities through evolution

---

Box 3.4 Suppose:

A gene for resistance to a parasite is identified in an indigenous breed of livestock by a bioengineering company. The local community has been aware of the breed's resistance to that particular parasite, through unfortunate experiences gained when non-indigenous (European) breeds were introduced in the past to upgrade their breeds. A bioengineering company isolates and patents the gene and introduces it into European breeds, which it sells in regions that have the parasite. The local community from which the gene originated has been told that their animals contain the patented gene, and that they may be expected to pay royalties on any animals they sell outside the region. The legal and ethical issues here are whether this is a case of plagiarism and, if not, who owns the genetic make-up of this trait. Is there a common heritage, or can it be appropriated for commercial gain? The common ownership of genetic resources is sometimes called 'biodemocracy', and the move towards gene patenting is directly opposed to this concept. An opposing outlook is that, in the absence of a competitive advantage from patenting, the wider community will not benefit from the gene.

---

and the exploitation of that diversity through:

- commercialization of useful characteristics
- non-profit-making use

Genetics in the Aquaculture Industry

Current developments in fish breeding deserve special consideration because of the changes that are occurring and because the fish breeding sector is particularly forward-thinking.

Almost all domesticated strains of fish exhibit faster growth than wild strains when compared alongside each other. In channel catfish (*Ictalurus punctatus*), growth rate has increased at a rate of 3 to 6% per generation since it was introduced to aquaculture. Part of this increase is genetic and part is due to changes in feed and holding conditions. Comparable improvements have occurred in the trout industry, and strains that have been selected for growth rate have developed high weight-to-length ratios. About 65% of Norwegian salmon production is with improved lines. Selection has been mainly directed at growth rate, but there have been improvements in body length, feed conversion efficiency, meat quality and disease resistance (Gjedrem, 2000).

Selection experiments show the gains that could be achieved in the near future. First, the h^2 of body weight in unimproved lines of rainbow trout can be as high as 0.89, but this quickly diminishes during successful selection for the trait. Typically, the h^2 for growth rate in commercial salmonid lines settles at between 0.21 to 0.40. Six generations of selection for growth rate in trout increased body weight by 30%, and in salmon selection for ten generations produced an increase of about 50%. In coho salmon, improvements have been made even in the presence of moderately high inbreeding. They do not seem to express some of the adverse effects of inbreeding, either because inbreeding depression has already been removed during evolution through the habit of returning to familial breeding grounds or because the innate high fecundity of these fish masks the inbreeding depression. Body weight in channel catfish has been improved by 12 to 20% in only two generations of mass selection, and in one strain it improved by 55% in four generations of selection.

Body conformation and composition can be modified by genetic selection, but attempts at increasing dressing percentage and reducing the number of intermuscular bones have not been very successful. The h^2 of body fat percentage in catfish and trout is about 0.50, and so selection for low-fat meat should be feasible in these species. In other species it is lower ($h^2 = 0.06$). There is a positive genetic correlation between growth rate and body fat percentage. In other words, selection for growth rate can result in an increase in fatness. Whether this becomes a problem depends on which market is being supplied. Compared with other species, salmonids produce fatty flesh, and, if the pressure for lower fat content increases, genetic manipulation may be considered as an option in future. Generally for salmonids, the fat content is too high if it exceeds 16 to 18% (Gjedrem, 1997). The cyprinids, on the other hand, produce low-fat meat, and boniness is a more important quality issue.

There is a genetic antagonism between growth rate and sexual maturity in some species, and this is a substantial problem for the salmon industry in some countries. Sexual maturity occurs earliest in the faster-growing larger fish, and it retards growth when the fish stop feeding during the summer in preparation for spawning. Some of the ways this is managed have welfare implications.

Not all strains of salmon show a high prevalence of **grilsing** (sexual maturity after the first winter at sea). Norwegian strains are less prone to grilsing than strains from other countries. In Scotland, sometimes a third of the crop is grilse, and this is managed by grading all the first-year sea fish during August and September to harvest or grade out the grilse. This is one of the busiest periods for salmon farmers, and the handling of large numbers of fish at high rates causes stress, injuries and bruising. Prophylactic drugs are sometimes used to control stress-induced disease outbreaks after this grading procedure.

The other approach to avoiding a high grilse rate is to feed-deprive the fish towards the end of the first winter in the sea cages. The setback in growth can be made up through compensatory growth when normal feeding is restored. Photoperiod manipulation has been a more common approach, and it has also been used to improve growth.

High grilsing rates are linked to downgrading and meat quality problems as well as arrested growth. Undersize plus early maturation is the most important cause of downgrading in Atlantic salmon (Table 3.7; Quigley, 2003). The effects that early maturation has on eating quality and appearance are as follows. When salmon reach sexual maturity, pigment carotenoids are transferred from the muscle to the gonads and skin. The muscle also contains less lipid and protein, which are replaced by water, and in extreme cases it is mushy. There are changes in external appearance, with the development of secondary sexual characteristics, such as the kype and a longer head. These features give away the fact that the fish is reaching sexual maturity and so the flesh is likely to have inferior eating quality. Another potential welfare issue that is linked to precocious sexual maturity is in sea trout, as they have poorer over-winter survival rates.

Hereditary skeletal **deformities** are quite common in some farmed finfish. This may be because there is less pressure from natural selection in the farmed environment, and deformed

Table 3.7. Causes of downgrading in an Atlantic salmon processing company in 2002.

Reason for downgrading	% of downgrades
Maturity	37.8
Deformities	24.6
Scale loss	15.9
Condition factor	11.9
Bruising	5.6
Wounds and lesions	3.9
Softness	0.3

Downgrading rate as a proportion of total production was 12.5%.
Condition factor is the ratio of body weight (g) to body length cubed (cm^3).

individuals survive for longer. The deformities in salmon include curvature of the lower jaw, short opercula, scoliosis, lordosis and gill deformities. Spinal fusion is a common deformity and is believed to be linked genetically to a fast growth rate. It makes accurate filleting more difficult. Cataracts occur in Atlantic salmon. They are partly genetic, and are linked to rapid growth whilst in seawater. In some cages, up to 90% of the fish are affected, with 30% being severely affected and 5% completely blind in at least one eye (Ersdal *et al.*, 2001). Blindness reduces the fish's ability to feed and thrive in competitive situations.

Some fish hybridize across species, but there have been few **hybrids** bred specifically for aquaculture. This was not for want of trying. The main shortcoming has been that hybrid vigour is not readily expressed in the interspecies hybrids. It is seen, however, in crosses between races, lines or varieties. In one case, intervarietal crosses of Chinese common carp expressed 20% greater growth than the fastest-growing parent. Cross-bred strains exhibiting heterosis are used in the carp industries in Israel and Vietnam. There can be unexpected behavioural repercussions when crossing different genotypes. For example, the blue catfish × channel catfish hybrid proved to be nervous and aggressive, to the extent that it was unsuitable for tank or cage culture and had to be grown in ponds.

Triploidy

Triploid fish were first promoted as a way of controlling sexual development, but their main use now is in producing sterile offspring. This has two benefits in the case of salmon. First, it maintains growth and meat quality by offsetting the effects of grilsing. Secondly, it is used in conjunction with other genetic manipulations for ensuring that genetically modified or non-native stock cannot pass on genes to wild stock should they escape from a fish farm. The first application has not found widespread use because, in practice, growth performance in salmon triploids has been inferior to that of diploids when transferred to marine cages. Prior to this, they have good growth, survival and smelting rates (survival to smelt stage) in freshwater. The likely reason for the poorer late performance is that triploid salmon develop a number of physical disorders as they mature. These include lower jaw deformity (Fig. 3.2),

Fig. 3.2. Jaw abnormality in triploid salmon.

cataracts, shortened opercula, reduced gill surface area and impaired oxygen carrying capacity by the blood due to the size and shape of the red blood cells. They can usually compensate for their reduced aerobic capacity, but when stressed they are more prone to hypoxia (Bernier *et al.*, 2004). For example, during the early 1990s when triploid Atlantic salmon were promoted in Scotland, it was found that they were susceptible to oxygen shortage in the water and were the first to die in plankton blooms and during routine handling procedures. The prevalence of the jaw deformity has been as high as 30% (Sadler *et al.*, 2001). Triploid rainbow trout have shown spinal deformities.

In spite of some of their physical limitations, triploid fish have a future in some niche markets. Sterility in sex-reversed trout stocks is achieved presently with triploidy. So far, triploid rainbow trout have performed well in aquaculture, unlike triploid Atlantic salmon. There is interest in commercializing triploid transgenic tilapia and coho salmon. They grow more slowly than their diploid counterparts, but they are sterile and grow faster than diploid non-transgenic stock. They could, however, be more prone to some infectious diseases, including vibriosis. Sterile triploid male salmon may be developed in future for ocean ranching, where return migration of superior types is wanted, without the ability to reproduce with domesticated or wild stock.

The benefits of sterile triploids in terms of meat quality depend on the season and the state of muscle lipid and water content. Triploid amago (*Oncorhynchus mason ishikawae*) can be softer between March and September, but they are similar to diploids during the rest of the year. As noted for salmon, at the time of sexual maturation, the labile lipid and protein stores in muscle are depleted and replaced by water, making the meat soft. In addition, pigments are mobilized from muscle for deposition in the eggs, and the skin becomes darker.

There are changes in head shape and the fish become more aggressive to each other.

Inducing triploidy is a relatively simple and innocuous procedure. A temperature or pressure shock is applied soon (e.g. 40 min) after fertilization of the eggs to block extrusion of the second polar body. Equipment is available commercially for applying hydrostatic pressure shocks to the eggs. Tetraploid individuals can be produced by disrupting the first mitotic cleavage using similar treatments, and these fish produce triploid offspring when crossed with diploids.

Monosex culture

One of the obstacles in tilapia farming is their small size. When regenerating naturally, tilapia often overbreed, producing large numbers of fingerlings that compete with each other for feed, resulting in a multitude of undersized fish. In addition they have early sexual maturity, and once this sets in further growth is restricted. In practice it has been found that as much as 30 to 50% of a harvest from a tilapia pond has been unmarketable because the fish were undersize. Male-only tilapia ponds overcome this problem. The traditional method of sexing tilapia was by visual inspection of the genital papilla whilst restraining the fingerling manually. This was stressful for the fish and time-consuming for staff. It was also wasteful, as 50% of the progeny (the females) were discarded. These problems have been resolved genetically with monosex YY tilapia males, which sire XY-only male progeny (Beardmore *et al.*, 2001). Monosex male tilapias are now used commercially and are gaining popularity. In species other than tilapia and catfish, female-only strains would be preferred where the need is to take advantage of the female's faster growth and avoid the precocious maturation and early deterioration in meat quality seen in males.

Genetic modification is a more appropriate way of manipulating gender than sex reversal through steroid treatment. It eliminates hazards with disposal of feed residues containing sex hormones. Nevertheless, sex reversal using 17-α-methyltestosterone (MT) delivered to fry through the feed is commonplace in the trout industry. Genotypic females are masculinized at the fry stage using MT, and their sperm is subsequently used to fertilize normal eggs, resulting in all-female offspring. This technique is used commercially in chinook salmon as well as trout (Johnstone, 1996). It is necessary because precocious maturation is a common problem in males. They change, almost overnight, from a silvery to black colour, and quickly develop a kype. Their ugly appearance renders them unmarketable as whole fish, but more importantly they are undersized for most portion markets.

Transgenics

Rapidly growing lines of fish with a large mature size have been developed using transgenesis. Two approaches have been used so far in salmon: insertion of the growth hormone (GH) gene, and insertion of an anti-freeze gene, which extends the period of endogenous GH production in cold conditions. In extreme cases, transgenic coho salmon have grown to over 30 times the size of normal siblings. Other GH transgenic lines have been developed in channel catfish, common carp and tilapia, and the growth advantages have been more moderate. With GH transgenesis, the growth enhancement depends on the growth potential of the recipient for the GH gene. In wild types, which have a relatively slow growth rate, the potential enhancement is large, but, in types that are already achieving high growth rates from genetic selection, the enhancement is smaller.

The potential welfare problems in GH transgenic lines include:

- Impaired immune function, including impaired T-cell function, reduced resistance to bacterial endotoxins and autoimmune activity.
- Over-expression of GH resulting in acromegaly. This takes the form of enlarged head, bulging operculum, excessive cartilage growth in the jaw and premature smoltification. The overgrowth in the operculum was sufficient in some individuals to disrupt normal ventilation. Excessive cartilaginous growth can also impair feeding and vision (Devlin *et al.*, 1995).

The meat from GH transgenic fish has higher protein, less fat and less moisture than that from non-transgenic siblings.

Transgenesis by incorporating an anti-freeze gene has been achieved with winter flounder (*Pseudopleuronectes americanus*) as well as Atlantic salmon. Atlantic salmon kept in sea cages off Newfoundland often die during the season when icebergs float south during winter. Introducing the anti-freeze protein into the salmon should improve their survival, but so far the concentration of the protein has been 100 times below that needed to prevent freezing under those conditions.

Salinity tolerance could be manipulated in the future by transgenesis. The risk in this case is that the genes may be under-expressed and salinity tolerance may be incomplete in a proportion of individuals. There would be a high probability of suffering in affected fish, especially if the under-expression resulted in a slow death.

4

Cattle

The welfare issues linked to beef production include:

- respiratory and digestive disease in feedlot cattle
- underfeeding
- handling problems
- dystocia and double muscling
- confinement in some veal units
- castration methods
- disbudding
- branding
- inadequate shelter
- transport stress

Dairy farming produces a substantial proportion of the world's beef as bobby calves and cull cows, and the welfare problems it faces include:

- underfeeding and poor body condition relative to production requirements
- cow–calf separation
- mastitis*
- lameness*
- metabolic and digestive disorders
- cull cow and downer cow management*
- discomfort and trauma
- inductions*
- calf management*
- social stressors
- heat stress

Topics marked with an asterisk have been included in this chapter because of their welfare importance and relevance in meat production. There are many other conditions in beef and dairy production that are less common but cause substantial

suffering in the affected animals, but space constraints do not allow their coverage.

Beef Production

Feedlots and cattle barn systems

The welfare issues connected with feedlots include:

- stress during introduction to the feedlot
- respiratory disease
- digestive disorders
- trauma
- routine inspection and veterinary care
- spaying heifers
- heat stress

In the USA, feeder cattle are introduced to feedlots either as 5- to 8-month-old weaners, or as 230 to 360 kg yearlings. Average daily live-weight gains are typically 1.4 kg, with a finishing weight range of 455 to 570 kg. Young calves have some difficulty adapting to feedlot conditions. They are susceptible to disease, and in particular respiratory disease that is linked to **stress-induced immune suppression**. They have had less opportunity to build up immune protection to respiratory pathogens compared with yearlings, and they are affected more by sudden changes in feed. The accumulated stresses associated with their introduction to the feedlot are also more severe. Not only are they weaned at the time they leave the farm of origin, but they are often transported long distances from the breeding ranches to the feedlots in the grain-growing regions. On top of this, shortly after

©N. Gregory 2007. *Animal Welfare and Meat Production*
(N. Gregory)

arrival they are put through a crush for vaccination, anthelmintic treatment, dehorning, castrating, branding and ear tagging. Together, these create a substantial stress, and it is not unusual to see calves fall over when they are let out of the crush. Some feedlots have installed rubber mats made from car tyres at the exit to the crush to reduce injuries during these falls. The calf also has to adapt to the feedlot ration during the following week. Calves react maximally to these combined stresses during the first 15 days at the feedlot, and this is when most morbidity is seen. If weaning coincides with transfer to the feedlot, the calves have suppressed appetite and growth for a period of 3 to 14 days. Various strategies have been tried to reduce this growth check (Table 4.1), but none has been sufficiently reliable to warrant widespread adoption.

There are contrasting views about whether it is less stressful to process the calves as soon as they arrive at the feedlot or hold them for 3 or more days before processing. Common advice is to delay the more stressful tasks such as castration and dehorning for 3 to 5 weeks, but to proceed with prophylactic treatments straight away if they were not performed before dispatch to the feedlot.

Failure to eat during this early period is due to unfamiliarity with the feed, but bruises and injuries during the journey to the feedlot may make the animals more reluctant to move, eat and drink, and this in turn may increase the risk of acquiring an infectious disease.

Stress-induced immunosuppression on entry to the feedlot is thought to be due to elevated cortisol levels in the circulation. Cortisol inhibits several aspects of immune function, including:

- primary antibody responses
- lymphocyte blastogenic responsiveness to antigens
- recruitment of neutrophils to sites of inflammation
- activity of the hydrogen peroxide–halide– myeloperoxidase antibacterial system
- macrophage phagocytosis and bactericidal capacity

It is not realistic to gauge immune competence from an animal's behaviour. Animals with a nervous temperament can have higher plasma cortisol concentrations and poorer growth rate whilst most measures of immune function are normal (Fell *et al.*, 1999). A possible exception is serum immunoglobulin M (IgM), which, paradoxically, can be lower in calm animals. A fast growth rate in the less flighty animals is thought to be due to a stronger appetite and feed intake.

Typically, the prevalence of disease in US feedlot cattle is 8%, and **mortality** is 1%. Respiratory disease accounts for over half these losses, and digestive disorders about one-third. Of the overall morbidity, 70% occurs in the first 45 days. At some feedlots over half the deaths occurred later than this when there were problems with bloat and enterotoxaemia.

The main disease risk during the introductory period is **bovine respiratory disease** (BRD). The suffering and signs associated with BRD

Table 4.1. Methods tested for reducing the growth check that occurs when weaner calves are introduced to feedlots.

Preconditioning at the farm of origin	Calves are weaned at least 3 weeks before sale, and trained to eat and drink from troughs, treated for parasites, vaccinated, castrated, dehorned and ear tagged at the farm of origin
Backgrounding at a holding site	Calves are gathered together from several sources and either fed a growing ration or grazed on pasture for 30 to 90 days at a location near their farm of origin
Pre-weaning at the farm of origin	Calves are managed in the same way as the preconditioning schedule except they are not treated for parasites or vaccinated
Trainer cows at the feedlot	On arrival at the feedlot, calves are penned with steers or mature cows that have been adapted to the feedlot, in the hope that this will encourage early feeding
Enhancing the immune system on arrival at the feedlot	Treating the calves with immuno-enhancing drugs (e.g. lipoidal amines, vitamin E)
Prophylactic treatment with antibiotics	On arrival at the feedlot, calves are given a course of subcutaneous injections of tilmicosin

are similar to acute respiratory diseases in other species, and include depression, inappetence, fever, nasal discharge, coughing and dyspnoea. It is initiated by viruses, which compromise immune defence mechanisms further and facilitate secondary infection by bacteria (Roth, 1984). There may also be a genetic component, at least in double-muscled breeds. Genetic selection resulting in either a larger birthweight or more muscular body could increase susceptibility to the disease, as the genetic correlations between birthweight and BRD are in the range 0.25 to 0.50 (Muggli-Cockett et al., 1992). Double-muscled Belgian Blue and Belgian White calves that have a low respiratory capacity are less tolerant of BRD, and the heritability for respiratory capacity in these breeds is about 0.48. Respiratory capacity was determined from the flow of respired gases when the calves were given an intravenous injection of the respiratory stimulant lobeline (Bureau et al., 1999). In other types the heritability of BRD is low (0.06 to 0.10).

When a case of BRD becomes advanced, a decision has to be made whether to treat with a drug or slaughter the affected animals. Drug treatment is not always effective and so it is usually used for a short period to see whether it is going to be some help. If it gives little benefit and it is decided to switch to a slaughter policy, there is a mandatory drug-withholding period, and this can reduce the chances of obtaining salvage value before the animals die. Decision-making can sometimes be made easier by measuring the animal's plasma lactate concentration (Coghe et al., 2000). In advanced cases, oxygenation and aerobic metabolism are compromised and plasma lactate concentrations are raised. If plasma lactate is > 4 mmol/l, prognosis is poor and it is advisable to slaughter without delay.

In Canada, there is a pre-sale vaccination programme for weaner calves that are put through auction markets. The aim is to encourage the development of protective immunity to pathogens associated with BRD before the calves are exposed to those pathogens in the market environment. There are special auctions that only offer calves that comply with the programme.

Besides BRD, calves are also prone to acquiring and excreting gastrointestinal pathogens when they are introduced to a feedlot. When calves were monitored from their farm of origin in Tennessee, through an auction market and then on to a Texas feedlot, the prevalence of Salmonella excreters was

0, 1.5 and 8% at the respective stages (Corrier et al., 1990).

Once the calves have settled down at the feedlot, a common aim is to minimize fluctuations in feed intake. Rumen health is optimized when feed is consumed as regular frequent meals, as rapid changes in intake can be linked to **acidosis** and bloat. Sometimes acidosis is difficult to control. If there is a turn in the weather and the cattle stop feeding for a period, they often return to the feed bunk with a strong appetite and overeat when the bad weather breaks. This results in a transient acidosis in the rumen, which may cause the cattle to go off their feed. Rumen pH rebounds upwards, appetite returns and the cycle is repeated. This condition is known as subacute rumen acidosis (SARA). SARA occurs when rumen pH falls to 5.5 and less, and it is thought to contribute to lameness.

A uniform, steady rate of feeding is also conducive to improved feed conversion efficiency (Prawl and Owens, 1998). When cattle have restricted access to feed they tend to take larger mouthfuls compared with ad libitum fed cattle, and they spend less time chewing their feed. Bolting the feed results in poorer digestive efficiency.

Feed is usually dispensed as a total mixed ration (TMR). Experience is needed in ensuring that there is adequate palatable feed in front of the animals whilst minimizing wastage and feed refusal. One study indicated that feeding TMRs can be linked to more aggression between animals at the feed bunk, compared with feeding roughage and grain separately (Hindhede et al., 1999).

Some feedlots have acquired a poor reputation for managing physical **injuries**. They are sometimes overlooked or misdiagnosed (Stokka et al., 2001). In the Pacific Northwest, 5.5% of feedlot deaths during the winter months of January to March 2000 were attributed to injuries. The majority of the injuries were in the hindlimbs, with the hip and stifle most frequently affected. Some injuries occur during mounting and riding behaviour, but the majority of hip injuries are thought to occur during transport and processing. Joint injuries result from poor handling or accidents on slippery surfaces. Shoulder injuries occur when cattle collide with the head bail of the crush or when downer cattle are dragged.

Osteochondrosis is an occasional problem in feedlot cattle in the USA. It is mainly seen in the stifle joint, and according to one estimate it explained 9% of the cases of lameness. Riding behaviour is responsible for some of these injuries, especially on hard floor surfaces. It is thought that high growth rates make the animals more prone to osteochondrosis, but this has not been proved.

When an animal acquires a serious injury, the most appropriate course of action is to promptly slaughter it at the feedlot. However, economic pressures may override that decision. Instead, it may be decided to salvage the value of the animal through an abattoir. Any hormone growth promoter will have to be removed and the injured animal retained at the feedlot for the required drug-withholding period before it is submitted for slaughter.

The prevalence of **liver abscesses** in feedlot cattle at slaughter in the USA is usually between 10 and 14%. Liver abscesses can develop as secondary infections during rumenitis. Bacterial emboli pass from the infected rumen wall via the hepatic portal vein and lodge in the liver, where they produce an abscess. The abscesses are not usually severe enough to cause clinical signs, but they are linked to poor growth performance and feed conversion. *Fusobacterium necrophorum* has been a common isolate.

Some abscessed livers in feedlot cattle rupture during evisceration at the abattoir. The tissue containing the abscess can be quite delicate, and it is suspected that in the live animal rupture can occur during handling and fighting. Abscess rupture could lead to systemic illness and even peritonitis.

Heat stress is a common hazard in feedlots. Shading, sprinkling, misting and fogging can all be used to keep cattle cool, and the method chosen depends in part on water availability. Providing shade or sprinklers has been more effective than providing misters, presumably because water from the misters was less effective at penetrating the coat to promote evaporative cooling. In Texas feedlots, introducing shade not only improved growth performance but also reduced the prevalence of dark-cutting beef at slaughter. It helped to reduce overt signs of heat stress as well as aggression. During heat waves it may be advisable to restrict feed intake or limit it to the cooler period of the day. This will help maintain body temperature within normal levels, but it is inevitably associated with lower overall feed consumption (Mader *et al.*, 2002).

When the beef market is depressed, fewer heifers are retained at farms as breeding replacements and more are put through feedlots for finishing. About 17% of heifers arriving at US feedlots are pregnant, and in December it can be as high as 64%. **Pregnant heifers** are often regarded as a menace by the feedlot industry. They incur additional care and labour and they are worth less when slaughtered. If they calve at the feedlot, the calf has to be either slaughtered as a calf or reared at the feedlot, as most feedlots have quarantine status and cannot dispatch live animals to another feedlot or farm. At one Kansas feedlot it was reported that 14% of pregnant heifers abort spontaneously within 3 weeks of arrival (Edwards and Landert, 1984). By the time they are slaughtered, about 3% of heifers are still pregnant (Lorenzen *et al.*, 1993). If the heifers are slaughtered whilst pregnant, they are penalized for their poorer killing-out per cent. At some feedlots, the heifers are palpated on arrival at the feedlot, and any that are found pregnant are given an abortifacient. This places additional stress on the animals and, if a corticosteroid is given, there is additional immunosuppression. When fenprostalene was used at one feedlot as the abortifacient, 43% of the heifers had retained placenta or uterine infection, and this incurred additional veterinary costs.

Buller steers are another menace for feedlot companies. There are two types of buller. Some are volunteers that are receptive to bulling activity. The other type is the victimized animal that may be ill or insufficiently fit to resist unwanted mounting behaviour. The incidence of this **buller syndrome** in the USA varies between 1 and 3%, but it has been as high as 11% in small feedlots. Buller behaviour often peaks in the autumn and winter, and, besides being disruptive to other cattle, it can lead to injuries. Early removal and separate penning of buller steers can help minimize injuries.

At some feedlots, heifers are spayed as a way of preventing bulling behaviour and unwanted pregnancies. Two surgical spaying methods are used. In the paralumbar fossa approach, the heifer is sedated with xylazine and given a paravertebral block at T 13 and L 1. In the vaginal approach, an epidural is used and the cervix is bypassed by puncturing the vagina and then transecting the ovary, which is positioned trans-rectally. Following the vaginal approach, the heifer may show slight stiffness in its gait, with an elevated tail, and there is a check in feed intake and growth. It is usually recommended that, when spaying is done, other surgical procedures, such as dehorning, should be avoided. Alternatives to spaying are either to install fencing, which allows the heifers to be run as a separate group, or to feed a masculinizing hormone such as melengestrol acetate (MGA) to counteract oestrus. MGA is commonly used, but it increases the heifers' susceptibility to **acute interstitial pneumonia** (AIP). This is a non-infectious condition, which is found almost exclusively in heifers and has become the second most important cause of death from respiratory disease in the feedlot industry. Affected animals develop an acute severe dyspnoea, which often ends in emergency slaughter or death (Woolums *et al.*, 2001).

Feedlots are best sited in low-rainfall regions. Otherwise the high density of cattle plus the rain can turn the pad (manure bedding) into a mire. Even so, **boggy conditions** occur in low-rainfall zones where rains are highly seasonal. Feedlots usually have earth floors, except for a concrete apron around the feed bunk and water trough. Concrete floors drain faster during the wet season, and this should also reduce the risk of odour problems. Small-scale outdoor kraaling is becoming popular in colder districts in Europe as a low-cost alternative to finishing cattle in a barn, and conditions underfoot are managed by using large amounts of bedding and by keeping the **stocking density** to within reasonable limits.

When barns are used, there is financial pressure to make good use of the building by stocking it intensively. There are hazards with high stocking rates. When 470 kg beef heifers were stocked in a barn at 1.5 or 3.0 m² per animal in slatted-floor pens for 104 days, the heifers with less individual space had lower plasma cortisol concentrations (Fisher *et al.*, 1997). Their immune responses to haemocyanin antigen and their white blood cell numbers were the same. The heifers that had more individual space spent more time sitting down and showed more head-resting behaviour. When dairy heifers were prevented from lying down because of overstocking, they developed lameness from solar (plantar) bruising (Leonard *et al.*, 1996). This was, however, an extreme case, as lying time was reduced to 5 h per 24 h.

Controlling **ammonia** emissions is becoming a more important feature of environmental management in cattle production. In cattle barns, long straw bedding is likely to release more ammonia into the atmosphere than either chopped straw, peat plus chopped straw (60 : 40) or no bedding on

a concrete floor (Jeppsson, 1999). However, cattle stayed cleaner with peat plus straw bedding, because peat absorbs more moisture.

The usual aims in providing cattle **bedding** are to give them a dry, clean, comfortable surface to sit on. Keeping the cattle clean is one of the main aims, and this depends on maintaining a dry upper surface. Bedding that has high water absorption and retaining properties will be less prone to surface ponding under pressure from the sitting animal. Balers that use high compression cause the stems of the straw to flatten and split and this increases absorptive capacity when the straw is used as bedding.

Cleanliness in slaughter cattle has become an increasingly important issue for the beef industry because of the implications for meat safety. A comparison of different barn systems stocked at different densities, concluded that:

- In straw-bedded pens stocked at high densities, the upper layer of straw was dirty regardless of the amount of fresh straw that was added.
- In slatted and sloped floor systems, low stocking densities allowed dung to build up, as did very high stocking densities. At very high densities some animals were obliged to lie in dirty areas, whilst at low densities there was insufficient animal traffic to disperse the dung or tread it through the slats.
- Slatted systems have low labour requirements but high capital cost (Scott and Kelly, 1989).

Long-haired cattle are more prone to getting dirty than short-haired animals. However, the animals become cleaner as the winter season progresses and hair begins to shed. Cattle get dirtier if the bedding is wet. This is a feature where rain projects into a barn. It also occurs where cattle are fed first-cut grass silage, because this results in wetter faeces.

Some feedlots specialize in putting condition on **cull cows**. These animals are often culled because of reproductive failure, and younger animals can produce good-quality carcasses. Cows in poor condition have low lean-to-bone ratios, and older cows can have low lean-to-bone ratios even when finished (Graham and Price, 1982). When put through a feedlot, there is preferential accumulation of fat, rather than lean, and much of the gain in fat is in the mesenteric and kidney knob and channel fat (KKCF) depots.

Barley beef

When the barley beef system was first developed, young cattle were fed ground barley. Serious losses were experienced from bloat and acidosis. When rolled barley was provided in place of ground barley, most of the problems were overcome, but even so there was a high incidence of rumenitis and some animals were affected by liver abscesses. Hyperkeratosis of the rumen wall has been a problem when cattle have been fed rolled–pelleted barley.

Barley beef is still produced from bulls in parts of Europe, but its popularity in any year depends on prevailing grain prices. The animals sometimes show oral stereotypic behaviours, and they are apt to eat the shavings used as bedding. At some markets and abattoirs the bulls are penned individually to reduce social stresses, injuries and dark-cutting beef. Liver abscesses are still a problem.

Veal calves

Confining veal calves for long periods in crates has been phased out in EU countries for animal welfare reasons, and veal production is presently less common in the USA because of a downturn in the market. In Europe, veal calves can be confined for the first 2 to 3 weeks, but thereafter they are held in groups in larger pens and are usually fed from troughs or buckets with floating teats. Keeping the calves in pens instead of raised stalls produces a cleaner calf by the time they are sent for slaughter and less knee-joint swelling on the front legs (Terosky et al., 1997). They are often fed a soya-based milk substitute, which may be supplemented with restricted amounts of maize silage, but the overall amount of dry feed is usually insufficient to prevent mutual sucking on each others' muzzles, ears or prepuces, or coat-licking and tongue-playing. Compulsive water drinking was a problem when water was available ad libitum, even though the calves were not dehydrated and had ample milk replacer (Gottardo et al., 2002).

Gastrointestinal problems that have been encountered in veal calf units include chronic tympanitis, chronic vomiting, abomasal erosions leading to ulcers, and failure of the oesophageal groove reflex. Providing the right cues before feeding is important in initiating the oesophageal groove reflex and avoiding ruminal drinking. Ruminal drinking occurs when a calf gulps the milk, licks it or bites it, instead of sucking it in. The oesophageal groove can be overloaded with milk if drinking is too rapid, and it floods into the rumen. The animal must learn to slow down and suck, and training it to anticipate a feed can improve the efficiency of the reflex. Ruminal drinking leads to fermentation of milk in the rumen to butyrate and lactate, and this causes a fall in overall digestibility of the feed. Inappetence sets in, and in extreme cases there is growth retardation, recurrent tympany, abdominal distension and clay-like faeces (Breukink et al., 1988).

Abomasal ulcers are usually seen in the pyloric region, and they are associated with the consumption of large volumes of milk. They can be avoided by decreasing the volume and increasing the concentration of the milk substitute (Welchman and Baust, 1987). These ulcers probably stem from overloading of the abomasums, which creates a risk of pyloric obstruction. There is mucosal damage from compression and ischaemia. At one time the prevalence of abomasal ulcers in Dutch veal calves was put at 23%, with a further 22% showing scarring from previous erosions (Wiepkema et al., 1987).

Anaemia occurs in white veal production from feeding an iron-deficient milk substitute. The intention is to produce a pale meat, because it attracts higher prices than pink veal. Growth is not adversely affected, but the animal may be clinically anaemic.

In the USA there have been problems with **stale calves**. These are very young calves that are taken back to a market to be resold because they are considered diseased or otherwise unsuitable by a calf buyer. It was estimated that calves in Wisconsin in the early 1990s were on average sold two to three times before they reached the final veal unit where they were retained (Friend et al., 1994). A law was introduced in 1992 requiring that calves up to 90 kg are marked each time they are sold. The usual marking method was to split the ear with a knife or notch it at an edge. Ear notching produces a mild startle response, but in other respects it does not appear to cause any immediate pain. Ear notching is no longer allowed, and the only tagging system that is permitted is a glue-on back tag. These are prone to alteration, and so the whole scheme has been largely ignored by livestock dealers because it has lost credibility. Clearly, this is a case where good intentions have not prevailed because of practical difficulties.

Dystocia and double muscling

Double muscling is controlled by a gene that inhibits the expression of the active form of myostatin. Myostatin normally inhibits muscle growth, but mutations have developed which remove that inhibition. The subsequent double muscling is present in both the homozygous and heterozygous derivatives of the mutation. There are at least six different mutations resulting in double muscling in cattle, but they all appear to operate by reducing the effect of myostatin, which in turn allows muscle fibre hyperplasia that is initiated prenatally.

One of the main concerns about the double muscling mutant is that pure-bred females have to reproduce by Caesarean in order to continue the homozygous double-muscled line. The pure-bred female is prone to dystocia because of the large birthweight of the pure-bred fetus. The risk of dystocia could, however, be avoided genetically using allele genotyping and when managed appropriately this could do away with the need for Caesarean sections. This would be achieved with a two-step backcross system as follows. Heterozygous heifers are bred to a normal (no double muscling) bull for the first generation. Then, as cows, they are put to a bull that is homozygous for double muscling. Half the calf crop from the heifers will be homozygous non-double-muscled and half will be heterozygotes. Half the crop from the cows would be heterozygous and half would be homozygous for double muscling. By genotyping the allele in the offspring, heterozygous females and homozygous males can be selected as replacements (Short et al., 2002). This approach does not involve the same level of dystocia risk associated with homozygous nucleus herds, and it does not expose the heifers to the inevitable need for Caesareans.

The double-muscling mutation involving the myostatin gene has occurred in a range of breeds, including Belgian Blue, Piedmontese, Pathenaise, Asturiana de los Valles and South Devon. The condition appeared in Belgian Blue cattle around 1960, and by 1980 nearly 90% of the breed within Belgium were double-muscled. The popularity of the double-muscled type is due to their:

- high proportion of muscle and low proportion of fat and bone
- higher killing out percentage

The increased muscle is particularly evident in the proximal parts of the legs and from the surface grooves between adjacent muscles in the hindleg.

Presently, in Belgium, about 90% of pure-bred Belgian Blue cattle require Caesarean section. In the UK, it is about 82%, and there is less need in cows with a large pelvic area and low body condition score (Murray et al., 2002). Dystocia can be a particular hazard in cows that are overfed (Maree, 1986). At one time it was thought that pure-bred females also had a smaller internal pelvic area, and that this contributed to the need for Caesareans, but it has been found that the Belgian Blue cow has a pelvic area that is 12% greater than that of the Charolais and Aberdeen Angus. Nevertheless, there is a fetus–dam mismatch in the pure-bred. In Belgian Blue heifers, the risk of dystocia can be almost halved by delaying calving until the heifer is 3 instead of 2 years old.

Caesarean sections are usually performed with local anaesthesia or a lumbar paravertebral block, plus appropriate physical restraint. A caudal epidural block is used if there is concern about prolapse of the ruminal wall or intestines through the incision.

Calves born to Caesareans are less susceptible to cold compared with calves born to severe dystocia. Calves can survive for up to 8 h during second-stage labour, but they are inevitably weakened when it lasts that long.

There are, of course, other reasons for dystocia besides inappropriate size of the fetus, but presently about 45% of all cases in Scotland are due to mismatching problems. An abundance of feed during late pregnancy can be associated with larger than normal calves, and can contribute to fat deposition in the pelvic cavity. This exacerbates the crowding of the pelvic canal, and it is associated with reduced elasticity and resilience in the vaginal wall (Sloss and Dufty, 1980). The most frequently observed complication in the cow following fetopelvic disproportion is paralysis of the obturator or ischiatic nerves. About 30% of cows suffering from this do not recover and have to be destroyed. The obturator nerve supplies the adductor muscles in the thigh, and dysfunction leads to splaying of the legs with the cow being unable to stand.

In rangeland and ranch conditions it can be difficult to attend to a beef cow with dystocia. In one study in Australia, it was reported that fewer than 10% of the cases that required calving assistance could be restrained quietly with a halter (Johnston and Sloss, 1967). The alternatives are moving them to a yard where they can either be held in a crush or roped, or lassoing and roping them where they are found. Once the animal is caught, sedatives or tranquillizers make management easier.

Various strategies have been tried for managing the risk of dystocia in double-muscled breeds. They are:

- delaying mating and first calving until the heifer is bigger
- ensuring that the cow is not over-fat when she calves
- limiting fetal growth by underfeeding the cow
- inducing early calving pharmacologically
- delivery by Caesarean

Each approach has its own hazards, and presently the preferred approach in many countries is to rely on Caesarean delivery as and when it is needed.

A balance has to be struck when trying to manage birthweight and subsequent calf mortality. The relationship between birthweight and calf mortality is curvilinear (Fig. 4.1). Undersized calves are immature and oversized calves experience dystocia, both of which reduce the chances of survival.

Calves that are born to cows with dystocia are at greater risk of dying soon after birth. The correlation between dystocia and perinatal mortality is high ($r = 0.86$; Smidt and Huth, 1979). Risks of injury to the calf during traction include broken ribs, dislocation of the head of the femur at the epiphysis, diaphragmatic tears and fractured vertebral column. Calves with dislocated hips may develop lameness from degeneration of the femoral nerve and the quadriceps femoris muscle, and this combination is more common in heavily muscled beef breeds (Hamilton et al., 1978).

Weaning

Weaning is an inevitable stress because it disrupts a social bond and comfort behaviours. The questions are when it should be done and whether there are ways of reducing the stress.

There is a fine balance between weaning early to allow return of body condition in the cow and good prospects of successful mating, and penalizing the growth of the calf. Early weaning works well if the calves are supplemented, and it provides a payback if the cows are in thin condition (Pordomingo, 2002).

If calves are weaned early (e.g. 4 months of age), there is a lot of vocalizing and pacing – more so than with weaning at 7 months of age. Although early-weaned calves may vocalize to begin with, they usually settle after 3 days if they are in visual contact with the cows, whereas calves that have been completely isolated from the cows still show signs of distress 6 days after separation. If weaning is too early it can check subsequent growth and reduce calf size when they are sold to feedlots at 7 months of age. In addition, weaning at the time of relocation to the feedlot suppresses overall growth more than relocating previously weaned calves.

Underfeeding

Underfeeding is probably the single most important welfare issue in beef production. The effects of seasonal underfeeding on cattle in developing countries are particularly relevant and they have been discussed in Chapter 2. More generally, underfeeding occurs in the following situations:

- lack of feed, e.g. unfavourable climatic conditions
- physiological situations where appetite alone cannot sustain metabolic needs, e.g. lactation
- when a farmer's financial strategy is to minimize feed costs
- neglect

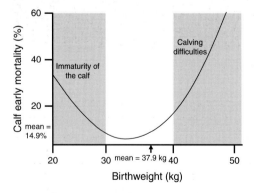

Fig. 4.1.   Relationship between calf birthweight and early mortality.

The risk is particularly high when these situations occur together.

In farming systems where underfeeding is a predictable event, farmers often rely on making the most out of re-alimentation and compensatory growth. There are production benefits in exploiting compensatory growth. The cycle of poor feeding followed by improved feeding often leads to enhanced lifetime feed conversion efficiency. This happens provided the level of underfeeding is within reasonable limits. In severe situations, when the period of underfeeding falls well below the maintenance requirements of the animal, the efficiency of digesting the feed is reduced and feed intake is inhibited (Grimaud *et al.*, 1998). This is linked to reduced rumen protozoa populations and increased rumen pH. If this state persists for too long, compensatory growth does not occur and the animal is stunted.

During the early stages of underfeeding, body fat is mobilized to provide the animal with energy for routine activities. Some muscle breakdown also occurs, and it accelerates as the fat reserves become depleted. Muscle catabolism during underfeeding is particularly damaging because it reduces the animal's strength. Research trials in Australia have shown that, during the early stages of underfeeding, muscle protein breakdown can be spared by giving the animal the β_2-adrenergic agonist clenbuterol, but this is at the expense of body fat and viscera (Sillence *et al.*, 2000). The wisdom or folly of this approach depends on long-term outcomes for the animals, which cannot always be predicted, and whether it is more cost-effective to use the money for feed instead of clenbuterol.

Underfeeding has negative effects on breeding performance. When breeding cows receive inadequate feed to the extent that body condition falls to 4 or less on the 1 to 9 point International Livestock Centre for Africa (ILCA) and US scales, the proportion of cows that express oestrus and become pregnant declines (Fig. 4.2, Table 4.2 Mukasa-Mugerwa *et al.*, 1997; Wikse *et al.*, 1994).

Cattle production in developing countries

Some of the welfare issues in cattle production in developing countries are discussed in Chapter 2. This section focuses on mortality rates and production levels in poorer countries, and the interaction between underfeeding and breeding performance.

Cattle production in poor rangeland systems can be disappointingly low. In normal years, the offtake rate in sub-Saharan Africa is usually 8% per year but it varies from 3% to 9%. Male cattle are usually sold when they are a little over 2 years old. Age at first calving is often about 4 years and the average lifespan of breeding cattle in Kenya is about 12 years (Roderick *et al.*, 1998). Calving rate is typically between 50 and 80%, calving interval is often a little over 600 days, and calf mortality varies between 6 and 40% (Homewood *et al.*, 1987).

Fig. 4.2.   Body condition scores 1, 4 and 7 on the ILCA scale.

Table 4.2.   Pregnancy rates in beef cows according to body condition score at weaning time.

	Body condition score				
	3	4	5	6	All
Number of cows	53	209	131	29	422
Number pregnant	27	159	121	29	336
% pregnant	51	76	92	100	80

Cow mortality rates are usually between 5 and 15% per year, excluding severe drought years when calf and cow mortalities can rise to 90% and 50 to 80%, respectively. In drought years calving rate has been known to fall to zero.

A long calving interval can be a sign that the cows are not recovering sufficient condition after the previous calving. If a cow does not ovulate and conceive during or soon after the rains, then it is unlikely that she will conceive until the following rainy season. Theoretically, a bimodal rainfall pattern through the year should result in a bimodal calving distribution. Two periods of rain improve the chances of achieving a calving interval of less than 2 years, in comparison with regions that have a single rainy season during the year. In practice, it does not always work that way. For example, in parts of Kenya where there are two rains, the calving pattern is compact and the average calving interval varies between 600 and 650 days. The condition of these cows needs more than one period of grass growth for recovery from lactation.

Wealthy cattle owners have larger herds, but the standard of cattle care is not always correspondingly higher. Calf mortality in Borana herds of Ethiopia has been put at 24%, 16% and 30% for wealthy, intermediate and poor households, respectively. The high calf mortality in the wealthy herds was probably due to reduced levels of management per calf, whereas in the poorer herds it was mainly due to malnutrition because of competition with the owner's household for the milk.

In poor and intermediate households, the aim is to limit consumption of milk by the calf, to allow offtake for the family. The emphasis is on calf survival rather than growth. Where calf feeding is well supervised, early calf mortality is low, but it can increase dramatically at the time of weaning when the calf depends on grazing at a time when the dry season has set in. In Benin, it is recognized that calf mortality is higher in transhumance herding systems than in sedentary herds (Aboagye et al., 1994). In transhumance herds, most of the calf deaths are from underfeeding in the first weeks of life.

Not all cattle pastoralists in developing countries are regular beef eaters. They prefer to sell or exchange cattle for goods or other animals they need. Cattle mortality is not always a complete waste. Dead animals are often eaten and sick animals are culled and consumed if they look as though they are about to die. Some live sales are made with itinerant cattle dealers or to ranches, and those animals eventually pass on to meat consumers in the larger towns or cities.

Draught and work animals

Draught animals make a contribution to the meat trade when they are slaughtered at the end of their working life or if they are maimed. This source of beef is particularly significant in China, which is the third largest cattle-owning country. Sixty-five per cent of these animals belong to the Yellow breeds, which traditionally are draught animals, but some have been upgraded to dual-purpose lines.

Branding

Branding is used to deter theft and to prevent unauthorized movement in disease control programmes, and it has been used as a way of identifying animals that are part of a subsidized destocking programme aimed at reducing production surpluses.

Branding is responsible for substantial devaluation of cattle hides. A survey of hide damage at an abattoir in Ethiopia showed that branding accounted for 24% of the pre-slaughter damage (Mersie and Bekele, 1994). In the USA, the butt and side are the most common branding sites, but at one time this was avoided by branding the cheaper cheek skin. For example, in the brucellosis control scheme, animals were branded with a 'B' on the jaw if they could only be sold for slaughter. A survey of opinions amongst US beef industry leaders in the mid-1990s showed that they felt that jaw branding should be phased out because of its poor public image (Schlink, 1995), and this move was hastened by a public outcry in the late 1990s.

When hot-iron branding has been compared with freeze-branding, the animals receiving the hot iron had:

- greater escape or avoidance reactions, such as vertical movement in the crush
- higher plasma adrenaline and noradrenaline levels 5 min after branding
- higher plasma cortisol levels 40 min after branding, and cortisol remained higher for longer

- greater amount of vocalization (Schwartzkopf-Genswein *et al.*, 1997)

It causes acute pain, but if nerves are sufficiently damaged by the procedure it is unlikely that there is chronic pain at the wound. There is likely to be some hyperalgesia around the wound, but this has not been tested systematically. Freeze-branding is not innocuous. It causes tail flicking and vocalization, and sometimes there is kicking. It probably causes a longer-lasting ache.

Castration

Over the past 20 years there has been a trend away from castrating cattle, because farmers are taking advantage of the higher growth rate of bulls. Bulls also produce leaner carcasses than steers. In parts of Asia castration is considered a sinful act and it is only practised when a person from a low caste can be recruited to perform the procedure. It is sinful because it is an act of violence and it changes the personality of the animal. It is also considered defiling amongst those who believe in the continued existence of the soul after death. In some communities, farmers who have their bull calves castrated experience social boycotts, such as difficulty in marrying off their children.

In other societies it is becoming common to use a local anaesthetic during calf castration. In the traditional method, the calf is restrained usually on its rump. The scrotum is incised, the testes exteriorized and the spermatic cord broken by traction. When a local anaesthetic is used, it is injected into the scrotum to eliminate pain during the incision. Injecting a non-steroidal anti-inflammatory drug would also help eliminate subsequent inflammatory pain (Stafford *et al.*, 2002). In large calves a calf-casting cradle is sometimes needed (Fig. 4.3).

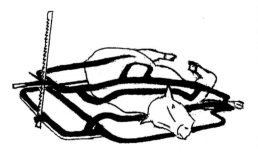

Fig. 4.3.   Casting cradle for calves.

Clamping each spermatic cord with a Burdizzo castrator is probably the least painful castration method. The inflammatory reactions and the plasma cortisol response shortly after Burdizzo castration are lower when it is performed at 1.5 instead of 5.5 months of age (Ting *et al.*, 2005). Our understanding of the pain that could occur during the weeks following castration is limited.

Calf disbudding

When fat cattle are sent to markets it is inevitable that horned animals will be mixed with unhorned animals when they have been sold and are allocated to the buyers' groups. The risk of injuries from horns is increased by the dispersal of horned animals. The only sure way of avoiding this hazard is to insist that all animals are unhorned if they are submitted for sale through a market. Horned animals can also inflict injuries on each other, but in groups of familiar animals this should be a lower risk.

The best time to remove the horns is when they are at the bud stage in young calves. It is a painful procedure and must be performed with a local anaesthetic. Applying local anaesthetic to the cornual nerves prevents the obvious behavioural responses to cautery disbudding and scoop dehorning, but calves may resist its administration. Lignocaine is the most commonly used local anaesthetic. It prevents pain for a 2 h period, but pain sets in as it wears off. This delayed pain can be prevented by injecting a non-steroidal anti-inflammatory drug, or by cauterizing the wound after scoop dehorning the lignocaine-treated calf (Stafford and Mellor, 2005). Cautery disbudding is to be preferred to scoop disbudding because there appears to be less risk of inflammatory pain, but breeding poll cattle would avoid pain and the disbudding procedure altogether. Premedication with xylazine is sometimes used to make restraint during disbudding easier. An alternative is to use a robust form of restraint, such as a calf-casting cradle (see Fig. 4.3).

Flies

Fly worry is resuming its former importance now that insecticide resistance is becoming more common and cattle are being kept in large units. The situation with the horn fly (*Haematobia irritans*) serves as an example. This fly gives cattle an irritating bite when it draws blood, and it focuses on the withers, shins and shoulder. In high numbers it can cause loss of growth. During the 1940s and 1950s it was controlled successfully with dichlorodiphenyltrichloroethane (DDT) and toxaphene, but resistance developed during the 1960s. In the 1970s, permethrin- and fenvalerate-impregnated ear tags were introduced, but fly resistance developed quite quickly. The approach now is to alternate between the organophosphate and pyrethroid insecticides whilst delaying use of either insecticide for as long as possible. In this way it is hoped to postpone or avoid combined resistance to both groups of insecticides.

Fly worrying can be seen when cattle stamp their feet, twitch their skin, tail flick and swing their heads towards their shoulders. These activities interfere with grazing behaviour, but fly-infested cattle compensate for this by increasing individual bite sizes (Dougherty *et al.*, 1993).

Stable flies (*Stomoxys calcitrans*) are also vicious biters. The irritation they cause can reduce weight gain and milk yield. They prefer to feed on the foreleg, where there is less interference from swishing tails, the coat is short and the skin is thinner with more sweat glands. In severe infestations, cattle will stop grazing and take refuge by standing in water if it is available, or they may bunch together to reduce the flies' access.

Mosquitoes are an underrated cattle nuisance. They are underrated because they are irritating shortly after dusk when discomfort in the cattle is less obvious. Cattle stamp their feet and ripple the skin in response to mosquito bites, but often the irritation develops after the mosquito has fed and left the host. Dairy cattle get obvious relief if their lower legs are hosed with cool water during the evening milking. In some parts of Texas and Louisiana cattle are not grazed near marshland during the summer months because of local mosquito problems. Brahmin cattle are less affected than European breeds of *Bos taurus*. When Angus and Hereford feedlot cattle were on a low plane of nutrition, mosquito attack suppressed growth (Steelman *et al.*, 1972), but this effect was not seen with high planes of nutrition.

Face flies (*Musca autumnalis*) do not bite, but they irritate stock when they congregate around the eyes. They favour dark coat colours, and are particularly numerous on animals that are sitting.

Electroejaculation

Electoejaculation is a common management practice in beef bulls and AI breeding stations. It is used for:

- obtaining a semen sample for evaluating breeding potential or semen abnormalities
- routine collection of semen in bulls with hindleg problems
- collecting semen from bulls with poor libido

Two types of electroejaculator are used. Programmed stimulators deliver a rhythmic current via a rectal probe with automatically increasing stimulation intensity for a period that does not usually exceed 1.5 min. Non-programmed stimulators apply the current continuously, and current duration is controlled by the operator. Typically, a 2-year-old bull would require about 15 impulses to stimulate ejaculation, whereas a 1-year-old may need about ten. Experienced operators regulate the current according to the intensity of the clonic contractions in the muscles of the hindleg. The aim is to minimize those contractions. Practical experience indicates that bulls do not develop aversion to the procedure, and most experts consider that it is no more distressing for the animal than other routine procedures such as vaccination and palpation whilst restrained.

Serving capacity test

Serving capacity tests have been a very controversial animal welfare issue. The concern has usually been about the way the test has been conducted rather than the justification for the test (Fraser, 1983; Mossman, 1983). Serving capacity tests are used for recognizing bulls with low libido or physical abnormalities, and the results are used as a marketing aid. Most physical abnormalities can be diagnosed by other means, and so the test is not essential in this respect. The test's main value is in gauging libido during a brief encounter with a cow, and in selecting bulls that will achieve prompt mating and hence a condensed calving period under range conditions. In large herds, a condensed calving period has the following welfare benefits. First, attendance at calving and calving assistance are more feasible when there is a tight calving pattern. Secondly, routine husbandry procedures such as disbudding and castration can be done at a uniform age, and this can be arranged to minimize pain and reduce handling in the calves. Thirdly, the timing of vaccine administration in calves can be managed more effectively when all the calves are at the appropriate age, and in some situations this could lead to reduced morbidity from disease.

Some bulls are shy breeders, and an active sex drive is needed especially under range conditions, where both the bulls and cows may have to travel some distance to ensure mating at the right stage of the oestrous cycle. However, the serving capacity test that has been developed has not always been a reliable method for predicting conception rate or calving rate, no doubt because there are many other factors that determine a successful pregnancy. This weakens the arguments supporting the reason for the test, but this is not the main issue that causes concern. It is the way the test is conducted.

In some versions of the test, the cow or heifer is confined in a service crate and made to stand to repeated mating by either a group of competing bulls or bulls presented one at a time. The cow may or may not be in oestrus, and confinement prevents her from swivelling to avoid the bull. Lubricants are usually used. Some people view the test as forced submission to mating. In human society this is not tolerated. Applying that value to this situation means that the test is ethically unacceptable, provided the same level of concern is given to cattle. Simply observing mating behaviour in a breeding paddock has failed to give repeatable scores for individual bulls.

Recommendations aimed at minimizing distress in the females include using placid females, using tranquillizers, limiting the number of consecutive tests on the same animal, and ensuring that the cow is in a sexually receptive condition. Separating aggressive bulls and ensuring that the test area is non-slippery are routine measures aimed at safeguarding bull welfare.

Serving capacity tests in sheep receive less criticism from a welfare perspective. The test in sheep does not involve any form of restraint, the ewes are in oestrus, and the ram is presented with more than one oestrus ewe at the same time (often three). In these respects, mating is consensual.

Inadequate shelter and exposure

Cattle are occasionally afflicted by snowstorms. In the 1967 snowstorms in New Zealand, it was found that cattle were the first stock to go down, and they experienced bigger initial losses than sheep. They succumbed within 2 days, but if they were fed during that period they had a better chance of survival. Sometimes, during severe cold, they forgo foraging. Instead, the emphasis is on heat conservation, and at these times hand-feeding is particularly beneficial, especially in late pregnant cows. Cows in the eighth month of pregnancy are very vulnerable. When immersed in snow they have a large conductive surface, which is not very well insulated. When ketosis sets in, appetite is suppressed and treating them can be difficult because they adopt an erratic, panicky behaviour.

In western Canada beef calves are usually born during winter but they are kept indoors. If the bedding is wet there is a risk of frostbite, especially in the hindlegs. This may only be discovered during physical examination for another complaint (Cruz and Naylor, 1993).

It is often thought that late-maturing beef breeds have lower cold tolerance than small 'hardy' breeds such as the Galloway. However, when comparisons have been made on the loss of condition under cold conditions (wintertime lows of −25°C), Limousin × Friesian cows were no different from Galloways (Wassmuth et al., 1999).

Feedlot cattle have limited protection from bad weather. This was a problem when a batch of cattle from a feedlot in Texas were slaughtered and it was found that they had extensive back bruising from a hail-storm. It took 49 days before the bruising disappeared and further animals could be slaughtered (Schmidt et al., 2000).

Dairy Cattle

Cull dairy cows and bobby calves usually produce poor-quality carcasses, but they are the main ingredients for low-value beef products such as ground-beef patties.

Bobby calf management

Bobby calves are of limited value to a dairy farmer and they are prone to being neglected and ill-treated. This starts from the time the calf is recognized as a bull calf and so has no value as a dairy replacement. In many countries it is now an offence to transport bobby calves to slaughter before they are 4 days old or before they have developed a dry navel. The intention is to ensure that they are fit to travel. This requirement imposes extra duties on dairy farm staff, or it may be decided to kill the calf on the farm without sending it to a slaughterhouse. The carcass may then be used as dog food.

Calf removal can be an upsetting time for both the calf and the cow. This can be reduced by removing the calf before bonding has occurred. In practice, this usually means that the calf is removed as soon as it is seen or as soon as it is standing. In this situation it is given colostrum by hand from a bulk supply. Bucket-reared calves often take to cross-sucking. This is worst where the milk is provided in an open bucket without a teat and where the calves quickly finish their meal. It may be due to residual hunger.

Scouring is the most common health disorder in calves. The primary cause is endotoxin produced by the infective microorganism. There are metabolic effects, and they also have to be treated. These are dehydration, electrolyte imbalance and acidosis. The calves are hypotonically dehydrated and suffering from excessive fluid loss through the faeces. Calves with a venous blood pH below 7.0 require immediate parenteral therapy because of their acidosis. In severe cases, hyperkalaemia may occur at a late stage. The elevated serum K^+ can be cardiotoxic, and it has been known for calves to die whilst being handled. The main problem in managing calf diarrhoea is in training the calf rearer to react appropriately to the different signs.

Respiratory disease can be common at some calf rearing units, but it would be unusual to see it in bobby calves because they are killed before it can develop. One of the main aims in ventilating a calf shed is to dilute the expired air from the calves. This helps control the transfer of airborne pathogens between animals. There are two components that are important: first, the air space per calf and, secondly, the rate of air changes in the shed. A deficiency in one of these components needs to be compensated by the other. Where the calves are healthy and the spatial volume for each calf is 5 to 7 m^3, the air exchange rate needs to be four or more air changes per hour (Anderson et al., 1978). Ventilation rate needs to be higher than this when one or more calves are exhaling pathogens. Alternatively, the calves need to be given more space to allow more time for the airborne

pathogens to die before being inhaled by another calf. Fortunately, most airborne pathogens in calf sheds have a short survival time in the atmosphere.

In heifer calves, supernumerary teats are usually removed with a pair of scissors whilst the calf is small. These teats have a teat canal and sphincter but lack a gland cistern complex. If the calf is not inspected early in life, larger supernumerary teats are removed by emasculator and blade.

It is generally accepted that more dairy calves are being wasted through stillbirths compared with 40 years ago. This could be due to the following trends:

- increased herd size with less attendance at calving
- decrease in the age and size of heifers at first calving
- better dairy cow feeding, but a greater risk of dystocia at calving
- greater use of large mature-body-size bulls, resulting in higher birthweights

In addition, more bull calves are probably being killed at dairy farms because of the low value of pure-bred Holsteins.

Cull cow and downer cow management

It is sometimes difficult to decide when to cull and when to persevere with a sick cow in the hope that it will recover. Take the case of a cow with a prolapsed uterus at calving. The veterinarian may obligingly, and with some effort, replace the uterus and vagina, and treat for milk fever if appropriate. However, the prospects for this type of cow are not always good, especially if the uterine artery has been damaged and the cow fails to stand within 6 h. Deciding when to euthanize can be difficult, and there is a tendency to persevere even with hopeless cases.

In a study on post-parturient downer cows in Australia during the 1960s, it was found that 14% of cases failed to respond to treatment for milk fever and had to be destroyed. The prevalence of failed treatment is now probably lower than this, and the overall risk of milk fever is reduced with better understanding of the mineral balance needed during transition feeding. Nevertheless, there are still treatment failures. Some of these failures arise because the cow does not have milk fever or any other metabolic disorder or a septicaemia. Instead, she seems to have 'given up'. She is a recalcitrant downer, and is probably experiencing a form of post-parturient depression. She refuses to move and does not respond to the usual stimuli that are given to encourage her to stand or bear weight, such as tail twisting and the use of hip lifters. This type of cow fails to react to threats made at her face and is unfazed by physical insults. On some farms aggressive approaches are used to get this type of cow to stand, including electric goading, setting a dog on her, a good kick in the ribs and intravenous injection of a central nervous system (CNS) stimulant.

Cows that develop acute septic or gangrenous mastitis deserve special consideration, and the following describes one of the unfortunate outcomes. If a quarter is gangrenous, there is no prospect for the cow being of further use as a milk producer. The best that can be done is to get salvage value from the carcass. However, the abattoir does not want this type of animal. There is a risk that bacteria from the infected udder will be transferred to the carcass during udder removal, and lymph nodes in the carcass will be harbouring these microbes regardless of how the carcass is dressed. If the carcass shows signs of being affected, it will be condemned and the farmer discouraged from submitting any cows in a similar condition. Instead, what some farmers do is to cut the teat from the affected quarter to allow the quarter to drain, and then place the cow in a paddock on a remote part of the farm. If it survives, the cow can be sold for meat consumption. If the cow dies, there has been no additional expense. Clearly, not all farmers would operate this way. Some would pursue the euthanasia option if the cow is a hopeless case, but attitudes inevitably vary.

Some cows are sent for slaughter with a health or physical defect, and this may make them more vulnerable to injury. In many countries it is an offence to send an animal for slaughter if it is non-ambulatory or not weight-bearing on all four feet. Those animals have to slaughtered on the farm of origin, or they have to be treated.

Sometimes attempts are made at salvaging downer cows that respond to treatment, but develop a dislocated hip that is not treated. These animals are not weight-bearing on all four feet but are sometimes transported. The prevalence of milk fever cases that develop a dislocated hip is about 1% (Fenwick, 1969). Similarly, cows that have limb dragging and persistent knuckling of the hindfeet, due to tibial or peroneal nerve compression or ischaemia whilst recumbent, are not fit for transport if they are at risk of falling over.

When cull cows are sent for slaughter, there can be delays before they are killed. There are fewer slaughterhouses that specialize in cow beef and so the transport distance can be longer, and sometimes the cows pass through more than one market. One concern from the welfare perspective is that dairy cows are more prone to metabolic disorders when feed is withheld. This risk does not apply to every cow, and the type of risk varies with stage of lactation and condition of the cow. Cows in early pregnancy are prone to hypocalcaemia, whereas those in late pregnancy may develop hypomagnesaemia, when subjected to starvation plus transport to an abattoir (Warnock *et al.*, 1978). In extreme cases they become 'downers', and this can be very disruptive to the orderly flow of cows to the slaughter point, as well as increasing the risk of abuse. Cows that are in late pregnancy are also more prone to develop high-pH meat.

In some countries it is recommended that, when cows are transported for more than 10 h, they should be separated with barriers between every two animals. One of the aims is to allow animals to lean on a barrier when tired. In practice, cows tend to lean against each other, and when kept in larger groups (eight per pen) there has been less bruising. This was partly due to less trampling when an animal sat down.

Mastitis

Mastitis can cause considerable discomfort and pain. The pain occurs during the unblocking of the cistern and ducts of the udder and the first flow of mastitic milk. There can also be pressure pains in the inflamed and swollen udder. Theoretically it should be possible to control this pain whilst the cow is receiving intramammary antibiotics. However, the pain-relieving drugs that are available have limited application:

- Opioids are not licensed for food-producing species.
- Anti-inflammatory corticosteroids can be immunosuppressive, and they should not be used in pregnant cows because of the risk of inducing calving.
- Xylazine has sedative effects and can cause a longer-lasting debilitation than is wanted.

- Phenylbutazone has been withdrawn from use in food-producing animals.
- Non-steroidal anti-inflammatory drugs (NSAIDs) could be used, but only for limited periods (between 3 and 5 consecutive days). However, this may be sufficient for cases where the cow needs short-term assistance in overcoming the infection.

Lameness

Lameness is the single most important welfare problem in the dairy industry. It can involve severe pain, it affects a large number of animals and it is one of the more important reasons for culling cows and sending them for slaughter. There has been considerable investment in research and development regarding dairy cow lameness over the past 15 years, but there is little sign that the prevalence is coming down. This might be because the emphasis so far has been on understanding the causes of lameness, which are difficult to control, rather than focusing on treatment schedules.

One of the present weaknesses in lameness management is in deciding when to treat a cow. One study showed that milk yield is depressed for at least 10 days before lameness is recognized by the stockperson. Daily walking distance whilst grazing was affected for considerably longer than this before the cows were considered lame. The implication is that subclinical lameness is causing sufficient pain or discomfort to affect production before it is identified as a problem. If lameness could be recognized earlier, it would be diagnosed and treated at a correspondingly earlier stage. Early treatment would help avoid the situation where lameness is not considered a problem until the condition has become advanced.

White line disease and solar (plantar) bruising or ulceration are two important causes of lameness in dairy herds (Murray et al., 1996). Risk analysis has shown that track maintenance and the degree of patience shown by farmers in bringing the cows in for milking are key features explaining differences between herds. Normally a cow walks with its head down looking at the track and accurately placing each front foot to avoid injury. The hindfoot is placed close to where the front foot was, and so it is usually safe as well. The number of congestion points in the track appears to be important. When a herd comes to a crowding point, the

flow of cows is interrupted, they jostle for position and there is unplanned foot placement and risk of injury. This effect is exaggerated if the herdperson is impatient. Farms with a high incidence of lameness also tend to have poorly drained tracks where the surface is broken up and the base material exposed. Farms with a low incidence of lameness tend to have tracks with intact surfaces that are stable and well drained.

In housed cows, laminitis and infectious foot diseases, such as interdigital dermatitis, feature strongly. Dirtiness of the barn floor is an important predisposing factor for infectious lameness. When cows stand for long periods in their own effluent, the hoof becomes softened. This in turn makes it more susceptible to damage and wear, and it becomes more prone to bacterial or spirochaete invasion. Slatted concrete and pens that are well bedded with straw help provide clean conditions underfoot.

Inductions

In countries that rely on seasonal milk production from fresh grass, the calving pattern may be controlled by synchronizing oestrus for insemination and by inducing calving in any cows that are likely to be late calvers. The welfare risks to cows from inducing calving include:

- higher prevalence of endometritis
- overuse of the procedure in individual animals
- immunosuppression at calving, when long-acting corticosteroids are used for inducing calving

If 5% of inductions lead to retained placentas and about 50% of these develop endometritis then the welfare risk from this cause is not particularly high. The risks associated with overuse are more difficult to evaluate. Some would argue that certain cows should not be induced because of their risk of developing metabolic disorders and becoming emaciated. In addition, some farmers plan the calving season with the inductions coming first. This introduces risks if these cows are in poor condition and cold weather delays the start of grass growth. Another issue is: should cows be induced if they look unhealthy, or should they be left to calve naturally?

Cows that are nominated for induction sometimes have a long-term breeding problem. For example, they may be the cow that has been underfed and allowed to become thin. Cows with a body condition of less than 4.9 on the Australasian scale at mating have lower submission rates during the first 3 weeks of the mating period, so they are likely to be the cows that the farmer wants to induce during the next season. Cows that are even leaner (body condition score of less than 4.2 at mating) are less likely to settle to insemination and so they have an additional risk of being late calvers next year. Inducing calving in these cows is used as a way of correcting previous mistakes.

Induced calving also raises welfare hazards for the calves. If calving is induced too early the calf is underdeveloped, it has poor vigour and it requires considerable patience and perseverance during rearing on the part of the stockperson. Some farmers take the view that these calves are not worth the additional effort, and so they are either killed at birth or left to die in the paddock. The degree of prematurity can be gauged from eruption of the teeth, but it is difficult to give precise recommendations as to when euthanasia is appropriate because the prospects for the calf depend on the diligence of the calf rearer.

5

Sheep

There are about 1.02 billion sheep in the world. The countries with the most sheep are China, Australia, India, Iran, Sudan, New Zealand, the UK, South Africa and Turkey, and the major wool-producing nations are Australia, New Zealand, CIS and China.

The welfare features in sheep include:

- exposure
- heat stress and water deprivation
- stress and fertility
- out-of-season lambing
- lamb mortality
- weaning
- longevity
- easy-care production systems
- endoparasites
- dystocia
- prolapses
- foot rot
- sheep scab
- stocking density
- holding sheep indoors
- shearing
- tail docking
- fly strike, mulesing and pizzle dropping
- castration
- tooth breaking and tooth trimming
- seed damage in skin
- slaughter of suckling lambs
- trauma
- subsistence farming systems
- transport and pre-slaughter management
- stunning and slaughter
- fear

Sheep production methods are continuously evolving. In parts of Europe there has been a change from outdoor to indoor lambing systems. In New Zealand, there is a trend away from single to twin lamb production, and in many countries there is a move away from wool towards breeds with a high yield of meat in the hindlegs. Whenever there is a change in direction there can be repercussions for conservation, the environment and animal welfare.

The imminent changes in the sheep industry in China are worth considering. About 42% of China is devoted to grassland production, and in the west and north some of the grassland is becoming degraded. Much of it is meadow steppe, but it also includes substantial tracts of desert steppe. Several provinces are overstocked and this is causing loss of plant cover and further desertification. When the winds blow at the end of the dry season, dust clouds are sent over central and eastern China. At the turn of this century, this reached a crisis when the capital, Beijing, received unusually disruptive clouds of red dust. In 2002, the State Council revised the Grassland Law, and now provinces in the west and north are faced with an order to reduce sheep numbers from the steppes by 50 million, and 67 million ha of degraded grassland will be taken out of free-grazing management. This will create a shortfall in sheep meat production, which will be compensated by feedlot production of sheep instead. In the short term, about 3 million sheep will be raised in fenced yards, and the households that previously herded sheep on the steppes will receive feed grain to support the change to an intensive sheep farming system. The impacts this will have on sheep welfare will

©N. Gregory 2007. *Animal Welfare and Meat Production*
(N. Gregory)

unfold, but there will inevitably be a change in disease patterns. Those involved in China's sheep industry will no doubt be giving this some thought as they change the husbandry methods.

Many of the issues in traditional sheep farming have been connected with underfeeding or disease. Underfeeding is common in rangeland systems, and seasonal underfeeding is a risk in lowland systems that have a short growing season and rely mainly on pasture. Underfeeding is considered in this chapter in the context of exposure, lamb mortality, parasitic disease and stocking density.

Exposure

Catastrophic climatic stress is a common cause of animal suffering affecting large numbers of sheep at a time. One of the worst incidents was in Iran in 1954, when 50% of the sheep population died from starvation and hypothermia during heavy snows. In New Zealand, heavy snowfalls and blizzards cause severe stock losses about once every 6 years, and in Australia night-time deaths from hypothermia at higher altitudes are common, especially after out-of-season shearing (Gregory, 1995).

Young lambs are particularly susceptible to death from hypothermia. Cold, wet weather increases the chance of a lamb dying from starvation. Hypothermia can quickly set in, and under extreme conditions a 10°C fall in body temperature within 30 min of birth has been noted (McCutcheon *et al.*, 1981). There are two periods when lambs are at most risk:

- Birth to 5 h of age – usually due to excessive heat loss from wet newborn lambs. Colostrum helps them through this period by increasing heat production.
- Twelve hours onwards – usually due to depressed heat production from starvation, lack of shelter and adverse weather.

Lambs in fine-woolled breeds, such as Merino and Corriedale, are prone to early mortality from the cold (Dalton *et al.*, 1980). In addition, sheep farming systems that strive for multiple births have a higher risk of lamb mortality during unfavourable weather at lambing time. The multiple lamb is smaller and more prone to hypothermia and rapid loss of vigour if it fails to get an early feed. This risk is exaggerated further in lambs born to older broken-mouth ewes because of the difficulty they experience in obtaining sufficient feed (Gunn, 1970).

The risk of lamb hypothermia can be reduced by:

- good ewe nutrition between joining and lambing
- providing shelter
- encouraging hungry lambs to get an early feed
- supplementing ewes where there is a high risk
- planning the lambing date so that it does not coincide with the risk of a snow blizzard

Snow poses dangers from underfeeding, hypothermia, exhaustion and casting. Sheep are poorly adapted to manoeuvring in snow. They have short legs and a low chest clearance, and so they depend on following trails and beating open paths rather than treading through the snow from above. This makes it exhausting to walk through deep snow. Snow shelter belts can relieve this by providing areas of limited snow cover in the lee of the trees, but they can introduce a smothering hazard for sheep trapped on the windward side (Gregory, 1995). Shelter belts also reduce the chilling effect of wind. For example, it was estimated that, at an air temperature of 17°C and a wind speed of 16 km/h, a 10-year-old pine shelter belt gave a 35% saving in heat loss for a newborn lamb (Gregory *et al.*, 1999).

Normally an animal living in its natural habitat would find its own shelter, but farmed animals may not be provided with sufficient options, especially if they are in a field of grass surrounded by only a wire fence. In that situation there would be limited relief from wind, rain and snow. Giving the sheep access to shelter and moving stock either to the sunny side of a hill or to lower ground are some of the few immediate steps that can be taken to prevent these losses.

The risk of lamb hypothermia depends on the date chosen for lambing. There has to be a balance between the risk of lamb mortality from hypothermia and maximizing reproductive success, growth rate according to feed availability and presenting the lambs for sale at a time when prices are favourable. There can be improved conception rates and lamb growth rate from delaying lambing (Knight *et al.*, 1980; Geenty, 1986). The change to more fecund breeds is increasing the risk of lamb mortality from hypothermia, and the multigravid ewe and its fetuses are also at risk during periods of underfeeding. If the ewes are in poor condition at the start of pregnancy and they receive inadequate feed, the fetuses will be lighter, and this raises the risk of lamb mortality at birth if the weather turns bad (McNeill *et al.*, 1999). The risk of pregnancy toxaemia in the ewe is also higher.

During winter, the sheep's appetite is suppressed by as much as 30% (Blaxter *et al.*, 1982). They also expend more energy in keeping warm. Together, these curtail growth in young animals and can jeopardize body condition in mature breeding sheep. This wintertime suppression in appetite and growth is more pronounced in flocks that have been inbred (Wiener *et al.*, 1992).

Heat stress

Heat is a seasonal stress for sheep farmed in semi-arid regions of the Middle East, Southern Asia, Africa and Australia. Under hot, sunny conditions ewes make good use of shade, partly to get out of the glare and partly to stay cool. In the absence of shade the ewes form tight groups and tend to hold their heads in the shade provided by the bodies of other sheep. For recently shorn sheep in hot climates, the heat load from the sun can approach the magnitude of the animal's metabolic heat production, and shade is a particularly important form of protection.

Underfeeding reduces heat tolerance in sheep (Robinson, 1969). The sheep also become more dependent on drinking, partly to replace the water lost through sweating and panting, but also, if water is available, there can be more drinking and urine output (Blaxter *et al.*, 1959). Drinking more is not an efficient way of cooling the body, but it provides prompt short-term relief. Panting reaches a maximum when rectal temperature is at 40.5°C, and at 42°C lambs are near their survival limit (Lowe *et al.*, 2002).

Severe heat stress in the last third of pregnancy inhibits the growth of the placenta and fetus, independently of reductions in feed intake. The lower birthweight does not necessarily affect subsequent size, as lambs can catch up provided the heat stress does not continue for too long, but viability during the first 5 days following birth may be reduced (Shelton, 1964). By the sixth day following lambing, both large and small lambs have usually established greater tolerance (Morgan

et al., 1972). Following this critical period, heat stress will suppress growth rate in lambs and wool growth in mature wethers.

Lamb mortality

In intensive outdoor farming systems lamb mortality rates are often between 15 and 20%. Most lamb deaths are due to dystocia and **starvation–mismothering–exposure** (SME). In indoor lambing systems, mortality from both these causes is lower, and overall mortality rates are also correspondingly lower.

Signs of death from SME in lambs include poor body condition, stomach and intestines are usually empty, dehydrated condition, liver is small and firmer than usual, there is very little internal body fat, and the normal pink fat reserves around the heart and kidneys are metabolized to a brown gelatinous substance.

Most lamb deaths occur within the first day of birth and over 75% have usually occurred by the end of the third day after lambing. Death from dystocia is more common in large single lambs than in twins, whereas deaths from SME are more common in low-birthweight lambs (Hall *et al.*, 1995). Dystocia presents a risk of fatigue in the ewe and loss of interest in the lamb, as well as liver rupture in the lamb during delivery.

Delivering a large single lamb involves more straining to expel the lamb. If the effort is excessive, there is a risk that the ewe will be exhausted and show no interest in the lamb after delivery. This type of ewe benefits most from assistance at lambing time.

Disturbance of a flock when providing lambing assistance can have two repercussions. First, those ewes that are in the process of lambing may develop uterine inertia if they are disturbed. The contractions are inhibited, and dystocia may develop if the inhibition is prolonged. Secondly, if there is a disturbance in the flock and a ewe leaves its newborn lamb before bonding has taken place, there is a risk that the lamb will be orphaned, particularly in outdoor lambing systems. Normally, ewes return to their birth site, which they recognize from the smell of their birth fluids on the ground, and so there is a second chance of mothering. The drive to return to the birth site dwindles with time, and the period for bonding is shortened when there is disturbance. Normally the bonding period lasts less than an hour from the end of lambing. The potential survival of twins is closely related to the length of time ewes remain at the birth site, and this again argues for minimizing disturbance of the lambing flock (Fig. 5.1).

Ewes can form a bond with a lamb before they actually lamb. Oestrogen released into the bloodstream before lambing can provoke lamb stealing from another ewe. If, after producing her own lamb, the thieving ewe loses interest in the first lamb, the first lamb may be left to die. Alternatively, she may neglect her own lamb. Mismothering and lamb rejection are seen most commonly when stocking density in the lambing paddock is too high.

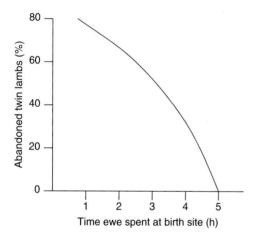

Fig. 5.1. Relationship between abandonment risk in lambs and time spent at the lambing site by the ewe.

A small internal pelvic size relative to size of the lamb often results in shoulder–elbow flexion during delivery. This is easily recognized from the nose presenting at the level of the fore hooves, but it is not necessarily linked to the need for assisted lambing (Grommers *et al.*, 1985). Instead it can be the passage of the lamb's hips through the ewe's pelvic opening that causes the problem. Some strains of Texel and Poll Dorset show extreme plumpness in the hindlegs and are prone to **dystocia**. These two breeds are popular as terminal sires in Europe and Australia because of their increased yield of lean and rounded conformation in the hindlegs. In other crosses, choice of sire breed influences the likelihood of dystocia largely through its effect on the overall size of the lamb at birth (Table 5.1). In spite of the breed differences in the risk of dystocia, the heritability of dystocia for most breeds is low (0.01 to 0.18). This means that breed substitution can be more effective at controlling dystocia than selection within breed.

The callipyge condition in sheep does not carry the same level of dystocia risk as double muscling in cattle. The first callipyge lamb was recognized in a flock of Dorset sheep in Oklahoma. Since then the gene has been introduced into strains of a number of breeds. Unlike cattle, the double muscling in callipyge strains of sheep does not express itself in the fetus. It only becomes obvious a few weeks after birth, and so there is no additional risk of dystocia. Callipyge lambs have superior killing out percentage, and as much as 28% more muscle in the carcass compared with non-carriers at the same slaughter weight. Growth

Table 5.1. Effect of sire breed used on Romney ewes on the size and cause of death of offspring (after Meyer and Clarke, 1978).

Breed of sire	Lamb birthweight (kg)	Lamb mortality (%)	Cause of lamb deaths (%)	
			Dystocia	SME
Oxford	4.8	17.1	45	27
German Whitehead	4.8	17.3	46	24
East Friesian	4.6	13.0	41	33
Border Leicester	4.6	14.9	31	34
Dorset	4.5	14.7	52	28
Cheviot	4.5	15.6	40	22
Booroola Merino	4.3	11.3	39	25
Finn	4.2	10.9	28	28

SME, starvation–mismothering–exposure.

rate is comparable, but callipyge lambs have a better feed conversion efficiency because of lower feed intake. Hypertrophy occurs in the fast-twitch fibres, especially in muscles of the hindlimb, and muscle calpastatin activity in the longissimus dorsi is higher.

Death from dystocia can be recognized from oedema of the neck and head, haemorrhage in the abdomen due to ruptured liver and petechial haemorrhages in the pleura, thymus and heart. Lambs that survive delivery may have experienced blood and oxygen deprivation during birth, and this can lead to weakness, failure in suckling and early postnatal death.

The genetics of lamb mortality, maternal behaviour and early growth rate, and the ways they interact can be summarized as follows:

- Genetic selection for high weaning weights could be compromising early lamb survival, but the reason for this association is not clear.
- Genetic selection for early growth rate in lambs is not likely to compromise maternal behaviour in a breeding flock. In fact, the converse is the case, as there is a positive genetic correlation between these two traits (0.40). In spite of this, the genetic correlation between weaning weight and number of lambs lost before weaning is moderately high (0.33; Conington *et al.*, 2001).
- Ewes that show very poor maternal behaviour can have higher lamb mortality. These ewes are potentially less productive but it is unlikely that they will pass on their poor maternal behaviour because it has a low heritability (0.13).

Out-of-season and indoor lambing

Out-of-season lambing became popular in Europe in the 1980s in response to the high prices offered for spring lambs. Oestrus is stimulated using hormonal induction, with mating occurring in July or August. The ewes are housed for 2 to 6 weeks before lambing, and fed *ad libitum* silage and a restricted amount of concentrates. When the lambs are born in December and January, they spend a day with their mother in a separate pen, and then they are transferred to a straw-bedded barn, where they are group-housed. The lambs

may remain in this barn until they are finished. They are weaned at 6 to 8 weeks, and may be fed *ad libitum* concentrates from about day 10 through to slaughter. Typically, they are slaughtered at 35 kg live-weight when 14 weeks old.

The advantages and disadvantages of **indoor lambing** are shown in Table 5.2. In general, indoor lambing is more labour-intensive in terms of catching and moving ewes, but fewer lambings require assistance. In large flocks (> 900 ewes), postnatal mortality can be high if the time available for flock supervision is limited (Binns *et al.*, 2002). As feed costs are high in indoor lambing systems, underfeeding, low lamb growth rate and loss of live-weight in ewes are hazards (Carson *et al.*, 2004).

Some managers prefer to shear the ewes before lambing indoors as this allows a higher stocking density. Recently shorn ewes prefer soft bedding that provides insulation, such as straw, whereas unshorn ewes show no distinct preference (Færevik *et al.*, 2005). It is said that some breeds (such as the Romney) adapt more readily to close confinement indoors than other breeds (e.g.

Table 5.2. Advantages and disadvantages of indoor lambing.

Advantages
- Reduction in pugging and preservation of pasture
- Protection from adverse weather
- Easier to supervise and assist lambing
- Lower lamb mortality rate
- Lower risk of stillbirths
- Close proximity can assist bond formation between ewe and lamb, especially in primiparous ewes or ewes experiencing dystocia

Disadvantages
- Greater labour requirement
- Additional feed costs
- Additional capital expenditure on buildings
- Increased risk of mastitis, especially when housed on straw at high stocking densities
- Increased risk of perinatal infections, especially in unhygienic sheds
- Increased risk of conjunctivitis and anaemia in lambs
- Ewes may not be well adapted to cold weather and lush grass when turned out with the lambs

Scottish Blackface), but there has been no critical evaluation of these differences. There is an increased risk of mastitis in lactating ewes that are housed on straw when they are kept at high stocking densities.

The main hazards with indoor lambing are mainly connected with health of the lambs. Diarrhoea is more common, setting in at about 1 month, but it is not necessarily connected with morbidity or a check in final live-weight. Affected lambs may, however, take longer to reach market weight. Other hazards are anaemia, which can be controlled with iron supplementation, and there can be outbreaks of listerial encephalitis and urolithiasis.

Weaning

Early weaning is an inevitable stress for lambs, but when performed at 16 weeks of age the stress is not severe, and it is not sufficient to cause immune suppression (Rhind et al., 1998). Bonding between lambs and ewes is strong up to about 100 days following lambing, and, even when there is limited suckling beyond that age, they maintain contact up to at least 190 days. It is the initial separation from the ewe that is stressful for the lamb. Familiarity with being briefly separated eliminates the plasma cortisol response to subsequent permanent separation (Sowińska et al., 2001). The elevations in cortisol are greater if the separation is total compared with situations where the ewe and lamb can still see each other. In parts of Africa, where it is not possible to keep sheep in separate groups, weaning is sometimes achieved by coating or plugging the teats with dung to reduce their appeal, or tying them with a cord.

Growing lambs indoors

Out-of-season breeding has made it profitable to raise lambs intensively on concentrate feeds, which may be provided for lambs in either feedlot corrals or covered barns (Robinson and Ørskov, 1975). Typical production targets are three or four lambs per ewe per year, with a daily milk yield of 3.25 kg during a 4-week lactation, supporting 0.6 kg daily growth in the lambs. Once the lambs reach 4 to 5 weeks of age, they can function as ruminants

independently of the ewe, and they are weaned. Thereafter they can grow at 400 g per day between 15 and 40 kg live-weight on a barley plus protein concentrate. The main welfare hazards in this type of system are respiratory disease when housed indoors, over-fatness in the ewes leading to inappetence and hypoglycaemia in late pregnancy, acidosis in the lambs if feed availability is erratic and urolithiasis in the male lambs. Respiratory disease is best controlled with adequate ventilation, and feeding whole barley grains can help control rumenitis in lambs without involving digestive inefficiency.

In corral feedlots, problems with clostridial enterotoxaemias have occurred when lambs have been exposed to soil or faeces carrying the pathogens. This can be avoided either by vaccinating or by reducing contact with soil and faeces by keeping the lambs on slats. Respiratory disease has been an important cause of losses in housed lambs, and often the lambs are found dead or die after a brief illness. Urolithiasis has been responsible for more prolonged discomfort, which is seen as straining during urination, foot stamping and kicking at the pizzle. It usually occurs when there is a mineral imbalance, but it has also occurred when the water troughs were not cleaned regularly. The lambs refused to drink **dirty water**, and this contributed to magnesium ammonium phosphate stone accumulation (Malone et al., 1985). Affected lambs can also have abdomens distended with urine from a ruptured bladder, or subcutaneous accumulation of urine from ruptured urethras. One approach used in preventing calculi has been to load the ration with salt to encourage drinking and flushing of the urinogenital system. Ensuring the correct balance of dietary minerals would, in the first instance, be a more appropriate remedy.

Wool biting and wood partition chewing sometimes develop in housed sheep. Wool biting usually starts when one animal picks on another's fleece. Providing adequate roughage does not stop or prevent the behaviour, but it disappears if the sheep are turned out to pasture. When severe, the condition can escalate, skin lesions may develop and losses may arise if wool balls accumulate in the rumen.

Coccidiosis is a problem in lambs that are born indoors and are turned on to ground that is carrying overwintered oocysts. It has also caused suffering and spectacular losses at some intensive indoor finishing units where the lambs were held on bedding that harboured the oocysts. Early sub-clinical infection in the lambing shed can provide resistance or tolerance to later infection in the fattening shed.

When attempts were first made at formulating complete pelleted rations for intensive lamb production, problems developed from **insufficient long fibre**. These included rumen parakeratosis and sleep deprivation from insufficient rumination. Providing small amounts of long hay have overcome these effects, and it is now usual to provide some hay or straw for all sheep kept indoors.

In many parts of the world, smallholders keep sheep under cover or indoors at particular times of year. This is usually on a small scale, but the problems encountered are similar to those experienced in large-scale intensive systems. The smallholders may, however, be in a different position when it comes to reacting to problems. For example, in 1992 in eastern Turkey there was heavy snow in late winter. The sheep farmers had to keep their animals indoors for longer than usual and did not have the means to buy in feed. The animals were grossly underfed during early lactation, and lamb mortality that year was exceptionally high. In some countries farmers always keep a reserve of conserved feed for this type of situation.

Overstocking

Overstocking is evident from:

- emaciation
- absence of feed, because either the pasture has been eaten out or none is provided

It can lead to increased susceptibility to parasitism and other diseases, and mortality is likely to be higher than usual. It is inevitable that the animals will have experienced hunger and discomfort in the course of becoming emaciated.

Overgrazing can be a sign of over-intensification or over-exploitation. It is not limited to high-input farming systems, as explained in the case of China at the beginning of this chapter, and it can impact on national agricultural policy. The situation in India is another example. In the past, the Indian government thought that the problems connected with overgrazing were attributable to goats. Goats were blamed for spoiling vegetation and plant composition. In 1976 a policy was initiated for promoting sheep production in place of goats. The aim was to limit the goat population to about 40 million and to increase the sheep population to around 70 million by the year 2000. At the same time, the Indian government was undertaking social reforms where common land was being distributed to farmers and people in lower income groups. However, many low-income farmers cashed in their allocation and sold their holdings to large cropping farmers. As a result many of them became landless. For landless people, it is more feasible to keep a goat than a sheep. Goats are more self-reliant, they are more enterprising in getting feed from wasteland, and they can be kept successfully whilst tethered on roadsides or confined in hutches. As a result the goat population in India rose from 47 million in 1950 to 123 million in 2000, whereas the sheep population rose to only 58 million.

Overstocking can have hidden risks. For example, it may force sheep to eat less palatable material, such as poisonous plants, or dead pasture carrying *Pithomyces chartarum* spores, which cause facial eczema. This and related fungi survive under moist conditions, and they are particularly prevalent in New Zealand and southern Australia. When sheep are forced to eat the spore-bearing litter because of overstocking, they develop liver damage, which can be fatal.

Sometimes overstocking is intentional, and in particular situations it may be considered the right thing to do. For example, over-fat breeding flocks are overstocked to reduce the subsequent risk of dystocia, and this is accepted as best practice for the longer-term interest of the sheep (Joyce *et al.*, 1976). Conversely, withholding feed, through overstocking, is a less acceptable way of reducing over-fatness in lambs before slaughter. It is sometimes done to ensure that the carcasses are not penalized for being over-fat.

In Europe, overstocking whilst underfeeding is probably the main reason for cruelty prosecutions involving sheep. There are four causes for this situation. They are:

- ignorance
- inexperience
- incompetence
- inconsideration

Ignorance arises from not knowing how to look after the animals. Inexperience occurs when the carer knows what he or she is aiming for, but does not know how to achieve it. Incompetence would be the case where the owner is abusing alcohol or has become mentally unstable, and inconsideration stems from callousness. Together, these are known as the four i's. They are not mutually exclusive. The remedy in a particular case will depend on identifying which of these causes is relevant. Similarly, the justification for raising a charge against the farmer for cruelty will depend on which of the four i's applies.

Shearing

Judging from the animals' plasma cortisol responses, shearing is one of the most stressful routine procedures that sheep experience. It involves unfamiliar noises, restraint and close contact with the shearer and shearing equipment. In sheep familiar with shearing, the sounds are sufficient to elicit behavioural signs of aversion. In addition, feed and water are withheld whilst the sheep are awaiting shearing, there can be cold stress following shearing, especially at night, shearing cuts can be painful and lead to caseous lymphadenitis, and in inexperienced hands the handling can be rough.

When shearing becomes harder for the shearer, it often happens that the sheep suffer as well. For example, in fleeces that present a strong penetration drag, the wool fibres can tangle round the ends of the comb teeth. This ties the skin to the comb for a short time and makes it harder to make a blow and increases the risk of a skin cut. In this situation some shearers change to a sharper-toothed comb, but this can increase the risk of skin scratches if the comb is not changed back again between sheep. In some developing countries, shearing methods are still rudimentary. For example, in

Mali, the Macina sheep, which is one of the few tropical long-wool breeds, is shorn with a hand-held double-edged knife. The sheep are shorn up to four times a year (Wilson, 1983).

Chemical defleecing can take much of the stress out of shearing for both the shearer and the sheep, provided sunburn can be avoided. Sunburn has resulted in the death of some sheep, and the risk has limited the application of this technique (Chapman *et al.*, 1984). One way of managing this has been to wrap the sheep's body in a stocking net after injection with the depilatory agent. The stocking and fleece are removed by hand once sufficient new growth of wool has come through to protect the animal's skin from the sun.

Out-of-season shearing, and in particular pre-lambing shearing, is practised on some farms to improve the yield of good-quality wool. It can improve a ewe's mobility, reduce ewe mortality from casting and encourage the ewe to lamb under shelter, which benefits the lamb, and it has been known to enhance birthweight in twin lambs, which can have survival advantages (Morris and McCutcheon, 1997). It does, however, have its risks, especially if a severe storm occurs shortly after shearing (Table 5.3). For example, in South Australia in 1960, 12% of the ewes that were shorn 1 to 5 days before a September storm died, whereas only 0.7% of the ewes shorn 8 to 12 days before the storm perished (Geytenbeek, 1962). Mortality in ewes and wethers under these conditions can be related to body condition. The greater the weight loss before shearing, the greater the risk of deaths (Hutchinson and McRae, 1969).

The risk of cold exposure following shearing can be controlled by using a cover comb, snow comb, blizzard comb or comb lifter (Fig. 5.2). These raise the cutting height of the shears and

Table 5.3. Advantages and disadvantages of pre-lambing shearing.

Advantages	Disadvantages
• Ewes tend to lamb under shelter • Lower risk of casting • Shearing through the wool break • Cleaner fleece, with fewer cotts • An extra shearing every second year • Better mobility in the ewes • No disturbance and check in lamb growth rate from a summer shearing • Eases labour requirements at haymaking time	• Risk of exposure and mortality in ewes if there is bad weather, no shelter, poor body condition and poor feed supply • Risk of loss of condition in the ewes, or a lower milk yield resulting in lower lamb growth rates

leave extra depth of wool on the animal, which helps reduce heat loss (Table 5.4). Theoretically they should also help reduce the prevalence of shearing cuts, but in practice this does not always work out because cover combs have nine teeth instead of 13 to limit the drag, and there is an increased hazard with slicing into skin near rigid tendons or through raised skin folds when using a nine-tooth comb.

Trauma

Common causes of trauma include injuries from shearing equipment, barbed wire, handling in yards, predation, dog worrying, road accidents and falls. Leg injuries can occur in lambs housed indoors on slatted floors that are not well maintained.

Most shearing cuts are partial-thickness rather than full-thickness cuts. Breeds with ribby pelts are more prone to shearing scars (Holst *et al.*, 1997). Severing teats, vulvas, pizzle sheaths and raised abscesses is a risk in all breeds, and some shearers place their fingers over these parts to reduce the risk of cutting into the sheep. Shearing cuts close to leg tendons can cause permanent walking disorders, and deep cuts in the neck have been known to result in excessive bleeding and death. Shearing cuts also

pose a risk of infestation with fly strike larvae and infection with the bacteria causing caseous lymphadenitis (CLA). The CLA bacteria can infect sheep through unbroken skin, so shearing cuts are not solely responsible. The bacteria can survive in sheep dip solutions and, where CLA is common, dipping should be delayed for a fortnight after shearing. Skin pieces left attached to the fleece are a problem for wool processors.

In parts of Australia, grass seed damage has emerged as the single most important fault in lamb pelts. Grass and weed seeds (such as barley grass, speargrass, silver grass and corkscrew geranium) are picked up in the wool and enter the skin, producing puncture marks. This causes irritation for the lamb, and as the seeds work deeper they produce scars on the surface of the carcass and set up abscesses. The pockmarks and embedded seed have to be trimmed from the carcass.

Tooth breaking has been used in some countries in the recent past, but now it would usually be disallowed as a cruel practice. It has been used in two situations. First, lambs are sometimes folded on turnips or swedes, but the intention is that they should only eat the foliage, leaving the bulbous stems for the ewes later in the season. To prevent the lambs eating the stems, they are run through the yards and their incisors are broken off with a pair of pliers. The second situation is where a farmer wishes to receive lamb prices for hoggets. The two permanent incisors are shortened to try and give the fraudulent impression that they are lambs.

Tooth trimming has been used in adult sheep to improve their bite and the strength of attachment to the jaw. This is said to prolong the productive life of the animal. Normally, the incisors become loose in their sockets as the sheep ages. The continuous jiggling that occurs during eating and the lateral forces, especially during cud chewing, accelerate enlargement of the socket and weakening of ligament attachment until the teeth are shed. Shortening the incisors with a powered

Fig. 5.2. Points of the teeth of a normal shearing comb and a snow comb.

Table 5.4. Saving in heat loss from using a cover comb at shearing (after Dabiri *et al.*, 1995).

	Standard comb	Cover comb
Remaining fleece depth (mm ± SE)	3.1^a ± 0.2	5.1^b ± 0.2
Heat loss (MJ/24 h): 7 km/h wind speed	13.4^a ± 0.4	11.0^b ± 0.4
25 km/h + 25 mm/h rain	19.7^a ± 0.6	14.3^b ± 0.6

Means in a row with a different superscript letter were significantly different at $P = 0.05$.

grinder reduces forward drift of the teeth beyond the dental pad and reduces the stresses on the incisors and their attachments during cud chewing. Although this would be a painful procedure in people, there is incomplete agreement as to whether or not it is painful for sheep.

Suffering associated with predation takes two forms. There is the suffering associated with the chase and being mauled, and there is protracted fear associated with the repeated threat of predation. Flocks that are regularly predated become apprehensive and flighty at the least provocation. This can make them more difficult to manage during routine procedures.

In Europe, the red fox is the single most important lamb predator, but there are few reliable estimates of the extent of the problem. The situation is similar with the coyote in North America, but it has been recognized that the level of predation can be high (Tigner and Larson, 1977). There are better estimates for some of the less common predators (Table 5.5). Often, the scale of the losses is minor compared with those caused by disease. This does not mean that the problem should be discredited as being trivial. It means that the problem needs to be put into perspective in terms of other priorities.

The situation in Norway has been closely studied using mortality transmitters, which are attached to lambs and are activated when the animal does not move for 2 to 3 h. Predation occurs after the lambs and ewes have been turned out to rangeland for the summer. Overall summer mortality is about 23% in the lambs and 4% in the ewes, and 75% of the lamb losses are from predation by red fox, wolverine, golden eagle and lynx (Warren et al., 2001). Mortality tends to be higher in male lambs as they seem to be less fearful. They have shorter fleeing distances, and they graze on their own, further from the ewes. Older ewes maintain closer contact with their lambs, and they are more likely to position themselves between their lambs and a predator.

Dog worrying can be a hazard on farms near towns and cities, and on occasion it acquires community nuisance status. For example, during 1980 to 1982, there were over 1000 livestock deaths per year from dogs around Adelaide in South Australia (Burley et al., 1983). Most of the losses were lambs, and on a number of occasions the dogs ran in packs of three or more, and these incidents accounted for 27% of the losses. When a dog catches a lamb it tends to maul it. Often, the lamb is grasped behind the withers and the chest is crushed, whereas, when a fox attacks a lamb, it often grabs it from below first at the sternum and then at the neck once it is dropped.

Sometimes it can be difficult to decide when an injured animal is fit for salvage and when it should be killed. However, the following minimum guidelines apply in most cases:

- There is little point in sending injured sheep for slaughter in the hope that there may be salvage value when it is unlikely that the carcasses will be fit for human consumption.
- If it is decided to slaughter the sheep, it is best to slaughter them where they are found rather than taking them to a slaughterhouse or other site.
- Animals that are suffering and need to be slaughtered should be killed promptly.
- Special consideration should be given to animals with serious mouth and feet injuries.

Table 5.5. Livestock losses from predation.

Predator	Conservation status	Region	Livestock prey	% of livestock killed
Snow leopard	Endangered	Nepal	Cattle	2.6
Lion	Vulnerable	Zimbabwe, Kenya	Cattle	1.7–2.4
Jaguar	Near threatened	Central and South America	Cattle	1.3
Wolverine	Vulnerable	Norway	Sheep	0.6
Lynx	Vulnerable	France, Scandinavia	Sheep	0.26–1.0
Leopard	Endangered	Africa	Cattle and sheep	0.2–0.6
Painted dog	Endangered	Zimbabwe	Cattle	0.17
Wolf	Not listed	North America	Cattle	0.07

Animals with these injuries are more likely to need to be killed because the prospects of future suffering can be greater.

Prolapses

Prolapsed vaginas are largely a result of increasing prolificacy in breeding ewes. Multigravid ewes are prone to abdominal overcrowding. The prolapses cause suffering from inability to void urine, irritation causing the ewe to bear down on the prolapsed vagina in an attempt to expel it, infestation with blowflies and pain and discomfort associated with treatment methods. The prevalence in any one flock is usually between 1 and 2%, but it can be as high as 17% when the ewes are consuming phyto-oestrogens in their feed. In global terms, a 1% prevalence would correspond to more than 5 million affected ewes a year.

This condition is usually treated by farmers, rather than veterinarians. The methods used for repositioning and retaining the vagina are often crude and do not involve any analgesia. They include a perineal purse-string suture, inserting a large safety pin through the vulva, harnesses and plastic spoons secured with a harness or wool ties along the rump. An empty bottle is sometimes secured in the repositioned vagina with the purse-string suture method.

Rectal prolapse has been a problem at some lamb feedlots.

Foot rot

The domestic sheep is thought to have evolved from the Asiatic mouflon, which today occupies the arid hill ranges in Western Asia. Its origins and early evolution were in a dry environment. Many sheep are now kept in wetter districts that favour grassland production and are too wet for arable cropping. It is not surprising that diseases such as foot rot that flourish in wet conditions have become the foremost complaints.

It is easy to recognize that this can be a very painful condition for sheep. The sheep limp and have obvious difficulty in keeping up with the rest of the flock when moved. Standing and walking whilst grazing are also painful experiences. Foot rot can lead to persistent hyperalgesia and depressed production (Ley *et al.*, 1995). The foot lesions are sometimes prone to fly strike.

Endoparasites and dags

Parasitized sheep are less active than worm-free sheep. They spend less time grazing and so they have lower herbage intake. Lambs born to ewes that have high worm burdens have shorter suckling bouts and reduced survival compared with lambs born to ewes with low worm burdens.

Feed intake, growth rate and weaning weights are often depressed during parasitism, even though there are no obvious outward signs of infestation. Sheep in general tend to avoid pasture that is contaminated with faeces. Subclinically parasitized sheep show stronger avoidance behaviour than non-infected animals.

The greatest risk of suffering in parasitized sheep is when they are underfed. For example, when lambs infected with *Haemonchus contortus* were underfed, they developed severe anorexia, weakness and anaemia, and some animals had to be culled (Wallace *et al.*, 1999). The appetite- and growth-suppressing effects of a worm burden are greater in lambs that are underfed (Table 5.6), and in the case of *Trichostrongylus colubriformis* they may in part be caused by the effects of the lamb's own immune response rather than direct damage caused by the parasite (Greer *et al.*, 2005). Ensuring that lambs have adequate feed will help improve their resilience to an established infection, but it will not necessarily improve their resistance to infection.

With *Ostertagia circumcincta* infestations in ewes, up to 200 ml of serum is lost through the faeces per day (Yakoob *et al.*, 1983). This may influence feeding behaviour. Infested animals often select clover from the pasture, presumably to make-up for this protein loss.

Endoparasite diarrhoea occurs with parasites that infest the small intestine, such as *Trichostrongylus* species. It can occur with abomasal parasites, but this probably depends on a change in microbial flora that is promoted by the elevated abomasal pH. Diarrhoea and subsequent faecal soiling of the breech (dags) are the predisposing cause of a number of sheep welfare problems. They include:

- increased susceptibility to fly strike attack

Table 5.6. Effect of feed restriction and parasite infestation on growth and appetite in lambs.

Feed allowance	Standard	Standard	Restricted	Restricted
H. contortus-infected	−	+	−	+
Weight gain (kg ± SE)	9.8 ± 1.19	9.6 ± 1.15	4.0 ± 0.71	2.2 ± 1.56
Feed intake (kg ± SE)	103.6 ± 1.70	101.9 ± 4.30	68.9 ± 0.0	60.4 ± 8.99

- extra handling associated with dagging and crutching sheep
- increased justification for mulesing and tail docking
- interference with suckling by young lambs

Dags also reduce the value of the fleece and they can interfere with successful mating. Ingesting fungal endophyte or lush pasture can also lead to dags, and forages such as Maku lotus (*Lotus pedunculata*) limit the build-up of dags, and this should reduce the need for crutching and the risk of fly strike.

The genetic correlation between *Ostertagia circumcincta* faecal egg count and live-weight for age in lambs varies between −0.63 and −0.90, whereas the phenotypic correlations are close to zero. This suggests that resistance to this gastrointestinal parasite may be an important genetic determinant of growth rate, and this probably applies to other parasites as well.

Sheep scab

Sheep scab causes intense irritation. The sheep are restless: they rub themselves on fence posts, bite at their flanks and scratch themselves with their hindfeet. They also produce unusual facial expressions when they scratch themselves with a hindfoot. There is mouthing with the tongue protruding during the scratching (Corke and Broom, 1999).

Fly strike

Fly strike is important because it is responsible for two amputation procedures that are performed without analgesia (mulesing and tail docking), and because of the suffering it causes when untreated. It often escapes treatment because the larvae emerge and invade the sheep rapidly and without warning. The larvae attach themselves to the host with two hooks near the mouthparts, which assist in opening the skin of the animal. They regurgitate acidic digestive fluids to digest the skin, muscle and fat of the sheep's body, and the digesta are then sucked up by the maggot. The damage to the skin and the inflammation are highly irritating for the sheep. They provoke a characteristic frenzied behaviour, where the sheep are restless, bite at their wounds, wag their tails and take short runs with the head held down. There can be substantial loss of fluid from the open wound. Untreated sheep that succumb to fly strike die from dehydration in combination with protein and electrolyte loss or toxaemia.

Mulesing is used in the southern hemisphere in wrinkly sheep as a general prophylactic treatment against fly strike. The skin fold in the breech is removed without analgesia. As the breech region is the main part affected by fly strike, focusing the treatment on this region is generally effective (Fig. 5.3). Breeding for wrinkle-free breeches would be a painless alternative (Tillyard and Seddon, 1933).

Pizzle dropping is severing of the sheath containing the pizzle from the underline of the belly without anaesthesia. This allows the pizzle to hang vertically facilitating drainage of urine. It is not a common procedure but is done for one or more of the following reasons, depending on the problem that is being managed:

- control of sheath rot
- reducing the risk of pizzle and belly strike
- reducing the risk of urinary calculi
- reducing urine stain on the belly wool

Tail docking

The advantages and disadvantages of tail-docking lambs are listed in Table 5.7. Overall, it is considered that the benefit from reducing the risk of fly strike outweighs the disadvantages. This, however, will depend on local circumstances. In cold arid regions where sheep are not kept in large numbers and the deep-burrowing blowfly *Lucilia cuprina* is absent, the risk of fly strike is low, and tail docking may not be necessary.

Fat-tailed sheep are not usually docked because the tail is a valued part of the carcass. However, an exception is the Turkish Karakaş breed, which is sometimes docked at 1 day old. This results in heavier lambs at slaughter and the fat that is absent from the tail is redistributed to other depots (Gökdal *et al.*, 2003).

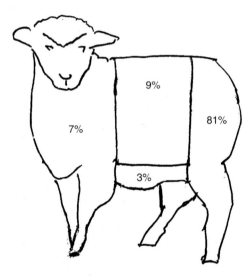

Fig. 5.3.   Distribution of fly strike on sheep.

Castration

The advantages and disadvantages of castrating lambs are shown in Table 5.8. Castration is a potentially painful procedure, but it is not usually done under local anaesthesia. Applying a local anaesthetic before castration would slow the procedure down, and there are mixed views about whether it is a necessary or practical procedure.

The methods are:

- rubber ring applied around the neck of the scrotum and above the testes (conventional ring method)
- rubber ring applied around the scrotum with the testes held above the ring and pressed against the abdomen (short scrotum method)
- cutting the scrotum and removing the testes
- crushing the cord without incising the scrotum

The rubber ring placed above the testes compresses the blood vessels, causing ischaemia and death of tissues below the ring. Ischaemia can be painful and, judging from the plasma cortisol response, this lasts for 3 to 3½ h. It can be abolished with a lignocaine injection. Cord crushing is done in a variety of ways. In parts of Africa it is performed between weaning and 1 year of age, by placing a stick either side of the scrotum at the position of the spermatic cords and beating the sticks together with a stone to crush the cords. In the Middle East it is sometimes done by the same

Table 5.7.   Advantages and disadvantages of tail docking.

Advantages	Disadvantages
Cleanliness	Pain and stress of docking
Reduces risk of fly strike	Risk of infection
Crutching is easier	Risk of neuromas
	Risk of sunburn and perineal cancer if tail is too short

Table 5.8.   Advantages and disadvantages of castration.

Advantages	Disadvantages
Prevents indiscriminate breeding	Wethers grow more slowly
Prevents stunting of younger ewe lambs	Check in growth rate
Prevents weight loss in ram lambs that reach puberty	Pain of castration
	Risk of infection

approach but tightening the sticks together with a rope cord. Small emasculators are available for lambs, which are similar in principle to the calf Burdizzo. The surgical method is considerably more painful than either the conventional rubber ring or the emasculator method, and the short scrotum ring method causes less pain and distress than the conventional ring method (Mellor and Stafford, 1999).

Stress and fertility

Ewes that are stressed just before or during the mating period may show delayed oestrus (Doney *et al.*, 1973; MacKenzie *et al.*, 1975). This could be due either to the subsequent heat being silent or to failure to ovulate. The effect can be greater in younger ewes, and exercise stress is thought to be more hazardous than stressors associated with pre-lambing shearing (Knight *et al.*, 1988). Embryo loss can occur if ewes are exercise-stressed or transport-stressed during the 3 weeks following mating (Doney *et al.*, 1976). Injection with adreno-corticotrophic hormone (ACTH) and severe underfeeding also cause embryo loss at this stage.

Longevity

The heritability of longevity is low (0.08) and it is positively correlated genetically with fleece weight, lamb carcass weight, muscularity and the number of lambs reared per ewe (Conington *et al.*, 2001). This indicates that genetic selection for these production traits is not likely to lead to poorer survival in breeding ewes.

Loss of teeth and reproductive failure are two of the main causes of culling. In the past in the UK, broken-mouth ewes have been kept on to earn a subsidy, even though their chances of surviving the winter were in some cases low. In addition, in many countries cull ewes from extensive farming conditions are often transferred to an intensive farming system for a final crop of lambs ('flying flock'). Their survival depends in part on the condition of their teeth. Feeding root crops and unchopped silage is inappropriate for broken-mouth ewes, and herbage intake can also be reduced. In one study, the consumption of long grass by ewes in poor tooth condition was 74% that of ewes with good tooth condition, and on

short grass it was only 61%. On short grass, intake was between 0.4 and 0.8 of maintenance level, respectively, whereas a ewe needs to consume 3.0 times maintenance during lactation.

Easy-care production systems

In easy-care systems, the sheep are left to themselves. They may be brought in once or twice a year for weaning the lambs and attending to routine needs such as shearing, culling ewes, drenching and foot trimming. Lambing is unsupervised, and no lambing assistance is provided. This and other disorders or diseases are resolved through survival of the fittest. Specific genetic lines have been developed for easy-care systems, such as the Marshall Romney. In some cases they have been developed after culling all ewes that have given problems during lambing.

The main justifications for easy-care farming are convenience and cost. When it is decided to farm sheep on an easy-care basis, the laissez-faire policy may need to be adhered to closely. For example, a study in Australia showed that, when separate flocks were confronted with a man with a dog, 15% of the ewes in the flock that had been handled only twice a year deserted their lambs compared with 5% in those that were used to being handled every month. Interference in wild types can be counterproductive. Where a non-interference policy is adopted, the sheep will inevitably be more prone to stress during transport and at the abattoir.

A misguided justification for easy-care lambing is that it encourages natural improvement of ease of lambing. It is said that, through survival of the fittest, dystocia is reduced, if not eliminated. That view is not consistent with the low heritability for dystocia in sheep.

Suckling lambs intended for slaughter

Young lambs from dairy sheep that are not required as replacements are sometimes sold for slaughter as suckling lambs. In addition, some meat-producing sheep farmers raise lambs for sale as 1-month-old suckling lambs. These lambs command a good price in Spain and Italy, where they are roasted for consumption on special occasions, and suckling lamb carcasses are imported from other countries. Up to 24% of the sheep slaughtered in Spain every year are sold as suckling lambs.

In the region lying between Iran and Kazakhstan, Karakul lambs are slaughtered for their pelts usually between 1 and 2 days of age. The pelts are used in hat making and for fashion wear, and the aim is to maintain a tight curl in the birth coat. This is achieved by removing the lamb at birth to prevent the ewe licking the coat, which tends to open the curl. If the curl is very tight, the lambs may be kept for 5 to 12 days to grow a bigger skin. At one time, when Persian lamb pelts were in strong demand, the newborn lamb was sewn into a coarse linen cloth so as to keep continuous pressure on the birth coat, and warm water was poured on to it every day to make the fibres soft and sleek. The bandage was let out periodically to allow for increases in lamb size, whilst keeping it tight enough to ensure the coat maintained its glossy ringlets. All these procedures required considerable handling and careful management of the lambs to avoid distress and injuries.

sheep on this island experience a population crash, with up to 70% mortality from starvation, which is exacerbated by endoparasites.

In many parts of Africa sheep are a supplementary enterprise to cattle. They reproduce more rapidly than cattle, even though they experience higher mortalities. When a drought strikes, sheep are usually slaughtered before goats, because they are less hardy, they tend to show poorer recovery following droughts and they are less valued for milk production. Sheep are suitable in saltbush regions. They are more salt-tolerant than other domesticated species, with the exception of the camel (Assad *et al.*, 1997).

Phosphorus deficiency can cause suffering, especially when it is severe and leads to broken bones. Sub-clinical forms are a particular problem in some parts of the world where subsistence cattle and sheep farming are practised. Subclinical deficiencies cause economic losses in production, especially when lactation coincides with a late dry season.

Subsistence farming systems

Semi-nomadic forms of sheep farming are still common in Asia, and the main welfare problem is underfeeding. Take Afghanistan as an example. Lambing is usually in late winter or early spring, which coincides with snow in the region. During the summer, the flocks range on highland pasture. Heavy losses occur when a severe winter follows a prolonged dry period. The ewes are in poor condition during the cold spell and cannot sustain themselves.

Survival during winter in cold climates can be determined by the parasite burden at the beginning of the winter and by the sheep's ability to feed. On the Island of Soay in Scotland, sheep with a broad span in their incisor teeth have a survival advantage as they can take larger bites of pasture (Illius *et al.*, 1995). Every 2 to 4 years, the

Dogs

One of the most fearful stimuli for free-ranging bighorn sheep, in terms of heart rate activation, is the presence of a man with a dog. Dogs seem to elicit an innate fear in sheep. In some countries dogs are used for controlling confined sheep, such as sheep in handling yards on farms and in abattoirs. These yard dogs are often muzzled to control biting, and they develop special skills for moving sheep. For example, 'backing dogs' at abattoirs are trained to empty a pen of sheep by jumping over the side railings and walking on the backs of the sheep to a corner where they drop to the floor and then start barking. Occasionally there is an unruly dog which tries to isolate the last sheep leaving a pen, which it then tries to attack. Injuries are avoided by ensuring that the dogs are muzzled whilst working.

6

Pigs

The countries with the largest pig populations are listed in Table 6.1. China is the single biggest pig producer, and presently it has about 100,000,000 people who own pigs. Most are backyard pig keepers with two or three animals, but there are 'specialist' producers who sell about 30 pigs for slaughter a year, and larger-scale growers who finish 200 to 330 pigs per year. This will change through investment in the infrastructure that supports large-scale farming systems. Over the next 7 years, demand for pigmeat in China is likely to increase by 22%, and the growth sector will be in the multi-thousand pig operations.

In China and elsewhere, increased production will be achieved partly through more efficient breeding. There will, however, be a biological limit to efficiency. In Denmark the top pig breeding units are now achieving over 30 weaners per sow per year, whereas 5 years ago it was between 28 and 29. It is estimated that the biological limit will be 41 weaners per sow per year, based on 2.52 litters per year, each with 16.4 weaners. It will take some time before this is reached and, until then, we shall not know all the impacts on animal health and welfare.

The main welfare issues in pig production at present are:

- confinement in breeding pigs
- piglet crushing
- hypothermia in outdoor piglets
- farrowing difficulties
- social pressures in gilts
- mixing and aggression
- hunger in the pregnant sow
- dry sow stalls
- heat stress
- weaning
- high stocking densities
- lameness
- gastric ulcers
- teeth clipping
- castration
- tail biting and tail docking
- nose ringing
- bedding and environmental enrichment
- dust
- ammonia
- cannibalism
- transport and lairage stress

Pigs in developing countries

Most developing countries have intensive forms of pig production as well as backyard pigs (Table 6.2). In non-intensive scavenging systems, communal ownership has developed as a way of managing disputes over damage to property, especially crops. The pigs are owned by the community, along with the problems they create. Those who share in managing the pigs and controlling the damage they can wreak are usually rewarded with an animal they can slaughter and consume.

This system often gives the scavenging pig a protected status with welfare safeguards. If a pig damages property it cannot be harmed. It can only be driven away. With the possible exception of roadkill cases, anyone who harms or kills a pig must make reparation to the community. For the pig, breeding and survival operate on a survival of the fittest basis. Pigs have limited road sense and

©N. Gregory 2007. *Animal Welfare and Meat Production*
(N. Gregory)

so, where there is fast traffic, accidents are common. There is no attendance at farrowing, and so the sow and piglets survive or perish by their own means. In communities where pigs are not protected, the damage they cause is a common source of frustration and they can be mistreated.

In semi-intensive systems, pigs are often confined at high stocking densities in pens that are not regularly mucked out. In some countries, they may be herded for short periods on aftermath, or they may be tethered in the fields with a rope secured through a hole in the ear.

It is helpful when evaluating pig production in a developing country to determine mortality levels, and compare them with expected values or values in other countries. Overall mortality rates between farrowing and slaughter in semi-intensive systems in West Africa are typically 30%, and diarrhoea is a common cause of morbidity. If performance in the USA is used as a

Table 6.1. Countries with the highest pig populations.

	Million pigs
World	956
China	470
USA	60
Brazil	33
Germany	26
Vietnam	25
Spain	24
Poland	19
India	18
Mexico	18

guide, the corresponding figure is 18%. The range between herds in the USA is, however, very variable.

Farrowing crates

Some sows are confined in pens for much of their breeding life. Farrowing crates are used whilst they are nursing piglets, and sow stalls or tether stalls are used for the greater part of the pregnancy period (Fig. 6.1). Dry sow stalls are being phased out in some countries because they are considered unfair for the sows, but farrowing crates are likely to remain in use.

Farrowing crates are designed to reduce piglet mortality from overlying by the sow. When they were first introduced they were used for only a brief time, from about 1 day before farrowing to about 2 days after farrowing. This is the period when the risk of crushing is greatest. However, it was quickly appreciated that farrowing crates helped to reduce the need for attending farrowings, and, as weaning age declined, it became convenient to leave the sow and litter in the crate for the whole of lactation. The main reason that sows are now kept in farrowing crates for extended periods is convenience for the producer.

The welfare problems associated with confinement in farrowing crates include:

- inability to exercise and forage
- thwarted interaction with the piglets
- the sow is unable to take a break from the piglets
- thwarted behaviour before farrowing (e.g. no nest building)
- greater likelihood of piglet savaging by gilts

Table 6.2. Pig production systems in developing countries.

Production system	Housing	Ownership	Feeding	Breeding
Scavenging	None	Often communal	None	Uncontrolled
Semi-intensive	Semi-permanent constructions from local materials	Individual smallholders	Household waste and some specially grown crops	Uncontrolled or use of local stud boar
Intensive	Covered pens made of concrete	Urban-based entrepreneurs	Agro-industrial co-products	Only selected boars used for stud

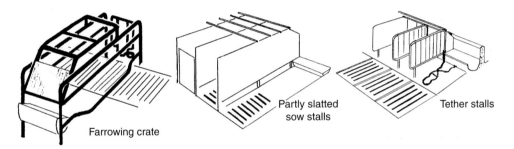

Fig. 6.1. Three confinement systems used for sows.

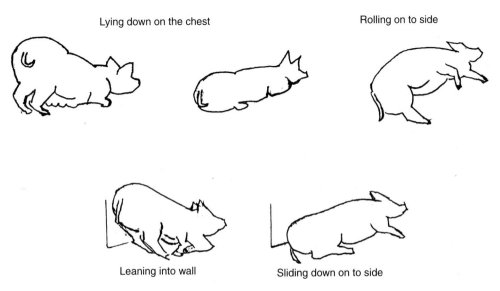

Fig. 6.2. Two sitting-down sequences in pigs.

- stress in some gilts when introduced to confinement
- skin lesions from contact or rubbing injuries

Skin lesions are seen on the lateral aspect of the knees, and they develop during sliding movements. They are worst in sows that are active, and these are often the sows that have thwarted behaviours.

Sow behaviour and piglet crushing

This section considers the ways in which sows sit down and accidentally crush their piglets.

When a sow is not confined in a small pen or in a crate, she would normally perform a regular behaviour sequence when sitting and lying down that helps minimize the chance of piglet crushing. First she makes snout contact with the piglets. This helps to initiate coordination of their behaviour. She then roots in the nest material, and the piglets mostly group on one side of the sow. She then goes down on to her chest by folding her forelegs under her, and then her weight goes backwards as her hindquarters come down (Fig. 6.2). At this stage she is lying on her midline. She then rolls on to her side that is opposite to the group of piglets.

If the sow is closely confined in a small pen, she does not use the same behaviour patterns. Instead, she is more likely to lean against the side of the pen and slide downwards on to her side. The sow's rapid descent makes it difficult for piglets to get out of the way in time.

The causes of crushing in farrowing pens are different from those in farrowing crates. The main risks of crushing in a strawed pen occur when the sow lies down from standing (52% of piglet crushings) and rolls over (36%). Getting up and manoeuvring between standing and sitting or between sitting and lying are lesser risks (Marchant *et al.*, 2001). In a crate, some crushing can occur when a sow slides down on to its side, but most occurs during farrowing and when the sows stand up from the lying position (Weary *et al.*, 1996). Crushing during rolling movements is unlikely in crates because they inhibit these movements, whereas sows can roll on to their bellies from their sides, and vice versa, more rapidly in pens and this poses a greater risk.

Five routine procedures are commonly used to help control crushing in farrowing crates:

- Draw the piglets away from the sow, using a creep heater. These are usually alongside the head and shoulders of the sow in the farrowing crate. One research study showed that it was helpful to use a movable lamp. At farrowing, the lamp was placed at the rear end of the sow and shortly afterwards moved to alongside the head and shoulders. This helped to draw the piglets away from the danger zone and it halved piglet crushing (Svendsen *et al.*, 1986).
- Provide a non-slip floor. Slippery crate floors create two problems. First, the sow cannot always control the way she lies down, and if she slides down quickly there is a greater risk of crushing a piglet. Secondly, the piglets need a reasonable grip to the floor to allow escape from the sow when she moves.
- Move the piglets during danger periods. Crushing sometimes occurs when the sow gets up, for example at feeding or mucking-out times. Moving the piglets away from the sow just before these times may help cut losses.
- When farrowing is supervised, give the piglets a feed immediately after farrowing and then shut the piglets in the creep area for half an hour. This helps train them to nest in the creep area away from the sow.
- Adjust the farrowing crate size for the individual sow. The average sow lying area in farrowing crates in 1.5 m², and it is 1.95 m long. The aim is to adjust the size to direct the sow's movement as she lies down. Hard fixation is necessary only for the first few days after farrowing. After that period the crate can be opened up, enlarging the lying area.

Various attempts have been made at reducing crushing in open farrowing systems, such as fitting a sloping buttress to the wall to provide protection for the piglets, but no single system has gained widespread acceptance. Maintaining a warm creep area can help. If the pen is cold, the piglets are likely to gather closer to the sow, and this can increase the risk of crushing. Providing bedding can improve the chances of piglet survival, presumably because it cushions a piglet when it gets trapped.

Sow's choice of farrowing crate design

Given a choice, young sows prefer a degree of enclosure in the farrowing crate (Phillips *et al.*, 1991). This provides privacy and a hide. Given the opportunity, they would root the ground to excavate a nest before they farrow (Hutson and Haskell, 1990) and they would build a nest of straw or other litter material, especially under cold conditions and as they get older.

When sows were offered the choice of 45, 60, 75 or 120 cm wide crates, they chose the wider crates (Phillips *et al.*, 1992). Piglets spend more time at the udder when the area inside the farrowing crate is large (2.16 vs. 1.54 m²), and they perform more sucking behaviour.

Sows in crates become considerably more active after the first week following farrowing, and in particular they root at the floor (Cronin and Smith, 1992). Some producers consider this is a good time to move them on to a pen. Providing straw or some other form of bedding in farrowing crates helps reduce sham chewing. In addition, the sows spend more time lying on their sides than on their bellies.

Attention needs to be given to the flooring that is used in farrowing crates. One survey showed that 41% of sows showed difficulty in lying down in the crates (Bonde *et al.*, 2004). This included slipping, interruptions in the sequence of actions during lying down and abnormal lying-down behaviours. Slipping was common where the floor was made of cast-iron slats or epoxy triangular slats, and it was more common in young sows that were unused to the crates. Sows that interrupted their lying down behaviour were also more likely to slip when lying down. Sows on slatted metal floors knelt and went from kneeling to lowering their hindquarters faster than sows that were on rubber mats. Evidently kneeling on the slats was uncomfortable. High-traction uneven surfaces may also be uncomfortable, but the higher friction causes the sow to take longer in placing and moving its knees and shoulders on the ground when attempting to lie down.

When expectant gilts are first confined in a farrowing crate, there is greater activation of the hypothalamic–pituitary–adrenal axis compared with moving them to a straw-bedded farrowing pen (Jarvis et al., 1997). Either the restriction imposed by the farrowing crate or the lack of nest building material or the harder floor surface or all three caused the additional stress.

In summary, the floor needs to be soft and comfortable and provide sufficient grip to allow the sow to lie down and stand up without slipping or scraping the skin on its legs. The lying area needs to be adjustable to allow for changing requirements, and it needs to be warm, especially during the first 3 days following farrowing.

Temperature needs of the piglets

Piglet hypothermia is a common hazard in outdoor breeding units. At birth the piglet has limited ability to regulate its body temperature and it is prone to hypothermia. Body temperature drops in the first half-hour by 2 to 7°C depending on air temperature and birthweight. Body temperature control develops in about 2 days if the piglets are kept at 15 to 24°C, but it may take up to 10 days if ambient temperature is 0°C.

When given a choice, newborn piglets prefer an ambient temperature of 30°C, with low air movement rates. Ideally, the ambient temperature for the piglets at birth should be 27 to 32°C. For the farrowing sow, an air temperature of 10 to 16°C is ideal, but more importantly she should have an opportunity to select the conditions that suit her best. This means providing nest material, which she can either burrow into if it is cold or lie on if it is hot.

The effect of temperatures lower than 27°C at birth on the piglet's metabolic rate is strongly influenced by the size of the piglet. It is usually the smaller piglets that are first affected by cold conditions. Those with low vigour are chilled, become lethargic and are prone to being crushed. Rapid chilling after birth and restlessness in nervous young sows both contribute to crushing.

Dystocia

If farrowing is protracted there is likely to be higher piglet mortality at birth and during the first 3 weeks of life (Janczak et al., 2003). Sows that are more fearful or anxious as young gilts and are then reared in stalls are more likely to experience problems with protracted labour.

Farrowing can be easier for the sow and piglets in pen systems compared with farrowing crates. Parry (1986) examined ease of farrowing in 71 sows that farrowed in crates compared with narrow pens (1.33 m wide). Sows in the crates made more straining movements and leg kicking movements and showed more quivering prior to and during farrowing. The newly born piglets from sows that farrowed in crates had higher blood lactate levels and were more stained with meconium. The time taken to farrow tended to be longer for the sows in crates. Taken together, this suggests that delivery had been easier for the sows and piglets in pens. In this particular study, subsequent piglet mortality was comparable in the two systems, but more of the deaths in the pen system were due to overlying.

Stress and reproductive performance

Sows normally lose between 20 and 30% of their embryos during a pregnancy period lasting 115 days. If early embryo loss could be reduced, there would be higher productivity. It is unlikely that this loss is due to any nutritional fault. It is only after a prolonged period of inanition (greater than 40 days) that embryo development is curtailed (Anderson, 1975), and it takes a severe nutritional insult for a sow to abort. Instead, it seems that emotional and exercise stressors are more important in influencing embryo survival.

Stress can cause embryo rejection, and it is thought that this has become more important because of the move to earlier weaning. There are three stressors that the breeding sow can experience. They are farrowing, weaning and mixing with other sows during the pregnancy period. Early-weaned sows are thought to be more prone to embryo loss (Varley *et al.*, 1984). Early-weaning brings two of the major stressors (farrowing and weaning) closer together. As these two stressors get closer, they add to each other and there is insufficient time for recovery. It is now recommended that females should not be moved from their service area or mixed with unfamiliar sows until 4 weeks after service. This helps prevent the third stressor making the situation even worse.

Moving recently mated sows into a group housing system where there are unfamiliar sows introduces stress at a critical period. It has been shown that if the sow is moved to a pen with a group of familiar sows she is more likely to produce a larger litter (Simmins, 1993). Breeding performance is better where there is a stable social group structure during early pregnancy. Continuously changing dynamic social groups result in smaller litters through early embryo loss (den Hartog *et al.*, 1993; Table 6.3).

Stress may also influence oestrus expression. Rearing gilts at high stocking densities has been associated with poorer reproductive performance, partly because oestrus was less obvious (Hemsworth *et al.*, 1986; Table 6.4).

Managing gilts

Gilts need special management because they are prone to being bullied by older sows, they farrow lighter litters that grow more slowly and their piglets can be more prone to disease. They also experience greater appetite suppression and weight loss immediately after farrowing, and so more attention needs to be given to feeding.

Introducing gilts to the main breeding herd can be a stressful experience. The gilts tend to stay on the edge of the pen as a subgroup, to avoid bullying and aggression from larger sows. Some farmers have changed to **parity segregation** systems partly as a way of managing this, but more often as a way of feeding growing gilts separately and applying prophylactic veterinary care for the gilt's piglets. In this system, introducing gilts to the main breeding herd is delayed until the second month of pregnancy ahead of the second farrowing. Delaying introduction this late also allows better control of the second mating. First-parity sows take about 2 days longer to come into oestrus than mature sows, and separate management allows less complicated scheduling.

Table 6.3. Effect of social stability during early pregnancy on litter size.

	Social stability		
	Stable	Dynamic	Significance
Proportion of sows producing ten or more piglets (%)	80	61	$P < 0.01$
Litter weight (kg)			
Born alive	16.1	14.3	ns
All piglets	16.9	14.6	$P < 0.05$

Table 6.4. Effect of space allowance during rearing on reproductive performance in gilts.

	Space allowance (m² per pig)		
	1	2	3
% gilts coming into heat during expected period	79	88	100
% gilts successfully mated	77	85	97

In some outdoor breeding units, the gilts are mated at their first heat (about 100 kg) instead of the third (120 kg) or second heat. This leads to a smaller first litter at farrowing, and it stunts the gilt, resulting in a less lazy, more athletic sow that is less prone to crushing her piglets.

Farrowing systems

Most intensive pig production systems use **farrowing crates**. The sow is confined in the crate until it is decided to let her out, which is usually at weaning time. Alternatives to the farrowing crate are as follows. In **open farrowing systems**, the sow is free to move within a communal or family pen. The **getaway farrowing system** is a type of open farrowing system. The piglets cannot leave the farrowing pen, but the sows can leave and enter a common area that is shared with other sows, where they feed and drink.

In **group farrowing systems** there are no farrowing crates. The sows are kept in a large communal pen that has nest areas for each sow. The sows have access to a communal feeding and lying area. During the first 7 or more days after farrowing the piglets are confined to the nest area. The barriers retaining the piglets are then removed and the litters are allowed to mix. This has the added advantage that the piglets are less likely to fight unfamiliar piglets at weaning time. There can be sow aggression during the nest building period before farrowing and at feeding time. The sows will defend their own nest area against other sows, but generally they do not enter each other's nest site. In **multi-suckling systems**, the sows usually farrow in crates, and then they and their litters are mixed with other sows and litters in a large pen, which is equipped with a covered creep area and brooder lamps.

Multi-suckling systems are used in Scandinavian countries. The aim is to imitate the natural social organization in a breeding herd. The sows are let out of their farrowing crates with their litters at about 11 days after farrowing and mixed in a large group pen. This provides social contact and exercise and frees up the farrowing crates for the next batch of sows. There can be difficulties with cross-suckling, which disrupts social order in established litters, throwing them all into competition and fighting at the udder, but this can be prevented by either using a getaway farrowing system

before the groups are introduced to the multi-suckling pen or delaying the stage at which they are transferred to the multi-suckling pen. Piglet mortality from crushing or starvation can be higher during the first days in some groups, but this varies considerably between groups.

One of the most comprehensive farm studies on indoor farrowing systems was done in Sweden (Gustafsson, 1983). Data were collected from 1500 herds and more than 128,000 litters. They showed that:

- sows farrowing in crates had more stillborn piglets and piglets that died at farrowing in comparison with sows that farrowed in pens
- sows farrowing in pens had more piglet deaths after farrowing
- sows confined in farrowing crates for 1 to 7 days following farrowing had fewer piglet deaths to 3 weeks than sows confined in farrowing crates for more than 7 days after farrowing
- overall piglet mortality was highest in sows (excluding first-parity sows) kept in dry sow stalls and then confined to farrowing crates for more than 7 days
- sows farrowing in pens, instead of farrowing crates, had longer parity intervals

The lower overall piglet mortality due to less piglet crushing is a major economic advantage for farrowing crates.

It takes about a week following farrowing before the sow reaches maximum *ad libitum* feed intake. In the intervening period she is losing weight, and this can be detrimental in lean sows and gilts nursing large litters. Excessive loss of condition and piglet underfeeding can be avoided by introducing the *ad libitum* feeding regime about 4 days before farrowing.

Stable vs. dynamic social groups

There are two approaches to managing sows that are group-housed:

- Stable groups – which have a fixed group composition from the time the group is formed. The groups are usually formed from sows weaned in the same week.
- Dynamic groups – in which every week sows are taken out of the dry sow house and new sows are brought into the group to replace them.

The stable social group system is less prone to uncontrolled aggression between sows, and so it is often preferred. Between three and 12 sows can form a stable social group or subgroup. The most common approach is to introduce a small number of sows to a boar or a group of boars each week for 3 to 4 weeks. This becomes a stable group that is kept together as a separate group, or animals may be mixed with sows of the same stage of pregnancy after the mating period.

In the dynamic system, sows are usually mixed in the mating pen. It is best to mix at the first signs of oestrus. This helps control aggression, as the sows are more interested in the signs of oestrus and in mating than in having a scrap. The sows are removed from the mating pen following service, and new sows that are in heat are introduced. Sow movement is perpetuated on a weekly basis, and so social stability is discouraged.

The optimum stocking density and group size for group-housed dry sows will vary with season, the way the manure is managed and the temperament of the sows. If they are given insufficient space they can become irritable and aggressive when trying to settle, and the lying area may become dirty. Excessive space can lead to chilling on cold nights. It is more difficult to locate and identify an individual sow when sows are in large groups. In particular, it can be difficult to draft out sows that are due to farrow.

Feeding the pregnant sow

Hunger in group-housed dry sows causes frustration, which leads to **aggression**. This can be managed either by regulating the way the feed is provided or by providing a bulky feed. In some countries, sugarbeet pulp is being promoted as a roughage for satisfying the sows' hunger, and it has been found to reduce stereotypies as well as aggression in comparison with conventional barley–soybean meal concentrate (Vestergaard and Danielsen, 1998). Presumably it provides a longer-lasting satiety either from greater gut distension or from longer-lasting hind-gut fermentation and absorption of nutrients. It is not known whether greater gut distension leads to an increased risk of prolapses at parturition, but there are reports that the prevalence of prolapses is increasing.

Individual feeding systems are sometimes preferred to group feeding systems because they help to limit fighting at feeding time. Individual feeding in stalls also allows the sows to be fed according to their individual condition, and it allows easier isolation for mating, routine inspection and health care procedures. In addition, some individual feeding systems can be operated automatically, in which case the sows do not get excited each time they see the stockperson.

The feeding systems that can be used in group housing systems are:

- floor feeding in a group pen
- trough feeding in a group pen
- individual dry sow feeding stalls
- trickle feeders in cubicles
- computer-controlled feeder stations

Aggression during group feeding is not a serious problem in outdoor units, but in sows that are loose-housed indoors it can get out of control. Floor feeding on straw can increase aggression in comparison with floor feeding on concrete.

With **trickle feed systems**, the feed is conveyed along a cable and disc tube and falls into a trough. The speed of the conveyor is regulated to control the rate of feed delivery. Where a dry meal is delivered, the trough must be provided with a nipple drinker; otherwise sows will leave the feed stall to have a drink before they have eaten their ration. They may then return to another sow's stall and a squabble starts. If the feed is pelleted, it is not essential to have a nipple drinker over the trough. In fact the pelleted feed can be dispensed on to the floor, and the trough can be done away with altogether. The feeding rate needs to be finely controlled when setting the individual trickle feeders. The effect of feed dispensing rate on aggression and on feed intake for a twice-daily feeding system is shown in Table 6.5. When dispensing speed increases, the number of aggressive interactions during feeding is reduced, but the amount of feed left over is increased (den Hartog *et al.*, 1993).

Electronic sow feeders (ESFs) are automatic dry sow feeders that have lockable gates. Typically there is one feeding station for every 30 sows. The feeders are activated by an electronic transponder in a collar or ear tag fitted to each sow. When a sow presents itself at the entry gate to the feeder, it is identified and allowed to enter. The entry gate is locked behind the sow and a programmed amount of feed is dispensed.

The design of the ESF and the way it is arranged determine whether there will be problems

Table 6.5.   The effect of dispensing speed from trickle feeders on aggression and feed left behind by dry sows.

	Dispensing speed (g per minute)					
	< 80	81–100	101–120	121–140	141–160	161–180
Number of changes in feeding place	4.2	2.5	0.9	1.6	0.6	0.6
Aggressive interactions during feeding	2.0	1.1	0.4	0.6	0.6	0.6
Aggressive interactions after feeding	0.8	1.2	1.0	1.7	1.9	2.4
Feed remaining						
None (%)	96	97	84	81	75	60
Little (%)	3	2	13	15	14	15
Much (%)	1	1	3	4	11	25

with feed stealing, queuing times and sow aggression. In some designs, when a sow is shut in the ESF, she has a pre-set time to finish the feed before the entry gate is unlocked. Intimidated, slow-feeding sows are prone to getting out before they have finished their allowance, because this ensures that they avoid a conflict with a sow queuing at the entry. This problem is overcome in designs where the entry gate is locked until the sow decides to leave, but the occasional sow may choose to sleep in the feeder stall, preventing access by other sows waiting to feed. A more common problem occurs if a sow manages to enter the ESF from the wrong direction. The sow in residence is then threatened from both front and rear.

Sows prefer to use walk-through ESFs in comparison with reverse-exit ESFs (Edwards *et al.*, 1988). However, in the past, some walk-through ESFs have led to more bullying and feed stealing. If a sow manages to enter the feeder behind a lead sow, she can steal the lead sow's feed if she manages to force the lead sow to get out of the way. Forward-exit ESFs can be more prone to this problem because a timid lead sow learns to escape without having to confront the follower. This situation is often associated with vulva biting.

The problems that have just been described are usually linked to queues forming, plus impatient sows needing their feed. Queuing can be reduced by siting the ESFs well away from the lying area of the group. Queuing, feed stealing and bullying can be reduced to controllable levels by

making a passage between the lying area and the feeder, and by designing a system that encourages a one-way flow of sows through the system, through strategic positioning of the drinkers. If bullying does develop, it can be helpful to change the routine: for example, by varying the time at which the feed cycle starts. Fighting at feeders is more common during the late afternoon and early evening, especially if this coincides with the start of the programmed feeding cycle. Confrontations at these times can be controlled by starting the feeding cycle late at night or early in the morning when the sows are normally resting. When all else fails, aggression can be controlled by floor-feeding silage twice daily in the lying area.

Low-ranking sows are disadvantaged in competitive, floor-feeding systems. Instead, if they have continuous access to an ESF or are fed *ad libitum*, they can modify their feeding strategy and achieve comparable intake to higher-ranking sows (Brouns and Edwards, 1994). They do this by visiting the feeder more frequently. There is no rigid sequence or order in which sows feed at an ESF when a feeding cycle starts. However, newly introduced sows that form subgroups tend to feed after the resident sows.

To summarize, ESFs have the advantage that:

- Sows can be fed an individual allowance.
- The computer can record the allowance, and with some systems it is easy to recognize which sows are not eating.

- The sows do not associate stockpeople with feeding, and so they are less agitated and are easier to handle.
- They allow feeding at any time. By feeding through the night the sows are quieter during the day.

Dry sow stalls (gestation crates)

Dry sow stalls are a simple and convenient way of housing and managing pregnant sows. Ease of management is their main advantage. However, from the sow's perspective, stalls are an unrewarding and uncomfortable environment.

Sows are prone to developing stereotypies when housed in dry sow stalls. The stereotypies often take the form of bar biting or vacuum chewing, and they are due to hunger and lack of suitable materials to investigate and work with the nose and mouth. Offering the sow more feed is not necessarily the right solution, as they could become over-fat and more prone to sitting on their piglets. Instead, farmers are now feeding the dry sows a high-fibre ration. This helps reduce oral stereotypies (Whittaker et al., 1998).

Sows kept in stalls take longer to lie down from the standing position. This is thought to be due to the discomfort of the floor, the greater difficulty in manoeuvring to lie down because of the limited space, and their poorer control over movements from not using their muscles (Marchant and Broom, 1996). Other physical hazards with stalls include skin lesions, vaginitis and foot disorders. In addition, gilts housed in dry sow stalls are more prone to developing gastric ulcers than group-housed gilts (Geverink et al., 2003). If a stall is too small for a sow, she spends more time in the recumbent position, instead of standing (Anil et al., 2002).

Pregnancy rate and litter size tend to be higher in individually stalled sows compared with those held in group housing (Kongsted, 2004). It is thought that the stress and fear associated with mixing unfamiliar sows, plus high stocking rates, were responsible for the poorer performance in group-housed herds.

Outdoor breeding herds

Over the past 15 years, a number of pig breeders in Europe have changed to outdoor systems.

This section describes some of the problems they have encountered, and how those problems can be managed.

Keeping pigs outdoors is not always appropriate in colder climates. One way of getting round this is to focus on a once-bred gilt system, where the gilts farrow in spring and early summer. They and their progeny are slaughtered from late summer through to late autumn except for replacement females, which are held over winter and mated to farrow the following spring. When this system was used in Sweden, the outdoor once-bred gilts were slaughtered at 446 days of age, weighing in at 146 kg live-weight. They were substantially leaner than their 121 kg maiden gilt counterparts because their fat reserves had been depleted by 9 weeks' lactation, but condition can be restored by feeding them for a period following weaning if needed. The meat from the once-bred gilts had a higher intramuscular fat content, and was juicier but stringier when used in cured meat products.

In France it has been reported that there is a pronounced increase in piglet mortality in outdoor units during the winter. This results in fewer piglets weaned per sow per year in comparison with indoor herds (21.2 vs. 23.5). Those outdoor units where farrowing was attended had the highest piglet mortalities, which is the opposite to findings for indoor units (Berger et al., 1998). The presence of people may be more of a problem with sows that are flighty. Flighty sows are more prone to making sudden movements when disturbed inside a hut (Herskin et al., 1998).

Control of endoparasites is important in outdoor pig units. The effects on the sow may not be obvious, but careful observation has shown that sows infected with nematodes are generally less active and spend more time resting inside the farrowing huts. This might be because they felt cold. This difference is less obvious in sows that are fed ad libitum.

In outdoor breeding units the huts are sometimes fitted with fenders or stepped entrances (Fig. 6.3). These reduce the chance of piglets leaving the hut and getting lost or experiencing hypothermia. The disadvantage with fenders is that they make it more difficult for piglets to leave the hut during summer, when the sow may prefer to feed the piglets outside. Those piglets that manage to get over the fender get a feed, whilst those remaining inside go hungry.

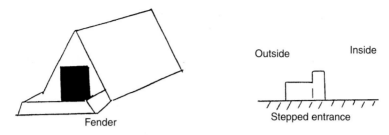

Fig. 6.3.   Fender and stepped entrance used for allowing piglets to re-enter a shed.

Damp litter can be a problem in outdoor huts. Rain can run off the roof and under the straw floor. Similarly, temporary straw-bale sheds are prone to becoming waterlogged in wet districts.

Lameness from bruising of the soles can occur on stony ground. This is best managed by maintaining pasture cover. Heavy soils that are continuously muddy can lead to softening of the feet, predisposing to injury and infection.

The stress of weaning is probably less in outdoor herds. Immediately following weaning, outdoor-reared piglets show less aggression and tail biting compared with indoor-reared piglets (Cox and Cooper, 2001). This may have been due to the greater opportunity to interact with pigs from other litters before weaning in outdoor units.

Outdoor pigs adapt more readily to unfamiliar conditions, including the lairage at abattoirs. They are calmer and fight less in comparison with pigs raised indoors under intensive conditions. In temperate and Mediterranean climates, outdoor pigs produce leaner carcasses, and this can be due to a lower feed intake as well as greater expenditure of energy during exercise (Hoffman *et al.*, 2003).

Heat stress in breeding pigs

Normally, a self-feeding lactating sow takes about five meals during the day and one at night. When she experiences heat stress, she skips some of the meals, and the first feed is often delayed to late in the morning. Overall feed intake drops off, but individual meal size often remains the same (Quiniou *et al.*, 2000). Her water requirement becomes more critical if she is to support lactation. For example, the ratio of water to feed consumption rises with ambient temperature from 4.2 l/kg at 18°C to 8.1 l/kg at 27°C. The upper critical temperature for the lactating sow is quite low (< 22°C).

Various methods are used by farmers to keep outdoor pregnant sows cool:

- providing shade
- providing wallows
- providing sprinklers
- providing straw under tall trees, to encourage the sow to lie outside on hot nights
- cooling the sheds by opening ventilation hatches, digging a hole at a corner of the hut or installing a hut sprinkler

Three factors contribute to **summer infertility**:

- adverse effects on libido and spermatozoa quality in the boar
- quieter oestrus and erratic cycling in the sows
- early loss of embryos

There are different causes of summer infertility in different herds. A common cause is failure of embryo survival in association with heat stress in the sow. If all the embryos are resorbed, the sow will return to heat anywhere between 30 and 90 days after service. If only a proportion of the embryos die, the sow continues her pregnancy but produces a smaller litter.

Besides keeping the sows and boars cool, the following approaches are sometimes used for managing summer infertility:

- Arranging for the gilts to farrow when the effects of summer infertility in the main herd are greatest.
- Keeping back cull sows and remating them to fill the farrowing gap created by summer infertility.
- Changing to indoor mating during hot weather, especially if it is suspected that the boars are being affected by the hot weather.
- Minimizing other stressors during the implantation period. Bullying by other sows

can contribute to summer infertility problems (Love, 1978), and ensuring there is plenty of feed to go round can reduce this effect.

- Withholding feed during periods of the day when mating normally occurs. Feed can be a distraction for the boar.

Sow longevity

Critics of modern pig farming methods claim that short life expectancy in sows is a sign that production is being pushed too far. The main reason for early culling is reproductive failure. Genetic selection for leanness can lead to earlier puberty in gilts and more piglets per litter during the early parities. However, leaner strains tend to be culled earlier and can have a lower lifetime productivity (Gaughan et al., 1995).

On some farms, gilts are mated at their first oestrus (100 kg live-weight) in order to stunt their growth and produce a sow with a small mature body size. The aims are to economize on feed requirements and produce a sow that is more active and less likely to crush its piglets. These sows are also less likely to develop lameness problems compared with sows that were first mated at their third oestrus (120 kg) when they were gilts (Le Cozler et al., 1999). The disadvantages of early breeding are that the sows are likely to have a shorter productive lifespan and produce smaller litters (Babot et al., 2003). A comparable reduction in sow longevity occurs if a gilt's size is limited by restricting its amount of feed (Le Cozler et al., 1999).

Weaning

At weaning time, the gut wall has to adapt to the sudden change from a milk-based diet to a grain-only feed. Initially there is a period of underfeeding before the pig learns to rely on solid feed. During this time the pH of the stomach contents rises and digestibility is suppressed until enzyme production by the pancreas and gut wall has adapted to the change in diet. Undigested solid feed passes along the small intestine, and this leads to a hind-gut fermentation, which may provoke an osmotic **post-weaning diarrhoea**. In addition, the consumption of dry feed causes abrasion and atrophy of villi on the small intestine.

The rise in pH has two important effects. It reduces the efficiency of protein digestion, and it can allow pathogenic *E. coli* to proliferate. The entry of *E. coli* in a protein-rich medium into the small intestine, which has a damaged surface from the switch to solid feed, at a time when the pig's immune response is normally low (the immunity gap), presents a high risk for a secretory diarrhoea. Normally this is managed with medication, but some nutritionists consider that a nutritional approach would be more appropriate. However, the ideal weaner feeding regime is undecided.

Post-weaning diarrhoea is a substantial welfare problem in the pig industry. Morbidity in affected herds can be greater than 80%, and in bad outbreaks mortality can exceed 25%. The number of *E. coli* in the faeces may increase 10^3–10^4-fold. In other respects, weaning would normally be a transient stress that the pigs get over. For example, play behaviour resumes after about 2 days (Donaldson et al., 2002).

Piglets do not start eating creep feed in substantial quantities before 21 days of age. In **early weaning systems** the piglets are weaned at less than 19 days. Early weaning relies on producing uniformly large and hungry piglets at weaning to ensure that the growth check is only slight. Selecting the right litter and the right conditions for early weaning can be important. Litter sizes of between six and ten piglets maximize pre-weaning growth whilst minimizing variation in pig size, and these litters have less risk of some piglets falling behind if weaning is early. Early weaning may need to be set later in hot climates, because milk yield during mid-lactation is likely to be lower and weaning weights correspondingly lower.

Early weaning (e.g. at 9 to 12 days of age) can lead to a more aggressive type of pig later in life, compared with weaning at 3 weeks (Yuan et al., 2004), and in the USA it has been associated with a greater risk of higher levels of mortality during the grower–finishing period (Losinger et al., 1998). The risk of aggression may be avoided by having communal creep areas pre-weaning, where different litters can mix (Weary et al., 1996). Early-weaned pigs show more belly nosing, escape behaviour and object chewing than 28-day-weaned pigs (Worobec et al., 1999).

In Sweden, weaning is often at 5 weeks of age. This is associated with a problem known as **disturbed milk production**, where the sow's udder dries up after 3 weeks. The cause is unknown and mastitis has been discounted. The prevalence in some herds has been as high as 10% and a genetic predisposition is suspected as it has a high heritability (0.70). Piglet mortality can be high from starvation if the situation is not managed appropriately.

Providing sows with a farrowing crate which they can leave whenever they like leads to less frequent nursing of the piglets. The piglets start eating solid feed earlier and, although they may be lighter at the start of weaning, they adapt to the weaning feed more quickly (Pajor *et al.*, 2002).

Group size and stocking density

There has been a pronounced change over the past 10 years towards keeping grower pigs in large groups. Sometimes as many as 1500 pigs are present in a pen. This economizes on housing costs and, when the pen is operated on an all-in–all-out basis, management is quite simple.

Often the pens are divided into a feeding area and a large common area. Where the whole pen is slatted, the centre often serves as the dunging area and the pigs lie against the walls. Practical experience with large group sizes has shown that it is more difficult to find and remove injured animals, inspection can be challenging if the stockperson is mobbed by the pigs and atmospheric ammonia is a hazard around the dunging area.

The welfare implications of group sizes as large as 1000 to 2000 have not been examined experimentally, and the following points come from studies on up to 205 pigs per group:

- Pigs kept in large groups can benefit from a greater degree of choice of micro-environment compared with pigs housed in a small pen.
- Social pressures are different. In large groups, recognizing social rank is difficult. At the start there is likely to be more aggression when social stability is being established. Thereafter, there is greater reliance on resolving rank disputes without aggression. There seems to be less social confidence, and this has been reflected in lower early weight gains compared with smaller groups, but the presence of extra feeders helps reduce this effect.
- With large groups, social pressure at the feeders gets worse as the pigs get bigger, and especially when *ad libitum* hoppers are available at

fewer than one per ten pigs (Spoolder *et al.*, 1999). Providing too many feeders can lead to feed wastage, and so a balance needs to be struck.

- Problems can also develop when a pig dominates a resource such as an *ad libitum* feed hopper or drinker, especially when the feeders and drinkers are situated too close together.

Aggression may also arise during social facilitation and overcrowding. This happens, for example, when a pig starts drinking and this encourages others to join in. In large group sizes this can become chaotic. Individual pigs that compete poorly are the ones likely to suffer. They have reduced growth rate either because they are more active or because of lower feed intake, and they may need to be culled or raised separately.

In over-competitive situations, the deprived pig is more likely to visit the feeder or drinker at night. Normally the average time per pig spent at a feeder or drinker decreases as group size increases. In the case of nipple drinkers, pigs in large groups use the water in a shorter period of time, whereas in small groups there is more non-drinking behaviour at the nipples. This situation is more critical for hot climates, and drinker availability needs to be more generous. In addition, the minimum stocking density in hot climates needs to be lower than in cool climates because of the pigs' need to stretch out to promote convective heat loss. In temperate climates, lying area requirements are based on the following equation (Ekkel *et al.*, 2003):

$$\text{Lying area} = 0.033 \times \text{body weight}^{0.66}$$

Within small groups (up to 20 pigs), stocking density has more effect on aggression than does group size (de Decker *et al.*, 2005). High stocking densities can create more activity around the feeders and reduce overall feed intake and growth rate (Bryant and Ewbank, 1974). Overall production per pen continues to increase with rising stocking density, and, provided the pigs can reach a satisfactory market weight and provided feed conversion efficiency is not compromised, there is a financial incentive to maximize returns from capital costs by overstocking. This dilemma is relevant in **wean-to-finish systems**.

In wean-to-finish systems, pigs are kept in the same groups and in the same pens from weaning to slaughter. This minimizes mixing stress and other disturbances. However, there is a strong incentive to stock at densities that maximize production per pen rather than production per pig.

Trauma

Whenever a pig is closely confined there is a risk that it will develop skin lesions. Shoulder lesions are quite common in sows confined in crates and stalls. For example, a Danish survey showed that the prevalence of shoulder wounds in sows held in farrowing crates was on average 12% (range 3 to 25%). Fully slatted crate floors and wet or soiled floors presented a greater risk, and thin sows were particularly affected.

When piglets are crushed by a sow, it is thought that many of them are killed by suffocation. Often the piglet is not immediately killed. For example, 16 out of 24 piglets survived being trapped for 4 min or longer, and they have been known to survive for 2 h under the sow.

Piglets are susceptible to sole bruising and foot tenderness if they are kept on a hard floor surface during the first 4 days after farrowing. After weaning, leg bone fractures occur mostly during falls on slippery concrete, and when limbs are trapped beneath pen rails or fixed feeding troughs. They also occur during loading and transport. The pig's front legs are more prone to sliding on slippery surfaces than the rear legs (Applegate et al., 1988). Slippery corners in corridors, slippery unloading ramps, smooth floors at feeders and drinkers and inappropriate gap widths in slatted floors are probably the greatest hazards for grower pigs.

Lameness

Sows confined in farrowing crates are prone to developing foot injuries. A Danish survey showed that the prevalence of foot wounds was 22% (range 2 to 43%). On average 12% of sows were slightly lame (range 3 to 18%) and 3% were visibly lame. Lameness was more common in thin and fat sows than in sows with normal condition and it was associated with skin lesions on the hocks and hindfeet (Bonde et al., 2004). Lameness can lead to uncontrolled lying-down behaviour in the crate, and this poses a greater risk for piglet crushing. Sows that slip when lying down are more likely to have hindfoot lesions.

Sows kept on concrete floors are more prone to lameness than those on straw bedding (Andersen and Bøe, 1999). Hock bursitis can develop in pigs housed on bare concrete, especially slatted concrete. Herds with a high prevalence of lameness can have a higher proportion of sows returning to oestrus than normal.

Sole erosion and cracks along the white line are the two most common foot lesions in grower pigs in the UK. Sole erosions are more common on concrete floors, whereas white-line lesions and wall separation feature in pigs kept on straw (Mouttotou et al., 1999). Partially slatted floors can be associated with both conditions. Slatted floors create an uneven distribution of pressure on the weight-bearing surface of the claws, and this poses a risk of claw and white-line damage. This may be made worse if dirt from sticky straw bedding is packed in the cavity at the apex of an overgrown claw. Repetitive pressure through this dirt pad probably acts as a wedge that splits the wall away from the sole.

Osteochondrosis is one of the main leg disorders causing lameness in breeding pigs. In the femur, it causes the hindlegs to turn out, and at the humerus the pig tends to be buck-kneed. Although rapid growth rate is often claimed to be an important contributor to osteochondrosis, it is not easily controlled by reducing growth rate through feed restriction. Instead, it may be a feature of lack of exercise. Confinement and lack of exercise can lead to cartilage lesions in joints. Exercise promotes penetration and release of synovial fluid by cartilage at the joint, and this helps the transport of nutrients into the cartilage and the removal of waste products released from chondrocytes into the joint space. This mechanical pumping is diminished by lack of exercise or prolonged standing in the same position, and the cartilage becomes weaker.

The heritability of lameness is not very high (0.16), but the pig industry looks to pig breeding companies to solve its lameness problems even though lack of exercise may be an important predisposing factor.

Gastric ulcers

There has been increasing emphasis on optimizing the digestibility of the grower pig's feed. One way of achieving this is to grind the feed more finely to produce a smaller particle size. This ensures more thorough acidification and digestion in the stomach. Fine grinding of the feed is, however, associated with a higher prevalence of stomach ulcers and higher counts of *Salmonella* in the digesta. It is thought that the prevalence of gastric ulcers in pigs has been increasing because of this change in feed milling practice (Nielsen and Ingvartsen, 2000).

Teeth clipping

Teeth clipping helps control facial scarring in piglets when they compete for a teat. It should only be necessary in large litters, or where there is a high risk of udder scarring in the sow. Udder discomfort from teeth gashes can cause the sow to sit for long periods in a dog-sitting position in an attempt to avoid nursing instead of lying on her side. When teeth clipping is performed, the teeth must not be clipped close to the gum, otherwise they may splinter. This creates a risk of tooth pulp infection.

Castration

Castration is used in grower pigs to reduce the risk of boar taint in meat. Some countries have stopped castrating pigs because it is a painful procedure, it is uneconomic because castrates grow slower, and avoiding boar taint by castrating the animal is unnecessary if the pigs are slaughtered before they reach puberty. In addition, castration can inhibit humoral immune responses. This may be important when castration is done at 10 days of age or later (Lessard *et al.*, 2002).

Some slaughter-generation pigs are now achieving puberty earlier and this has increased the need for castration. Introduction of the Meishan breed, as a way of improving early puberty and prolificacy in the female lines, has increased the risk of boar taint in the cross-breds. They are prone to producing meat with boar taint and this could warrant castration. An alternative to castration might be to include sugarbeet pulp in the grower ration, as this helps reduce the accumulation of skatole in carcass fat, skatole being one of the contributors to boar taint odour (Whittington *et al.*, 2004).

When boar taint is absent, the eating quality of fresh meat from castrates and boars is similar.

There are some small differences in eating quality in some of the drier processed meat products (Smith *et al.*, 1983; Bañón *et al.*, 2003a,b). Castrate meat often has a higher intramuscular fat content and this can improve juiciness and tenderness in processed products.

There is no doubt that castration without an anaesthetic is a painful procedure. Pulling and severing the spermatic cord may be the most painful part of the procedure (Taylor and Weary, 2000). Injecting a local anaesthetic into the testes or the funiculus spermaticus before incising the scrotum and removing the testes is now recognized as the best pain preventive in many European countries. It may not, however, provide any relief from subsequent inflammatory pain at the wound.

Castrating gilts that are not destined to be replacements in the breeding herd is not as common as it used to be. According to legend, this practice developed following a successful emergency Caesarean performed on a sow in Switzerland in about 1500. The main spaying method that evolved from this achievement is to restrain the gilt, which may be up to 30 kg live-weight, on its side, incise the left flank without sedative or analgesic, and then exteriorize and remove the ovaries plus oviduct. In Canada gilts used to be suspended by their hindlegs for the operation. The procedure is still practised in outdoor herds in parts of Spain, Sicily and Tuscany, especially where there are wild boars, but the practice has been stopped in France because it is considered cruel. Meat from spayed gilts is valued for producing traditional processed meats.

Tail biting and tail docking

In the UK, about two-thirds of pig farmers recognize tail biting as a problem in their herds (Chambers *et al.*, 1995). It is worse on farms with large herds and on farms holding pigs on slats instead of straw bedding. Other predisposing factors are automatic feeding instead of manual feeding, wet feed instead of dry feed, high stocking densities, and castrates compared with gilts.

The prevalence of tail biting in Denmark is much lower (10%). When an outbreak does occur, it is in larger pigs, weighing more than 60 kg, and during the colder periods of the year, when ventilation is poorer. Survey work in Danish abattoirs shows that 1.2% of pigs experience tail biting and in Sweden the prevalence is similar (1.9%). Differences between countries could be due to genetic differences besides standards of husbandry. In Landrace pigs the heritability for tail biting is 0.27.

Besides causing pain in victimized pigs, tail biting carries a risk of abscess formation in the animal's spine. Abscesses usually develop in the caudal half of the vertebral column, and posterior paralysis occurs if there is compression on the spinal cord. The abscesses cause carcass condemnation.

Tail docking is used to prevent tail biting, but in some countries it is prohibited on the grounds that it is an unnecessary or avoidable procedure. However, it is less painful than castration, which is often allowed (Prunier *et al.*, 2005).

Nose ringing

In outdoor breeding herds, controlling a sow's rooting behaviour can be difficult. At worst, the paddock becomes completely ripped, and when it rains it turns into a bog. The usual remedy is to use nose rings.

Nose ringing is a job that neither the operator nor the pig enjoys. Inserting the ring is stressful and painful, and it damages the trust the pig has in the stockperson. Nevertheless, nose ringing has its purposes:

- It prevents sows from uprooting a paddock and, at worst, turning it into a mire that is difficult to walk and work in.
- It allows herbage growth, which contributes to the pigs' feed. In ripped ground, productive species are replaced by broadleaved weeds.
- It stops sows from scraping a hollow when making a nest within the hut. This increases the risk of crushing if a piglet does not get out of the hollow quickly enough as the sow lies down.
- It can be used to control bullying behaviour in boars.
- It reduces fence breaking.

Nose ringing stops rooting behaviour by causing discomfort when the pig tries to root. Rooting is a normal exploratory and foraging behaviour, and some consider that it is inappropriate to prevent this behaviour. Nevertheless, nose ringing is commonplace in outdoor pig production. Sows with rings in their noses exhibit more chewing behaviour in place of rooting.

Pig producers use a wide range of materials as nose rings. They include soft wire with sharpened tips, purpose-manufactured C-clips and steel or nickel-plated rings. C-clips and soft wire rings are inserted in the top of the nose, and the steel rings are placed in the septum between the nostrils (Fig. 6.4). Rings in the top of the nose are prone to wearing or tearing out, and the temptation is to put in more than one ring to increase the chance that one will stay in longer. There is a fine balance between putting the rings in too deeply and too superficially. Superficial rings are prone to tearing out whilst rings that are too close to the nasal bone are uncomfortable for the pig.

Various restraining methods are used whilst inserting the ring, and all are stressful. A common way is to snub the nose with a rope, and lash the free end to a post, whilst the sow is confined in a race. She is then ringed and released as quickly as possible.

Alternatives to nose ringing include:

- Providing excessive amounts of pasture. Pigs are less inclined to root if good grazing is available.
- Encourage alternative interests or activities, e.g. provide forage crops, silage.
- Allow sacrifice areas where the sow can root up the ground.
- Turn the sows on to harder ground when damage starts to appear. The ground can be made firmer and more difficult to dig up by running a flat roller over it.
- Reduce the stocking density.
- Keep the sows that are notorious rooters in paddocks that have well-established trees. Normally pigs choose damp patches well away from large tree roots, which obstruct their tunnelling behaviour.

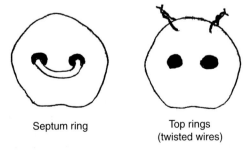

Septum ring   Top rings
                               (twisted wires)

Fig. 6.4.   Pig nose rings.

Ear notching

Ear notching is used on some farms as a way of marking and identifying grower pigs. It is done when they are piglets, which often shake their heads and less commonly scratch their ears with a raised hindleg, indicating irritation or pain.

Bedding and environmental enrichment

Variety provides pigs with a rewarding environment. It also encourages a pig to be less fearful of the unfamiliar (Beattie *et al.*, 2000). It leads to a less aggressive pig, and one that is less intimidated by the presence of people. Bedding and complex feed sources contribute to enrichment, but in their natural muddy environment they have a predilection for unearthing roots and other feedstuffs. In the piggery environment, they would prefer to manipulate low-dry-matter materials that are structurally complex, such as moist mixed-crop silage and fodder beets. A mixed-crop silage containing oats, vetch and lupin seeds gives the pigs something to search out (Olsen *et al.*, 2000). In the absence of these sorts of material, pigs direct their oral manipulation at pen equipment and other pigs. This can lead to restlessness, aggression and tail biting.

Straw bedding is a practical but not ideal alternative. It encourages activity in pigs, and this increases the number of visits they make to feeders through the day. It gives grower pigs something to work with their mouths and it helps reduce the risk of tail biting, but it can increase the risk of respiratory disease.

Floors

Slippery-surfaced and poorly maintained floors can be an important cause of injuries in pigs. When slats are used, they should give adequate grip for the feet whilst minimizing foot abrasion. Slat edges should be rounded, not sharp or chipped. The slat width and gap width for farrowing sows plus piglets are 18–25 mm and 8–11 mm. For dry sows and boars, 80–100 mm and 10–25 mm are appropriate. Problems can occur when metal slatted floors get broken and develop sharp edges. This is not an issue with cast-iron slats, but it is a problem with mesh floors. Plastic slats tend to be kinder on the feet than metal, but they are slippery and noisy. If plastic is preferred, it is best to choose a design that has a surface profile that provides a good grip.

Preference studies have shown that piglets prefer plastic-coated expanded metal to woven wire. Perforated metal slats are intermediate in choice. The piglets in these studies started to show these preferences between 5 and 48 h after birth. Surface grip was the single most important factor. Other useful observations were:

- Knee damage in piglets tends to be lower on old cement screed.
- In sows, leg and teat damage is unusual on straw-bedded floors, but more common on fully perforated floors. Leg damage in sows in farrowing crates can be reduced by using perforated floors in only the rear third of the crate floor.
- Slatted floors help to maintain pen hygiene in grower accommodation, and in the post-weaning period this can reduce diarrhoea and morbidity due to gastrointestinal disease (Rantzer and Svendsen, 2001b).

Preference studies have also examined the hardness of the floor surface which sows will choose to farrow on. In one trial all the substrates were covered with a layer of a synthetic rubber that is commonly used on the floor of cow cubicles. The findings were not surprising; the sows chose either a generous (18 kg) straw mattress or a profiled hollow in preference to a flat concrete floor. When different amounts of straw were tested, all the sows farrowed on straw when at least 4 to 5 kg was provided. Less than this was inadequate.

In the farrowing accommodation, inadequate cleanliness of the floor or nest can lead to mastitis. Part-slatted farrowing crate floors are more hygienic, and when they reduce slipperiness there can be an added advantage of lower piglet mortality (Rantzer and Svendsen, 2001a). It has also been found that Enterobacteriaceae on the teats and infections with *E. coli* mastitis were less frequent when sows were kept in farrowing pens with separate lying and dunging areas, in comparison with farrowing crates.

Straw bedding can reduce abnormal behaviours. For example, when sows and gilts have been given chains or wooden bars to help reduce irritability, some played with them obsessively. This behaviour was controlled by providing straw bedding along with the chains and logs (Spoolder *et al.*, 1995).

Dust

High levels of atmospheric dust can create an uncomfortable environment and may increase the risk of respiratory disease. Dust is a particular hazard where pigs are floor-fed a meal containing low levels of fat. Activity amongst the pigs stirs up dust, and it stays suspended for longer in poorly ventilated sheds and if the humidity is low.

Human safety standards require dust levels to be kept below 15 mg/m^3, as an average for an 8-h work period. The corresponding recommendation for piggeries is 4 mg/m^3, but levels sometimes rise to 100 mg/m^3. The majority of airborne bacteria adhere to dust particles larger than 4 μm in diameter. Most of the dust in grower sheds comes from the feed, and 80% of the particles are between 0.5 and 2.5 μm diameter. These smaller particles are more readily extracted by automatic ventilation systems than the larger, heavier particles, which mainly come from starch components in the feed.

Ammonia

Ammonia can be a problem in indoor enclosed livestock farming systems where the stock are fed high-protein feeds. Between 20 and 25% of nitrogen consumed in the feed is emitted as ammonia-nitrogen in dry sow accommodation. Ammonia emissions tend to be high where urine is allowed to stand as puddles, instead of draining away. The recommended upper limit for ammonia in piggeries is usually between 20 and 25 p.p.m., but grower pigs show aversion to levels as low as 10 p.p.m. (Jones *et al.*, 1996). Chronic exposure to about 40 p.p.m. ammonia can interfere with some pigs' sense of smell (Jones *et al.*, 2001). Dust particles can carry and release ammonia. As much as 3.9 mg can be adsorbed on to 1 g of dust. Sawdust bedding is better aerated than straw bedding and so it emits less ammonia and more N_2O. It should be possible to limit ammonia emissions by limiting the protein content of feeds, but this will be at the cost of slower growth rates and fatter carcasses (Ferguson and Gous, 1997).

Drinkers and water quality

In temperate climates the usual water requirement for a dry sow is 5 to 10 l per day, and for a lactating sow 15 to 30 l per day. When given the opportunity, sows will often mix water with their feed just before they start eating. Otherwise, they may stop feeding intermittently to take a drink.

The sow's behaviour can be managed by the siting of the drinker relative to the feeder. If the drinker is near the sows' normal feeding or loafing area, they may play with it and it can become a focus for competition between sows. This can be avoided either by providing more drinkers or by placing them so that the sows have to move to them when they are thirsty. The dunging area is an ideal place for drinkers in this respect, and it also makes management of spilt water easier.

Growing pigs consume about three times as much water as the weight of dry matter consumed as feed. However, this relationship breaks down when feed intake is severely restricted and hunger-induced polydipsia sets in.

Poor water quality is an occasional problem in low-rainfall regions (McLeese *et al.*, 1991). It can reduce performance and cause scouring, especially in young pigs. Sulphate anions are probably more detrimental than chlorides. Scours could develop from osmotic or more direct laxative effects.

Mixing and aggression

In group housing systems the contact between stock handler and sows is different from that in individual sow housing systems. Sows in groups seem to be less fearful of people. There is, however, more opportunity for aggression between sows, and this has to be managed in the following ways:

- Remove sows that are being bullied and put them in a pen with sows of the same size or temperament. Some farmers, however, feel that it is more important to take out the bully and pen her separately.
- Provide ample space. This allows submissive animals to escape from a bully or aggressive animal.
- Run the first-parity sows in separate pens from the main herd.
- Mix animals at a time and under circumstances when sows are less likely to fight. For example, when weaning begins, the sows can be let out of the farrowing crates to a paddock for one night. The novelty of the paddock,

the absence of piglets and tightness of the udder all help to distract the sows from fighting.

- Mix sows only with other familiar sows. This can be organized by using colour-coded ear tags, each colour indicating the age of the sows. Sows are mixed only with other sows that have the same colour tag.
- Train the animals. Some farmers equip themselves with a pig-board and Alkathene tubing when mixing sows or boars. A sharp smack on the snout is used to control outbreaks of fighting. This has to be done in confined quarters (e.g. pens or yards).
- Mix large numbers at a time (nine or more). This can be associated with less aggression, but it is not invariably effective.
- Mix unfamiliar pigs after dark.
- Provide feed *ad libitum* during the early period following mixing.
- In the case of boars, some producers run the boars successfully in pairs outdoors. Rearing them in pairs helps reduce aggression when they are kept together in paddocks or pens as adults. It can also be helpful to match dissimilar boars; otherwise, if they are not used to each other and there is no clear social hierarchy they will probably fight. Putting a new (young) boar in with an older boar can work well, and it often acts as an incentive for the older boar to work harder. Boars that have a poorly defined territorial area may be more prone to being aggressive to sows. This is sometimes seen in boars that are confined in stalls and are let into a corridor for mating.
- If all else fails, it may be necessary to cull bullies. It may also be necessary to consider culling sows that are aggressive towards people, especially if there are young children on the farm. Aggression is not heritable ($h^2 = 0.08$), and so culling will not have longer-term

benefits in breeding herds. Some producers claim that sows that are aggressive to other sows are also aggressive to people.

When grower pigs are mixed and two unacquainted pigs meet and fight each other, they push each other shoulder to shoulder and intermittently bite at each other, slashing the other's shoulders with an open mouth. These contests can last uninterrupted for 5 min or more, and frequently end with one pig biting and chasing the other. Mixing stress can be associated with poorer growth performance. When sibling groups have been raised separately without any mixing, they grew faster and had better feed conversion efficiency than groups of mixed pigs (Lund *et al.*, 1998; Table 6.6).

When sows and gilts are mixed with a pen of dry sows there is likely to be some fighting. This may last for about 3 h before things settle down. The newly introduced animals may form a subgroup, and they may lie in the dunging area instead of the main lying area. The subgroup can persist for a fortnight or more before it becomes integrated with the main group. Premixing the introduced subgroup before it is introduced to the main group can help to reduce the amount of fighting (Durrell *et al.*, 2003). Providing straw is not a reliable way of reducing aggression.

Fighting during mixing is less likely to occur if the pen is rectangular rather than square (Barnett *et al.*, 1993). This works if the rectangular pen provides greater opportunity for subgroup formation.

If sows are mixed during the mating period, reproductive performance can be affected. However, this is not an inevitable outcome. There are exceptions to every rule, and one study found that when sows were mixed after weaning at 5 weeks, they had 10% fewer reproductive failures than sows that were stalled individually after weaning (Fahmy and Dufour, 1976). It was also found that

Table 6.6. Effects of raising pigs between 25 and 100 kg either as sibling groups or in mixed groups.

	Sibling groups	Mixed groups	*P* <
Number of pens (pigs)	5 (48)	5 (49)	
Growth rate (g/day)	889	865	0.05
kg feed/kg live-weight	2.66	2.76	0.01
Prevalence of agonistic behaviours (%)	43.4	56.6	0.001

flushing helped to bring the sows on earlier, but embryo survival rate was poorer.

Cannibalism

Sometimes a family line crops up in a herd where the sows are aggressive towards their piglets. It usually occurs in the gilts, and they often grow out of it as they get older. The heritability of this type of aggression is between 0.12 and 0.25, and it should respond to culling. Gilts reared in stalls are more prone to this problem than those reared in straw pens. Farrowing crates can help to reduce piglet savaging by reducing access to the piglets, but gilts held in farrowing crates are more inclined to savage their piglets compared with gilts in pens (Cronin and Smith, 1992).

7

Poultry and Rabbits

During the second half of the 20th century, the broiler chicken and turkey industries were transformed from minor contributors to world meat supply into major producers. There are six reasons for their success, and each contributed to cost competitiveness with the red meat sector. The reasons are:

- Changes in production methods allowed poultry production to become less seasonal.
- Wholesale and retail markets developed rapidly through the frozen food trade and at the outset freezing enabled distribution of poultry carcasses.
- Intensification of production and processing enhanced their competitiveness through economies of scale.
- Birds were genetically selected for a high meat content in the carcass, in line with consumer needs.
- High feed conversion efficiency when fed grains.
- Low labour requirement plus rapid turnover suited some farming systems, and helped keep costs down.

Duck, goose and rabbit meats enjoy a smaller market. The duck and goose industry has focused on some of the lower-volume and more lucrative trades, and their carcasses are probably too fat for high-volume markets. Rabbit production has been less successful because the carcasses do not have ideal portion-size or conformation with easy-to-eat muscle blocs, and it has been difficult to recover labour and housing costs from the sale of fryers alone.

Chickens

The potential welfare problems in the chicken meat industry include:

- chick mortality in subsistence farming systems
- chick management before placement at the broiler farms
- lameness
- ascites
- hunger and associated problems in broiler breeders
- ammonia, dust and high temperatures in broiler sheds
- overstocking
- dehydration
- inadequate inspection and culling
- pre-slaughter stress
- slaughter methods
- caponizing methods
- broken bones in spent hens

Developing countries

In developing countries, chicken meat is often a co-product of egg production. For example, in Africa, male birds and hens that are no longer wanted are taken by traders from villages to town markets, and they are usually killed and consumed on the day they are sold. The birds are often transported in cane baskets attached to the backs of lorries. Mortality during some of these journeys is high (Anon., 1992).

Free-ranging chickens in the villages usually have to find most of their own feed. Productivity is low.

©N. Gregory 2007. *Animal Welfare and Meat Production* (N. Gregory)

For example, growth rate in male birds is between one-sixth and one-twelfth that of intensively reared broilers (Kuit et al., 1986; Wilson et al., 1987). The local breeds in West Africa typically produce three to four clutches of ten to 14 eggs a year. Hatch rate is often high, but so is chick mortality and theft. Mortality is mainly from dehydration, predation by raptors, cats and dogs, infections, and accidents from being run over or trampled by other animals. Often only one or two females survive to maturity. They are often housed at night, but where this does not happen they avoid predation by roosting. They are often confined for the whole day when crops are sown until germination has occurred, and at this time they depend on feed provided by the owners. In Bangladesh, mortality rate is lower when broilers and layers are kept in cages compared with deep litter (Solaiman and Azad, 1995).

In hot climates, a cool resting area is necessary even for chicks. Corrugated iron roofs help protect the birds from the sun's direct heat, but the roof needs to be insulated underneath: otherwise it acts as an oven.

In some poor countries there is consumer resistance to intensively grown broilers. This is due to lack of familiarity and trust in buying a bird that is frozen, and because broiler meat lacks flavour compared with hen meat. In other countries, and especially in South America, intensively produced broilers have become the single most important source of meat. Here, lack of flavour in the modern broiler is no longer considered a problem.

Lameness

Lameness is recognized as the single most important welfare issue in the broiler industry. Poor walking ability is common and in some birds it is painful. It is linked to high growth rates in modern broiler strains.

The most effective way of avoiding lameness is by reducing the growth rate during the second week of life. This has been known for many years in the broiler industry, and, in the past, the usual approach has been to reduce the protein content of the starter ration when there was a problem in a previous crop. This limits early growth rate and produces birds with better walking ability. Reducing the lysine content of the starter feed can also limit early growth rate and improve the birds' gait

score (Bizeray et al., 2002b). Other approaches include modifying the lighting pattern to reduce voluntary feed intake and limiting the amount of feed available for a short period early in life. Feed restriction early in life can also reduce the prevalence of ascites and sudden death syndrome (SDS), and improve the birds' subsequent resistance to heat stress (Zulkifli et al., 2000; Urdaneta-Rincon and Leeson, 2002). Feeding a mash instead of a pelleted ration will reduce feed intake and improve walking ability, but it has increased deaths from ascites and SDS (Bennett et al., 2002). Encouraging the birds to walk more as chicks has not always been effective in reducing lameness later on (Bizeray et al., 2002a). This has been tried unsuccessfully by scatter-feeding wheat, and by installing barriers between the feed and water sources, but some studies where the distance between the trough and drinkers was increased showed that this helped reduce the prevalence of twisted legs (Reiter, 2004).

The growth restriction needs to be applied early to capture the benefits of compensatory growth, and to realize the potential advantage in feed conversion efficiency. It might be hoped that the feed restriction will limit fat deposition in the carcass, but this has not always been the case (Lippens et al., 2000; Urdaneta-Rincon and Leeson, 2002).

In the 1990s tibial dyschondroplasia (TD) was one of the main causes of broiler leg disorders. TD occurs when there is a plug of avascular cartilage below the growth plate of the tibiotarsus. The chondrocytes in this plug do not seem to differentiate at an appropriate rate, and in very severe cases this results in bending at the head of the tibiotarsus and angulation of the limb. The present view is that TD is not such a common problem as it was, and this may be because the condition has been corrected by genetic selection. The majority of the TD cases that are seen today are not lame (McNamee et al., 1998; Garner et al., 2002). Instead, limb deformities, spondylolisthesis, scoliosis, osteomyelitis of the hock or hip and physical instability are collective contributors to abnormal gait. Physical instability is thought to be due to the birds' rapid breast muscle growth and realignment of the centre of gravity (Corr et al., 2003). The centre of gravity has been moved forwards, and the birds now have a balance problem whilst walking. This is apparent in the low frequency at which they take steps and their shorter wider steps, which throw the centre of gravity into lateral excursions. They develop outward-pointed toes, which improves lateral support, and they take a longer period during each stride, when both feet are supporting the bird's weight, in combination with a shorter swing period.

Two views are emerging about whether or not there is suffering associated with leg disorders in broilers. One view is that birds with a gait score of 3 or more (on a 0 to 5 point scale) experience pain when walking. This is based on the findings that gait score 3 birds walk faster when given an analgesic, and they voluntarily select an analgesic when given the opportunity (Danbury et al., 2000). The alternative view is that some birds may experience pain when walking, but others walk badly simply because they are physically unstable. The prevalence of pain and the pathophysiological conditions that cause pain have not been clearly defined. Once a bird is lame, it visits the feeder less frequently and it often eats whilst in the sitting position (Weeks et al., 2000).

Ascites and SDS

Ascites is recognized as a production disorder in broilers. Like lameness, it has developed as a result of intensification of broiler production and it can be worse in fast-growing broiler lines. Three theories have been put forward about the primary cause. They are:

- cardiac disorder
- pulmonary incompetence
- excessive metabolic demand relative to cardiopulmonary capacity

For many years, the third reason was the favoured explanation, but it has recently been shown that the oxygen demand on a metabolic body weight basis in fast-growing broiler lines is less than that in slow-growing lines (Malan et al., 2003). This means that rapid growth does not necessarily imply high oxygen requirement as a causal factor for ascites. Instead, it is suspected that the hypoxaemia that leads to ascites is due to circulatory insufficiency associated with a progressive bradycardia (Olkowski and Classen, 1998). This may be due to valvular insufficiency, with pulmonary hypertension developing as a secondary effect.

The prevalence of ascites is higher in cold climates, but it is controlled at tolerable levels. In Canada, the overall prevalence of post-mortem condemnation is 1.9% and, of those, 8.5% are due to ascites. Cold conditions encourage farmers to save heat by limiting ventilation, and this can exacerbate the ascitic condition.

The heritability of ascites is 0.12 to 0.41 depending on breed and gender, and the heritability of right ventricular hypertrophy in broilers is 0.45 to 0.54. Selection for growth rate has probably been responsible for the predisposition to ascites. The r_g between live-weight for age and ascites in males is 0.22 to 0.35. Feed restriction early in life helps reduce subsequent mortality from ascites (Camacho et al., 2004).

The heritability of SDS is 0.25 to 0.30. These deaths are due to acute heart failure, whereas death from ascites occurs after a protracted illness. When birds die from SDS, they lose balance, show vigorous wing flapping and often flip on to their backs when they give a kick. The whole process is over in less than a minute, and it is not necessarily preceded by any particular event or behaviour. It is seen in heavyweight males, and overall mortality from SDS in a flock has been as high as 10%, but more often it would be less than 1% (Gardiner et al., 1988). These deaths are sometimes called 'flip-overs'.

Chicks

The welfare issues in newly hatched chicks include:

- handling methods used during processing at the hatchery and placement at the farm
- dehydration during prolonged holding and transport
- disposal methods for undersized, deformed, injured, surplus and unwanted chicks
- temperature, humidity and noise conditions during delivery and following placement

The conditions used in the incubator and hatcher at the hatchery are critical for producing healthy chicks. If the temperature and humidity are too high, the navel may fail to close properly and the yolk sac protrudes from the abdomen. Of lesser concern, the chicks may have red beaks and red hocks. Together, these features, plus the volume of the yolk sac and overall vitality, are sometimes combined in a chick condition score, which is used to assess batch quality.

At the time of hatching the yolk sac accounts for 15 to 25% of the weight of the chick. It provides energy and water, which keep the chick alive during the critical early stage. In the first few days following hatching there is rapid development of the intestine, which is supported by nutrients derived from the yolk sac. Once gut function is established and the chick begins to eat, survival rate should be high.

Problems can stem from the way chicks are managed before they reach the grower farm. The majority of chicks hatch within a 2-day period, and the hatcher is usually emptied at about 48 h after the first chick has emerged. There is little point in extending the hatching period beyond this time when the aim is to maximize the number of viable chicks (Kingston, 1979). Cropping the hatcher at frequent intervals will have no benefit if the chicks are simply held in trays in a different room until enough chicks have hatched to make up an order or delivery. In addition, repeatedly opening and closing the hatcher door to crop the hatched chicks will probably slow down the hatching rate because of fluctuations in temperature and humidity. Those hatching towards the end of the hatching period have a lower viability and a greater risk of leg disorders. The early-hatched chicks remain without feed if the hatching period is extended and this will result in dehydration and compromise their chance of survival. Access to feed and water is delayed further by processing, which includes sexing, vaccination, beak trimming, packaging and transfer from the hatchery to the broiler sheds.

Hatcheries are sometimes criticized for the fast and rough handling methods used when sorting, grading and packaging the chicks. Automatic chick counters are particularly rough because of the change in speed the chicks experience (Knowles et al., 2004). In terms of the time the chicks take to right themselves, this procedure seems to be the most disorienting experience. Vaccination is often done with a spray at the hatchery. It relies on the chicks ingesting the vaccine when preening and inhaling some of the aerosol. The only welfare issues in this procedure are inadequate coverage when hand-spraying the vaccine and chilling in those hatcheries where a blower is used to dry the vaccine on to the plumage.

The longer the delay between dispatch from the hatchery and feeding at the farm, the greater the weight loss from dehydration (Fanguy et al., 1980). If the delay is more than 72 h, there is likely to be raised mortality and poorer growth to 4 weeks of age. In Taiwan, it has been estimated that for each 10 km increase in journey distance, cumulative mortality in the first week increases by 0.5% (Chou et al., 2004). A feed supplement has been developed which is provided during transport. It is in the form of a soft bead, containing carbohydrate that has been gelatinized and has a moisture content of 30% or more. Chicks that eat it benefit from early gut development and faster initial growth during the first weeks in the grower shed (Batal and Parson, 2002).

The time without feed and water is likely to be longest when chicks are transported between countries. Sudden changes in temperature can also create problems. Temperature control in the freight compartment is difficult during intercontinental flights. When a plane starts to descend as it approaches its destination, engine speed is reduced to save fuel. This limits ventilation in the hold, and temperature starts to rise. When the plane is standing at the terminal there is very limited ventilation, and in tropical regions temperature increases very rapidly. It has been known for temperature to rise from 20 to 40°C and relative humidity (RH) to rise from less than 40 to over 95%, between descent and unloading. The risk to chick survival is greatest when they are waiting at the airport for collection and during subsequent road transport to the farm. Mortality rates of 20% can be expected if the chicks are exposed to 41°C for more than 4 h. Clearly it is in no one's interest to allow mortality to reach this level, but sometimes conditions get out of control, especially on busy, congested roads.

In cooler climates, chilling can occasionally lead to high mortality following placement, especially when the shed is not warmed sufficiently before the chicks arrive. In addition, late-feathering males in feather-sexed strains sometimes need a longer brooding period to avoid chilling effects.

Undersized, deformed and unwanted day-old broiler chicks are usually euthanized either with CO_2 or by maceration in a double-roller mill. Carbon dioxide causes a period of gasping before collapse, indicating that there is breathlessness before loss of consciousness. Maceration is instantaneous provided the contra-rotating rollers are set properly. An advantage with maceration is that it can also be used for unhatched eggs. The eggs need to be destroyed: otherwise there is a risk that chicks will hatch after the eggs have been dumped at a disposal site.

Housing conditions and stocking density

Ammonia concentration in a broiler shed increases as the flock gets older and excreta accumulate in the bedding (Table 7.1). Typically, the concentration starts at about 2 p.p.m. and usually it does not exceed 20 p.p.m. If it reaches the 25 to 50 p.p.m. range, broiler growth starts to decline and mortality rate can increase (Miles *et al.*, 2004). Concentrations as high as 50 p.p.m. would be very unusual in broiler sheds, but the risk of reaching this level increases when the litter is wet. Wet litter occurs when there are leaking drinkers, when stocking density is too high and hinders drying of the litter, and if the shed is poorly ventilated. Litter moisture content should be maintained at less than 30% to keep the ammonia level below 20 p.p.m. Humid air can encourage ammonia production if it hinders drying of the litter, and it may in itself be a sign of damp litter. Gas brooders are prone to producing a humid atmosphere, but provided the chicks are placed on fresh dry litter they should not lead to ammonia problems.

Dust can become a problem when:

- birds are active, for example because of natural daylight, restricted feeding or supplementation with a scatter feed
- birds are being weighed, caught or handled
- shed ventilation is set low because it is cold
- ventilation is inadequate in naturally ventilated sheds

When dust levels get out of control, the birds' nostrils can become blocked. This would not be common on broiler farms, but it can occur with broiler breeders. It adds to the risk of respiratory disease. Dust can transmit microorganisms that are responsible for respiratory tract infections. Some of the respirable dust is from the litter. Inhaling this dust and pecking at the litter are two routes of transmission.

In humid environments, water vapour adsorbs on to the dust particles, making them heavier, and they settle more readily on the shed floor. This dust can be re-suspended as it dries and when the birds are active. There are ways of reducing this dust lifting. Adding fat to the feed can help, but if it leads to capping of the litter there may be a greater risk of hock burn.

Birds use a number of strategies when they get heat-stressed. Many of the strategies are directed at cooling their heads. They drink more, and if the water is cold this lowers the **temperature** of the crop, which is positioned close to the carotid arteries. This provides a cooling effect for blood passing directly to the head. The birds also extend the head and neck, spread their wings to promote convective heat loss and start panting. The high air speeds used in tunnel-ventilated sheds promote convective heat loss, and this is a useful strategy in hot climates where it is costly to lower overall air temperature in the shed. However, above 35°C an evaporative cooling system is also needed, as the convective cooling effect of the wind tunnel starts to change into a heating effect at about this temperature.

Broiler managers assess overheating from the air temperature and from the prevalence of panting. When air temperature exceeds 32°C, feed intake will decline and grower birds will feel uncomfortable. Water intake increases, bird activity declines and panting is more obvious. At high stocking rates, the risks are magnified. Stocking densities need to be lower in tropical and subtropical climates.

Controlling shed **humidity** can be difficult in wet tropical environments. If humidity outside the shed is high, the chances of lowering humidity inside are low. There are some simple safeguards such as ensuring that there is no wet vegetation alongside the air inlet to the shed, but in general the hazards have to be managed through regulating temperature rather than atmospheric humidity.

Under humid tropical conditions, exposing the chicks to a brief period (1 day) of high temperature during the first week of life can improve their survival during subsequent heat exposure when they are particularly susceptible to heat stress (Zulkifli *et al.*, 2004). Early feed restriction has a similar benefit.

Table 7.1. Ammonia emission from broilers at different ages.

Age (days)	Range in ammonia emission rate (mg/bird/day)
0–10	0–57
15–30	10–97
30–42	32–171
> 48	71–234

It is thought that modern broiler strains are less heat-tolerant than older broiler strains. Slow-growing strains of broiler survive heat stress better than fast-growing strains, and continuous genetic selection for high growth rate is negatively correlated with heat-related growth depression and with ascites (Yalcin *et al.*, 2001; Deeb *et al.*, 2002).

In **cold conditions** the sheds have to be heated, and there is reluctance to increase ventilation rate if this raises heating costs. Air and litter quality problems are worst in winter because of this cost-saving strategy, and ascites may be more common in the birds.

Rearing birds in dim lighting reduces their activity and this could benefit feed conversion efficiency. Continuous dim (6 lux in 24L : 0D) **lighting** regimes can make them more fearful and more reactive to stressful situations in comparison with a 12L : 12D regime (Zulkifli *et al.*, 1998). As broilers get older they prefer these darker conditions, but at 2 weeks of age about 200 lux is preferred (Davis *et al.*, 1999).

The **stocking densities** used in different countries depend on climate and the costs of housing. In general, birds kept in hot climates are stocked less densely to avoid overheating (Table 7.2). When stocking density is expressed as kg per m^2, it is a notional estimate of density at the end of the growing period. Stocking density reaches that level for a very short period, and, if the sheds are cropped or depopulated at set times, sometimes the target weight is not achieved and the final stocking density is lower. Conversely, if a flock performs particularly well or if a slaughter date has to be rescheduled, the anticipated final stocking density may be exceeded.

Table 7.2. Examples of broiler stocking densities used in different countries.

	Typical stocking densities (birds per m^2)
Thailand/Libya	8–12
Brazil	10–12
USA	14
China	15–16
France	16–25
Netherlands	23

Keeping birds at high stocking densities for too long can be associated with:

- deteriorating litter quality
- more breast blisters, hock burn, pododermatitis, scabby hips
- rising daily mortality
- more culling for leg disorders
- increased disturbance as birds try to move around, and less activity directed at ground pecking

These are not inevitable effects. They are risks, and the outcome depends on individual circumstances. The risk of deteriorating conditions will increase with litter moisture. In one study it was found that litter moisture started to rise when final density was in the 35 to 40 kg/m^2 range (Dozier *et al.*, 2005). Below this range, the risks were low. When high densities lead to high shed humidity and moist litter, the prevalence of pododermatitis, abnormal walking and mortality is also likely to be high (Jones *et al.*, 2005).

In terms of economy of production, feed conversion efficiency is not influenced by stocking density. However, yield per unit of shed area rises with stocking density within the ranges shown in Table 7.3 (Feddes *et al.*, 2002). This is a substantial benefit in terms of revenue over building costs. The coefficient of variation for liveweight is higher at the lower densities, presumably because the greater floor space allows the faster-growing birds to reach their full potential. As birds become more crowded, feed intake declines, particularly during the last week before slaughter (McLean *et al.*, 2002).

Newly hatched chicks are poorly feathered and need to be protected from draughts. This can be achieved with cardboard brooder circles. As the birds mature, feathering develops and in hot climates they benefit from moderately high air speeds. In recent years, there has been a trend towards increasing the air speed in tunnel-ventilated broiler sheds. Under high ambient temperatures (35°C, 60% RH), the optimum air speed is usually 2 m/s during the last 2 weeks of the growing period. At higher air speeds than this, the birds may not be evenly distributed in the shed and they eat less and grow more slowly (Yahav *et al.*, 2001).

High stocking densities can increase the risk of heat stress and, when this occurs, feed intake declines (McLean *et al.*, 2002). Mortality from SDS can increase as birds are stocked more heavily

Table 7.3. Effect of stocking density on broiler performance.

	Stocking density (kg/m²)			
	45	35	29	23
LFA (kg @ 38 days)	1.898[b]	1.931[b]	1.995[a]	1.915[b]
CV in LFA (%)	13.0[b]	13.6[b]	13.4[b]	15.3[a]
Shed yield (kg/m²)	46.9[a]	34.6[b]	28.6[c]	22.9[d]
FCR	1.72	1.72	1.73	1.70
% condemned	2.3	1.3	0.70	1.3
% birds with scratches	31	31	26	30

LFA, live-weight for age; CV, coefficient of variation; FCR, feed conversion ratio.

(Imaeda, 2000). Wet litter may encourage the survival of coccidia, *Salmonella* and *Campylobacter*.

broiler units, and occurs occasionally in free-range layer hens (Haslam *et al.*, 2006).

Pododermatitis and hock burn

Pododermatitis is used as an indicator of problems arising from overstocking. Like hock burn, it is a type of ammonia burn, but in severe cases the birds are less active and some can be lame. Ammonia builds up in the litter, especially if it is wet and capped. When the birds' feet tread through the capped layer or when the hocks press against an ammonia-rich patch of litter, the ammonia saponifies the fats in the epidermis and this allows penetration of NH_3 and NH_4^+ into the deeper tissues, which are denatured, setting up an inflammatory response. Long-term exposure leads to blackening of the outer tissue, which becomes leathery in texture (Fig. 7.1). Sheds that use misting systems can have high levels of pododermatitis, because the litter is wetter.

In Sweden, the prevalence of pododermatitis is used as a way of judging the optimum stocking density for a farm. Flocks that have high levels of pododermatitis are subsequently limited to a stocking density of 20 kg/m². This is a regulatory standard. Flocks that do not have a problem can stock at up to 36 kg/m², and there are gradations between the two limits. The prevalence of pododermatitis is assessed at the processing plant.

Hock burn is a similar condition to pododermatitis. In temperate climates, the prevalence of hock burn is directly related to the number of drinkers per unit area in the sheds. It has been a problem in heavily stocked free-range and organic

Drinkers

Dusty and very dry litter can be a sign that the birds do not have access to sufficient water. For example, if the litter below the drinkers is completely dry, the water pressure may need to be increased. When water is restricted the birds eat less. Their water requirement increases as ambient temperature increases above 20°C.

The height of the drinker line needs to be adjusted as the birds grow. In principle it would be helpful to have the occasional nipple set at a lower level, to allow short birds and severely lame birds to drink. Birds with poor walking ability have problems reaching the drinker, and sometimes lose balance when stretching to reach a nipple drinker.

In birds that can stand properly, the height should allow them to stretch their necks to drink using the end of the beak. This corresponds to a 55° angle when drawing a line from the feet to the beak. In this position, almost all the water should go into the mouth and not on to the litter. Birds that have abnormal joints may find it challenging to maintain this position to achieve a satisfactory drink.

Nipples save time spent on cleaning the drinker equipment, and when properly managed they are less prone to water wastage and damp litter. They can be associated with leg rotation, but this is less pronounced where nipple drinkers with cups are used. There have been some reports where birds with access to open trough water have

Normal   Pododermatitis

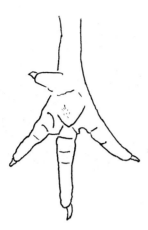 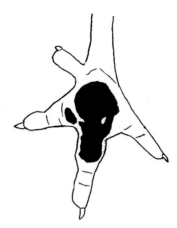

Fig. 7.1. Pododermatitis of the foot.

grown faster than birds with nipple drinkers, and it is suspected that this might have been in flocks with leg disorders (Lott *et al.*, 2001).

Flock inspection

Flock inspection is usually the first job that is omitted on busy farms when labour is in short supply. Birds need to be inspected regularly. Omitting flock inspection allows unforeseen problems to get out of control, and it reduces the opportunity for responsive management according to changes in bird behaviour. Inspection at an early age also helps reduce fearfulness and stress responses later on during catching.

Before inspection, the lighting is usually increased to encourage movement. The aim is to walk within 3 m of every bird in the shed once a day, inspecting at up to 10,000 birds per hour. At this rate most birds will stand up and move out of the way without panicking. Failure to move and abnormal movement should be spotted. Leaking drinkers should also be identified. This is also the time when culls and dead birds are gathered for removal. As the flock matures, the inspector will have to make more passes up and down the shed to inspect the whole flock.

Monitoring daily water use in each shed can help identify problems such as decreased drinking, which is an early sign of ill health or leaking drinkers.

Free-range and organic broilers

The market for free-range and organic broiler meat is growing slowly in Europe. The birds have access to outdoors, usually from a conventional broiler shed fitted with pop-holes in a side wall. They are managed intensively but they are usually slower-growing and have a poorer overall feed conversion efficiency in comparison with standard broilers. They can be leaner, from depositing less abdominal fat, and they are prone to developing breast blisters, especially if they have perches (Nielsen, 2004). The welfare advantages of this system are implicit from the more rewarding environment, but there are risks of predation.

Broiler breeders

The welfare issues in broiler breeders include:

- frustration from feed restriction
- prolapses
- cannibalism
- entrapment in separate sex feeders
- culling methods
- handling methods used during test weighing

Hunger, prolapses and cannibalism have arisen from genetic selection for growth rate and appetite. They are not problems in the slaughter generation, but have emerged as problems further up the breeding pyramid. If breeder birds are allowed to eat to appetite, they have suppressed fertility (in both females and males), excessive deposition of body fat and greater risk of vaginal prolapse, lameness and heart failure (Savory *et al.*, 1993). Instead, the feed has to be restricted to control those problems. As broiler lines have been selected for growth rate, they have increased the size of their abdominal fat pad ($r_g = 0.37$; Singh and Trehan, 1994), as well as their appetite. A large abdominal fat pad, plus a high intake when feed is not restricted, plus an enlarged reproductive tract in the females, leads to abdominal overcrowding and an increased risk of vaginal prolapse. There has also been reduced fertility, especially in the cockerels.

Limiting the birds' feed intake increases the risk of cannibalism through hunger and frustration. The frustration is probably worst during the growing phase. Early signs are voracious feeding, polydipsia and a trough-pecking stereotypy (Hocking *et al.*, 1996). There are three theoretical approaches to trying to limit hunger in these birds. They are:

- providing relatively small amounts of a high-energy feed
- feeding material that encourages hind-gut digestion and produces delayed and possibly protracted absorption of nutrients
- offering a poorly digestible bulky feed that provides gut fill but limited nutrient value

None of these approaches is completely satisfactory, and frustration will occur regardless of how the limited nutrients are delivered (Savory and Lariviere, 2000; de Jong *et al.*, 2005). However, diluting the broiler breeder's rearing ration with oat hulls, to allow higher dry-matter intake, has been successful in reducing polydipsia and litter dampness (Hocking, 2006). There may be ways of masking the expression of the frustration. For example, stereotypic pecking at the feed trough is less pronounced at lower stocking densities, and providing a brooder lamp can also reduce this pecking (Spinu *et al.*, 2003).

The risk of cannibalism may be reduced in broiler breeders by trimming their beaks (Fig. 7.2). This can be a painful procedure for the bird, depending on how much of the beak is taken off, and it can cause a lasting pain if neuromas form in the stump.

Male broiler breeders often show aberrant sexual behaviour, and in particular forced copulation. This is a feature of the genotype rather than feed restriction (Millman *et al.*, 2000). They have their

Fig. 7.2.   Beak-trimmed chicken.

sharp claws removed without analgesia to prevent back scratching of females during mating. The birds are kept on litter, and so sharp points on the claws do not get worn down naturally. Despurring is normally done with a hot wire and declawing with scissors. Trimming the tips of the claws would not be painful for the cockerels, but it is painful when the amputation is through bone or a joint.

Capons

In the USA, about 1 million capons are consumed every year, and production is based around Iowa. The birds are grown mainly for the gourmet and ethnic markets. They are killed at higher weights than standard broilers, and between 15 and 18 weeks of age. Absence of leg disorders in US capons is more critical than with standard broilers, because breast skin disorders are more common in heavyweight lame birds and because capon products are sold with skin on. In East Asia, the birds may be caponized at 10 weeks of age and grown on for a further 18 weeks. Capons in general have large amounts of abdominal fat, and the breast and thigh meat has a reputation for being tender.

An experienced operator can surgically caponize 200 2- to 4-week-old birds an hour. Anaesthesia is not used. The birds are fasted for 12 to 24 h to allow easier access to the testes. The bird is fastened on a board on its left side with the wings held together above the body. The legs are also fastened together, and the bird is stretched out to its full length in order to expose the rib cage area.

Feathers over the last ribs are removed and the skin disinfected. Using a sharp scalpel or knife, a 2- to 3-cm incision is made through the skin and other tissues between the last two ribs. The skin is pulled to one side before making the incision so that skin cut and muscle cut are not aligned afterwards. The incision should be deep enough to expose the abdominal air sac covering the intestines and other abdominal organs. Care must be used to avoid cutting a vein in the skin that runs diagonally towards the back of the bird. The abdominal air sac is punctured with a sharp hook or probe to expose the internal organs. The testes are located on the dorsal wall at the anterior end of the kidneys, posterior to the lungs. The first testis is grasped with special forceps and then twisted free from connective tissue while slowly pulling it from the attachments. Care needs to be taken not to rupture large blood vessels located between the two testes. The upper, right testis is then similarly removed. Electrically heated cautery equipment is sometimes used for incising the skin and removing the testes. It prevents excessive bleeding and may reduce the incidence of 'slips'. The rib spreader is then removed and tension on the bird released, allowing the skin and thigh muscle to move back into place. Once the bird is released, the incision should close without need for sutures or bandage.

'Windpuffs' may develop within a few days due to a build-up of air under the skin that escapes from air sacs cut during surgery. Puncturing the skin with a sharp instrument will release the trapped air, and is repeated as necessary. Removal of both testes is necessary and any fragments that remain will grow and produce enough male hormone to create a 'slip'. While a 'slip' will not normally function as a cockerel, it will not yield the desirable meat qualities of a good capon.

Layer hens

End-of-lay hens are one of the cheapest sources of meat. In some countries, the price received for the birds does not cover the catching and freight costs incurred in sending them for slaughter, and so instead they are killed and disposed of on the farms. In other countries, the egg producers pay the hen processing companies to take the birds away.

In some flocks, the yield of meat may be low because of emaciation by the time the birds reach the end of lay (Gregory and Devine, 1999).

In North America, about 14% of layer hens submitted for slaughter are condemned *post mortem*, and 28% of those condemnations are for emaciation (Anon., 2004). Hen meat can also be tough. Both defects are managed by cooking the carcasses before the meat is recovered, as this improves recovery of the meat, and by using the meat in comminuted products such as pie fillers, pastes and soups.

The welfare issues associated with catching, transport and killing end-of-lay hens are different from those of broilers. They are:

* risk of broken bones during catching
* risk of hypothermia during transport
* less effective electrical stunning

Cage layer hens have brittle bones by the time they reach the end of lay, and with some cage front designs it can be difficult to extract the bird from the cage without causing injuries. The keel bone and the ischium are prone to breaking when the former strikes the trough at the cage entrance and the latter hits the upper lip at the front of the cage. The average prevalence of broken bones in cull layer hens in North America is probably between 8 and 10% (Budgell and Silversides, 2004). In Europe it varies considerably between flocks (0 to 53%), and the way the birds are caught and removed from the cages contributes to this variation (Gregory and Wilkins, 1992). Birds kept in barn and free-range systems have stronger bones, but they acquire broken bones during the laying period from flight accidents (Gregory and Wilkins, 1996). Banning battery cages will not solve the broken-bone problem, which is considered by some as the single most serious welfare problem in the egg industry.

End-of-lay hens are often poorly feathered. Mortality rates as high as 11% have been recorded from hypothermia when transporting them in cold wet conditions and holding overnight without any cover over the vehicle (Gregory and Devine, 1999). Distinguishing between birds that are dead and those that are hypothermic but still alive requires close inspection, as the hens are cold and stiff from hypothermia before they die.

Turkeys

The welfare issues in the turkey industry include:

* early mortality
* pecking injuries and cannibalism

- leg disorders
- respiratory disease secondary to poor air quality in sheds
- toe clipping and skin scratches
- vaginal prolapse in breeder birds
- trauma
- killing methods used for cull breeder birds
- catching methods and transport

Turkeys need regular inspection, especially when raised in backyards. They are not renowned for protecting themselves from misadventure, or for figuring out how to overcome or cope with the physical problems they get themselves into. They also need to be inspected regularly for the first signs of cannibalism. If a bird is sick, it may be set upon by other birds and eaten. Self-defence is mainly by avoidance, and sickness reduces a bird's ability and inclination to move away from attack. Sick birds need to be housed separately as soon as they are identified.

Early mortality

Early mortality in grower females is usually about 1.5%, and for males it is about 2.5%. One of five causes is usually suspected when it is higher than this. It might be chilling or overheating during transport from the hatchery, an outbreak of an infectious disease, trauma during vaccination, delay in delivering the poults or 'starve-out'. Female poults are susceptible to cold conditions during delivery to the farm, whilst toms are at greater risk when it is hot (Carver et al., 2002). Turkeys are often vaccinated in the neck, and problems can develop if the needle is inserted too deep. Starve-out occurs in non-starters (poults that never start feeding) and stalled birds (poults that start feeding but stop). Causes of starve-out include:

- chilling
- overheating
- excessive beak trimming
- poor access to feeders or drinkers
- unpalatable water (e.g. residual disinfectant)
- inadequate lighting
- overcrowding at the brooders

If newly hatched poults do not get access to feed and water within the first 2 days of hatching, the development of their gut wall can be retarded. There are delays in villus and goblet cell development in the ileum and increased villus apoptosis throughout the small intestine (Potturi et al., 2005). This may not lead to deaths, but it can limit nutrient uptake and get the poults off to a slow start. If they have not learned how to eat by the third day, their outlook is poor. Mortality in non-starters generally occurs from the third to fifth day of age, after a period of progressive weakness. Post-mortems often show a loose slimy inner lining to the gizzard, which may plug the entry to the duodenum.

The risk of dehydration is higher in serviced poults than in non-serviced poults (Donaldson et al., 1994; Table 7.4). Serviced poults are sexed (by cloacal inspection), desnooded, toe-clipped and beak-trimmed at 1 day old. It is presumed that serviced poults experience greater fluid loss through urine, panting and bleeding from the snood, toe and beak wounds.

Beak trimming, feather pecking and cannibalism

Beak trimming is often a necessity for breeder and grower turkeys kept in natural daylight. Otherwise they are prone to killing each other. When needed, it should be performed early in life (between 11 and 21 days of age) to minimize neuroma formation, and secateurs are preferred to other methods

Table 7.4. Body weight and body water changes post-servicing in turkey poults.

Time post-servicing (h)	Body weight (g ± SE)		Body water (g ± SE)	
	Serviced	Unserviced	Serviced	Unserviced
16	56.3 ± 1.8	58.6 ± 3.2	43.6 ± 1.3	44.1 ± 2.4
40	50.7 ± 1.7	54.9 ± 1.6	38.7 ± 1.2	41.2 ± 1.4
64	43.2 ± 2.0	51.0 ± 1.9	31.7 ± 1.7	38.5 ± 1.6

because they are considered less painful. Beak-trimming poults at 1 day of age with a Bio-Beaker, as would be done in hatcheries, can lead to higher mortality and poorer growth rate during the first 20 weeks of life compared with trimming with a hot blade at 11 days (Renner *et al.*, 1989). Mortality can be minimized by ensuring that the hole created by the Bio-Beaker is not close to the nostrils. The Bio-Beaker produces the hole with a 1500 V current, causing the front portion to eventually drop off. It has two advantages. First, it can be applied at 1 day old without interfering with the chicks' ability to feed and drink during the first few days whilst the beak is intact. Secondly, there is less chance of the beak regrowing, and so there is less need to trim a second time.

Housing conditions and stocking density

Turkeys can be reared successfully in open-sided pole barns as well as environmentally controlled sheds. When given a choice the 2-week-old bird will select a lighting level of at least 20 lux (Barber *et al.*, 2004). Rearing in the dark helps control exploratory pecking, which can lead to cannibalism, but providing rewarding pecking substrates could be considered as an alternative.

When turkeys are kept at high stocking densities there is more interference between birds. For example, one bird moving about a closely stocked shed will cause those that are sitting to either change their position or stand up and move when disturbed. High stocking densities can also lead to a higher prevalence of lameness and breast buttons (Martrenchar *et al.*, 1999). Breast buttons are an unsightly blemish over the keel bone. The prevalence in North America has usually been less than 50% in heavy toms, but it has been as high as 67% (Newberry, 1992). It is a pressure injury that develops in heavy birds when they are sitting on the floor, and the prevalence can be influenced by the lighting regime in the shed. Increasing day-length patterns from 8 to 23 h between 4 and 16 weeks of age has been associated with fewer breast buttons than a constant 23 light (L) : 1 dark (D) pattern.

When sheds are heavily stocked there is a greater risk of one bird inflicting skin scratches with its claws on other birds, especially when they are clustered around a feeder or drinker. Toe clipping is used on some farms to control skin scratches.

Mortality to 4 weeks of age can, however, be higher in toe-clipped flocks (Newberry, 1992).

Leg and foot disorders

Modern turkey strains have been genetically selected for large mature body size and impressive breast muscle development. This is putting unnatural stresses on the hindlimbs, and valgus deformity of the distal aspect of the hindlimb is common in mature birds.

Hip joint degeneration is a problem in male breeder turkeys especially as they get older. It can also occur in females in the sire lines. The defect is an antitrochanteric degeneration and can be controlled to some extent by feed restriction during rearing (Hocking *et al.*, 1998). Providing turkeys that had hip joint degeneration with a steroidal anti-inflammatory agent has had contrasting effects on the amount of walking (Hocking *et al.*, 1999). This suggested that walking is not necessarily painful for these birds, at least in the mild forms of hip degeneration. In grower birds, the prevalence of leg disorders is said to be between 4 and 13%, and mortality from lameness is between 2.7 and 4% (Martrenchar, 1999).

Turkeys are more prone to pododermatitis than broiler chickens, presumably because they live longer and their foot pads are in contact with dirty litter for longer. This condition starts as an inflammation of the foot pads and progresses to erosions, hyperkeratosis, cracks and ulceration. Lesions can be seen as early as 3 weeks of age and, when the condition is not controlled, by 6 weeks over 66% of the flock can be affected. It is suspected that, when these lesions are severe, the wound acts as a portal for entry of bacteria that cause synovitis and lameness (Clark *et al.*, 2002). It is worst in sheds with poor ventilation systems, and is least in birds kept on concrete floors with a thin layer of wood shavings (Martrenchar *et al.*, 2002).

Disease

After the poult stage, two important causes of mortality are sudden death syndrome (SDS) and cardiovascular disease. SDS is probably linked to cardiovascular disease as there is usually hypertrophy of the ventricles. It can account for about 25% of losses, and spontaneous cardiomyopathy, congestive heart failure (CHF) and aortic plus atrial rupture contribute a further 28%. Most of the losses are seen in the heavyweight birds (Boulianne *et al.*, 1992).

Air sacculitis can affect a high proportion of a flock when it occurs. High levels of dust can make the severity and prevalence worse (Martrenchar, 1999).

Trauma

Breeder males are prone to injuries when semen is collected for artificial insemination. Normally, semen is collected twice weekly. A stag is caught and restrained by grasping one leg at the hock and the opposite wing at the elbow or shoulder. The bird is then repositioned and held by both legs. The femur can break if it struggles during handling or is held inappropriately. The break follows the line of a pre-existing stress fracture, which is thought to develop from repeated handling (Crespo *et al.*, 1999). Affected birds usually die in 1 to 2 days if they are not culled (Crespo *et al.*, 2002b). They die because they cannot reach the feeder or drinkers, or they may be killed by their pen-mates. Stags that take limited exercise can develop porous cortical bones in the femur and they seem to be particularly prone to developing these fractures.

In breeder hens, trauma can develop from an everted or prolapsed vagina. Prolapse of the vagina is a particular risk in the females of sire-line breeder turkeys that have been selected for high meat yield. Up to 10% of the sire-line females may be affected (Buchanan *et al.*, 2000). It usually occurs at the onset of lay when plasma oestradiol concentration is rising, vaginal collagen content is falling and the vagina is becoming more elastic. It results in losses of about 5% of the grandparent sire-line birds (Buchanan *et al.*, 1999).

Pecking injuries are a common reason for culling grower toms. They are usually avoided or controlled by beak trimming and by keeping the birds in subdued lighting. Research trials have shown that supplementary UV_a lighting, visual barriers within the pens, providing straw, providing sheets of reflective material and using fluorescent lighting at 10 lux instead of incandescent lighting can help reduce pecking injuries (Martrenchar *et al.*, 2001; Moinard *et al.*, 2001). Intermittent lighting patterns (8L : 16D) have been linked to less pecking injury compared with one 23L : 1D

and with eight 1L : 2D episodes per day (Lewis *et al.*, 1998). Growth rates and feed conversion efficiency were, however, better with the eight 1L : 2D lighting pattern. Turkeys are more likely to fight with unfamiliar pen-mates, and so mixing birds from different pens after thinning is inadvisable (Buchwalder and Huber-Eicher, 2003).

In 2000, a processing plant in the USA reported that it was condemning turkeys at a rate of 10 to 30% per day. The legs were bruised and some had oedema at the hock joint. The injury was linked to avulsion of the ligament that bridges the front of the hock joint. The exact cause was not established but the condition was thought to be provoked by slippery flooring when there was high humidity in the grower shed and the litter became wet, and from spontaneous avulsion when the birds stood up (Crespo *et al.*, 2002a).

Another company had a problem with back scratching when the birds were being caught prior to slaughter. Smaller turkeys were prone to climbing on each other if they were frightened by the catching team. One solution has been to clip the toes, but greater care on the part of the catchers would be more appropriate.

An integrated turkey company experienced an outbreak of breast bruising when it increased the slaughter weight of its birds to satisfy the growing demand for processed turkey meat products in place of whole birds. The bruising occurred when the larger birds were placed in the side-loading transport crates on the transport vehicles. The birds were too big for the entrance and their breasts struck the entrance as they were being loaded.

Ducks and Geese

The leading duck-producing countries are China, India, Vietnam, Indonesia, France, Thailand and Ukraine. In warm climates much of the production is outdoors, whereas in temperate climates indoor rearing on shavings is more common.

In Japan, Korea, China, Taiwan, Vietnam and Cuba, ducks are allowed to graze rice fields that have just been planted. They contribute nutrients through their droppings and help control weeds and pests, and they mix sediment with water, which is said to improve the soil. They start grazing from about 21 days of age, when they can withstand cold and wet conditions. They are grazed for 12 h in the day, and then brought in at night to avoid predation. When stocking density does not exceed 150 birds per ha, they do not usually need supplementary feed. They can be finished whilst grazing. Otherwise, they are penned and given a high-energy feed for the last 7 to 10 days to produce a finished carcass. When given cereal-based feeds, they prefer a wet rather than dry mash, but in warm climates a wet mash is more prone to souring. They can benefit from shade and death from overexposure to the sun can occur (Emmel, 1930).

Ducklings use up their yolk sac faster than chickens, and so the intervals between hatching and first feed and drinking are more critical. Growing ducks drink and excrete more water than chickens and turkeys, and their droppings consist of over 90% water. In indoor systems, extra care is needed in ensuring that litter is kept dry. This means that ventilation must be adequate and bedding needs to be topped up regularly. Drinkers are best placed outside to reduce tracking of water through the litter, provided they are not likely to freeze over. Alternatively they can be placed on a slatted or wire mesh floor. Duck skin is not as strong as broiler skin, and is more prone to injury, especially on abrasive floors. Slats, wire floors and cage bottoms pose risks if they are not in good condition. Ducklings are commonly reared on wire floors in India.

The main goose-producing countries are China, Ukraine, Hungary, Egypt and Taiwan. In East Asia, geese are allowed to forage in rice paddies in a similar way to that described for ducks.

The welfare issues in geese production include:

- stress associated with force-feeding
- stress associated with deplumation
- disease

The countries producing foie gras include Hungary, France, Madagascar, Lithuania and Taiwan. In France much of the production is with Mule ducks, but in the south-west of the country the Muscovy has been the traditional foie gras bird. Force-feeding is used in producing **pâté de foie gras**. The birds are force-fed for a period of 14 to 21 days, when they are between 9 and 25 weeks old. The geese are fed through a funnel or plastic tube introduced into the oesophagus. Feed is forced in until the crop is fully distended. The amount of feed that can be delivered increases after the first week, once the crop becomes stretched. During the force-feeding period, the liver increases in weight from 80 g to between 600 and 1000 g. The preferred feed is maize that has been cooked briefly in water. Goose fat may be added at up to 2% to act as a lubricant. With semi-automated cramming equipment, a wet mix is used.

The geese may be handled up to six times a day, but usually they are fed only three times a day. A quiet, non-aggressive strain is essential, and the Landes and Toulouse breeds are preferred in France because of their temperament. Force-feeding is followed by a period of torpor lasting from 3 to 5 h. Breathing may be laboured. Research trials have not revealed any signs of aversion in geese, while ducks can show aversion but not full avoidance (Faure *et al.*, 2001). They did, however, show stronger aversion to the presence of an unknown person than towards the person who did the force-feeding.

Feathers are an important co-product in geese breeder flocks, and in the goose meat and foie gras industries. Within Europe, **deplumation** (live plucking) is practised in Poland, France, Hungary and Germany. The down and feathers on the breast are removed as the birds come into moult. The first crop is taken at about 9 weeks of age, and subsequent pluckings are taken at about 50-day intervals, coinciding with each successive moult, until the birds are about 2 years old, when they are slaughtered for meat consumption. The number of pluckings taken from a bird depends on the market value at the time for feathers and down relative to meat and liver. Before deplumation, the birds are kept under shelter to keep them dry, and they are transferred to a closed shed or vehicle that contains the feathers as they are removed. Gangs of up to 15 people do the plucking. Following deplumation, the geese are transferred to a dry area that is protected from wind and rain. Breeding geese are normally plucked whilst they are out of lay.

Rabbits

Historically, rabbit meat consumption has been associated with poverty. Commercial rabbit farming has helped to change that image by increasing the availability of rabbit meat in the main retail market. Between 50 and 60% of the world's rabbit meat supply is now produced on commercial rabbit farms. The main countries with commercial systems are Italy, France, Spain, China and the CIS. The remaining production is in backyards at individual households, and this system is particularly common in the CIS.

A large farm holds up to 12,000 breeding does, but more often there are fewer than 1000. The primary product is meat, but Angoras are grown for their wool, which in France is plucked by hand, and in China it is also harvested at slaughter. There used to be strong demand for rabbit hair from meat-producing strains for making bowler hats, but this market has now declined.

The potential welfare problems in rabbit production include:

- atmospheric ammonia
- doe aggression
- hypothermia in kits
- hyperthermia in kits
- inappropriate nesting facilities

In the wild, the doe has a strong territorial and maternal instinct. Intruders at the nest site will be repelled aggressively. The only way to manage this in commercial rabbit farming is to cage each doe separately, where she rears her litter. Hutches or flat-deck wire-floor cages are used, and a nest box is provided for the litter.

This system entails substantial animal handling. The doe has to be:

- Taken from her cage for mating in the buck's cage (or the buck may be introduced for a short period into the doe's cage). The buck may eat the kits if left permanently with the doe.
- Palpated 14 days later to check that she is pregnant.
- Given a clean nest box before kindling.
- Inspected with her litter after kindling and kits are cross-fostered if necessary.
- Inspected regularly, especially for litter requirement in the nest box.
- Rebred 7 to 14 days after kindling.

All this makes heavy demands on staff time, which is one of the main reasons why commercial rabbit farming has been less attractive than broiler chicken farming.

One of the hazards of close confinement is trampling of the kits by the doe. Wild does normally visit the nest once a day to allow suckling, and close the entrance of the nest between visits. The behaviour pattern is different in commercial farms. The nest boxes used in commercial rabbit farms do not allow nest site protection in the same way, and this, plus the confined size of the pen, causes the doe to visit the nest many times each day (Baumann et al., 2005). Repeated visits increase the risk of kit trampling and crushing.

Doe replacement rate is high (about 120% per year), and so welfare issues associated with long-term confinement in cages are not common. The main reasons for their short productive life are sickness and death. Doe longevity has a low heritability (0.05 to 0.24), and so there will be little benefit in selecting for it genetically. Postponing the age at which maiden does are first mated beyond the usual age of 150 days can reduce the risk of early culling and so it may help increase longevity (Sánchez et al., 2004). Young does are prone to excessive mounting before they are separated into adult cages, and this can result in false pregnancies.

Rabbits are susceptible to hyperthermia. Normally, when it is hot, a rabbit moves into its burrow and stays underground to keep cool. This is not possible in rabbit farms. Instead, the stockperson has to keep close observation for signs of heat stress, and modify shed ventilation and temperature accordingly. Ventilation is also critical for controlling ammonia in the air. Rabbits produce large amounts of urine containing urea, and so high atmospheric ammonia concentrations are a common problem.

The advantages and disadvantages of keeping rabbits in pens compared with cages or wire-floor hutches depends on the standard of care that is provided. If the rabbits receive limited attention, those in pens become very flighty and difficult to handle. Handling injuries become a problem, along with skin sores from wet bedding. These issues are lesser hazards when the animals are in hutches or cages because caged animals can be more easily inspected, caught and handled, and they are separated from their own urine, which can dry on the underfloor or be directed out of the shed. Needless to say, the cage or hutch system is more expensive to install. In spite of the potential advantages for cage systems, under Nigerian conditions pen systems proved to be more productive (Mobolaji-Bukola et al., 2002). Breeding was more successful in pens because there was less need for heat detection and introducing the buck at the right time. Pre-weaning performance was also better in pens because of higher mortality in hutches. This may have been due to inability of the rabbits in hutches to burrow near the ground to keep cool, and the lower levels of fibre in their diet. Rabbits kept in cages sometimes develop excessive grooming and bar gnawing. This can be checked by enriching the environment such as access to a raised shelf where they can shelter.

Important features in hutch design include:

- adequate insulation from heat
- mesh floors to prevent accumulation of manure
- correct gauge of mesh to prevent foot injuries
- shallow depth to allow easier catching

Too much litter in the nest box can be a problem, as well as insufficient litter. The kits need to be kept warm, but they must not get smothered. If a kit leaves the nest box there can be a risk of death from hypothermia, as does rarely pick up escapees and return them to the box. Death from hypothermia and starvation is also a hazard for weak kits that do not manage to feed adequately. The doe usually suckles the young only once a day and, if a weaker member of the litter misses out on a feed, it can be life-threatening.

If maiden does are mated too early in life, there can be high mortality in the offspring. This occurs if the doe fails to produce enough milk, because of underfeeding. Similarly, the breeding life of the doe may decline if mating is too frequent and she is not allowed to regain condition between litters.

The litter is usually weaned at 4 weeks, and placed in a grower cage. Pens have been tried, but ammonia has been a problem and they do not grow as fast as fryers kept in cages. In addition, they spend less time resting and grooming when kept in pens in large social groups.

The overall mortality in rabbit farms is high compared with other intensive mammalian farming systems. In the slaughter generation it varies between 25 and 40%, including stillbirths. Mortality between live birth and weaning can be more than 20%. The main causes are crushing by the doe, enteritis, hyperthermia, hypothermia and starvation from lack of milk.

In the future, keeping rabbits will probably become concentrated in backyard operations in poorer communities, and commercial rabbit farming may become focused in countries where low labour costs allow the industry to compete with broiler chickens. Presently, urban rabbit production is commonplace in Indonesia, Mexico, Ghana and Egypt. China produces about 400,000 t of rabbit meat plus 10,000 t of rabbit hair, and this could increase now that it has established a presence as a rabbit meat exporter. In southern Europe, the rabbit processing industry has changed from a large number of small abattoirs killing a range of species including rabbits, to large plants that specialize in rabbit slaughter.

8

Aquaculture

About 133,000,000 t of fish are produced in the world every year, of which 76% is destined for human consumption. About 30% comes from the aquaculture industry and 77% is produced in developing countries.

Aquaculture is the fastest-growing sector of the meat industry. Between 1996 and 2003 aquaculture production increased by 58%. It is replacing shortfalls in the marine fishery industry that have arisen from overfishing during the 20th century. There is limited overlap in the markets served by the fisheries and aquaculture industries, and so the two are coexisting reasonably comfortably in the marketplace. The only exceptions to this are the shrimp and prawn markets (Table 8.1). The group of fish that contributes most in aquaculture production is the cyprinids (41% of total production). Salmonids and cichlids both contribute 4%. New enterprises that became established during the 1990s were tuna, *Portunus* crabs, *Penaeus indicus* prawn and turtle farming. Previously established enterprises that showed impressive growth during this period were fresh- and saltwater crustacea, flounder, halibut and tilapia farming.

Table 8.1. Main fish species harvested by the fisheries and aquaculture industries (% of total production from each industry).[a]

Species or group	Aquaculture	Fisheries
Important aquaculture species		
Carp, barbel, other cyprinids	40.7	0.7
Oysters, mussels, scallops, clams	26.2	3.0
Salmon, trout, smelts	4.3	1.1
Tilapia and other cichlids	4.0	0.8
Shrimps, prawns	4.2	3.9
Freshwater crustaceans	1.6	0.4
Important fisheries species		
Herring, sardines, anchovies	0	11.8
Tunas, bonitos, billfish	< 0.1	7.0
Cod, hake, haddock	< 0.1	6.7
Squids, cuttlefishes, octopuses	0	3.9

[a]Fish categories not represented in Table 8.1 that make a significant contribution in the fisheries sector include fish used for reduction (16.4% of total fisheries catch) and a range of unclassified pelagic, coastal and freshwater fishes (22.0%).

©N. Gregory 2007. *Animal Welfare and Meat Production*
(N. Gregory)

Farmed Fish

To appreciate the welfare issues in the aquaculture industry one has to understand the farming systems. The systems can be grouped into two main categories:

- A natural water resource is stocked with fish, which are extracted as and when required by the local community. Stock numbers may be replenished occasionally if they do not regenerate quickly enough, either with wild-caught stock or with stock raised on a breeder farm.
- Fish are bred and reared in dedicated tanks, ponds, concrete raceways, cages or sections of river with a controlled cropping schedule.

There are many systems within each category and they are based on the ways the fish are held and water quality is managed, as shown in Table 8.2. Ponds are the main system used in carp production. Tanks are used in hatcheries and for growing high-value species that require exacting conditions. Raceways are used at some trout farms, and raceway tanks are preferred for turbot and flounder juveniles. Fish towers are not common, but have been used at some salmon breeding farms for raising juveniles. Cages are common in large lakes, coastal waters and estuaries, and can be either floating or attached to the bottom. Floating cages are widely used by the salmon industry, and they are also becoming popular for farming common carp in the larger rivers of South-east Asia. Unfortunately, there have been mass mortalities

Table 8.2. Aquaculture production systems.

Holding system
Natural reserves
Managed ponds
Paddy fields
Tanks
Open raceway
Tower systems
Cages
Pens
Water management system
Natural discharge and recharge
Multiple reuse system
Closed system
Recycle system

at some of these carp farms from temperature-related deterioration in water quality. Pens are large floating nets and are used in large lakes, bays and open ocean (Fig. 8.1). The floating bag is a relatively new system, and although not established yet is attracting interest. It is made of woven polyester fabric, which excludes outside water, light and predators, and is fitted with a pump that forces the fish to swim against a current. The enhanced flow results in fitter fish with a greater disease- and lice-resistant barrier to the skin.

About 70% of world aquaculture production is in China. Of China's production, 59% comes from freshwater farms, where the main system is pond culture, but paddy–fish culture in rice fields is growing in importance. China has 6.75 million ha of inland water that is suitable for keeping fish, but not all would support intensive farming methods. Extensive natural reserves are used where it is not possible to control water throughput, and the lake or reserve is stocked lightly with fish. No feed or fertilizer is used. In semi-intensive systems, limited amounts of fertilizer or organic manure are used, and feed is provided to supplement natural food. In intensive aquaculture, the fish are stocked densely and they depend on additional feed. Water has to be replenished at a high rate to ensure adequate oxygenation and waste disposal, and oxygenation may be enhanced by mechanical power.

In rice–fish systems, carp serve other functions besides providing human food. They are effective at controlling weeds and mosquito and fly larvae, and they contribute fertilizer for the rice plants (Gomiero *et al.*, 1999). Pig dung is commonly used as a fish feed. Undigested components provide energy, protein and minerals for the fish, at the risk of off flavours in the fish meat. The manure from one finished pig can help produce about 40 kg of fish. Animal manure also supports grass growth, which serves as a feed for some farmed fish species.

In multiple reuse systems, water flows through a series of tanks or raceways containing fish. In closed systems, the water is reconditioned using biological and mechanical filters and recirculated through the system. New water is added to replace evaporative loss. If the percentage of the water that is replaced with fresh water each day is less than about 20%, it is usually known as a recycle system. In recycle systems water temperature, oxygen, carbon dioxide, alkalinity, calcium hardness and

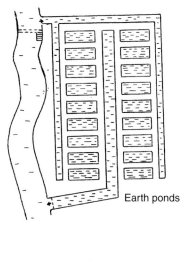

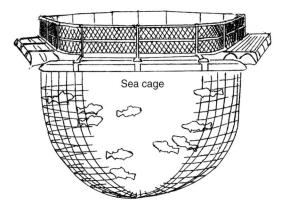

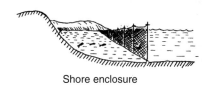

Fig. 8.1.   Four types of fish-rearing enclosures.

ammonia have to be managed mechanically or biologically.

The key to most welfare problems in aquaculture is in managing the condition and quality of the water. This and the other stressors in aquaculture are listed in Table 8.3.

At breeding units, stress can influence production through its effects on reproductive performance. In general, stress occurring during early ovarian development delays ovulation, the eggs tend to be smaller and the progeny may be less viable, whereas stress during late vitellogenesis causes the fish to spawn immediately. In carp farms, spawning is sometimes synchronized by lowering the water in the pond and then further stressing the fish by walking around the pond whilst beating the water surface with a cane.

Organic aquaculture is being developed in some European countries. The main focus presently is on carp farming, and the standards aim at controlling the purity of the product and at preventing disease using natural methods. There are few provisions for fish welfare in current standards.

Table 8.3.   Main welfare problems in the aquaculture industry.

Chronic stressors
- Poor water quality
- Inappropriate water temperature
- Overcrowding
- Disease
- Insufficient hiding facilities, cover or protection

Acute stressors
- Handling
- Grading
- Transporting
- Holding in confinement
- Vaccinating

Stocking natural water reserves

Stocking natural ponds and waterways is a common fish farming system in developing countries.

In Africa, about 300,000 ponds were stocked shortly after the Second World War. Many ponds were successful but some failed from overpopulation, especially when stocked with tilapia. Others failed from lack of continuity in the supply of subsidized fingerlings or from dwindling water supplies (de Graaf and Janssen, 1996). In addition, mortality rates were high if the fingerlings or fry were too small when they were introduced to their new reserve.

Stocking Lake Victoria with Nile perch (*Lates niloticus*) has been a notable commercial success for the local community. When food safety standards permit, this fish has a good export trade because of the meat's low bone content, succulence and firm texture. The lake was first stocked with this species in the 1960s. The original aim was to use the fish as a predator for controlling the population of unwanted indigenous species, but it gradually took over the lake. A Nile perch can grow to a large size, sometimes in excess of 230 kg, and it is a prolific breeder. At first the lakeside communities disliked the meat because of its high fat content, and this allowed the fish to get established and thrive. It eventually eliminated about 350 of the native lake species. It is caught by gill nets, drift nets and longlining, but presently the lake is being overfished and numbers are in decline.

In developing countries, fish ponds have sometimes developed from opportunistic sites. For example, cattle watering holes have been seeded with fish, and fish ponds have developed from excavations left behind by road menders. In mangrove regions, natural or bunded ponds are used for keeping fish trapped during falling floods. Some sites have been tried out of curiosity, and had little chance of success. Some large-scale failures have been in Cuba and India.

Several authorities on aquaculture consider that natural reserves have now reached their maximum production potential. From now on, if the reserves are to support greater production, they will either be integrated into intensive aquaculture or production will have to be enhanced by simple means. One of the simple enhancements is to add fertilizer whilst maintaining the existing stocking density. For example, in one case in China, the addition of fertilizer doubled the production of fish per hectare.

Other simple technical developments that have helped natural water reserve fish production are:

- separation of tilapia breeding units, allowing periodic restocking of the natural reserves and managed ponds
- monosex tilapia strains
- introduction of large-scale African catfish (*Clarias gariepinus*) production

Tilapia are becoming an important warm-water farmed finfish. They have been introduced into 150 countries and are being cultivated by more than 75 nations. Several species are farmed in fresh water, and there is also a hybrid species that can be used in brackish or salt waters. The hybrids have been produced by crossing *Oreochromis mossambicus* with *Oreochromis niloticus* and *Oreochromis hornorum*. They are more sensitive to stress and to secondary infections, and more care is needed during routine procedures such as crowding and handling; they also require protection from light, noise, fluctuating water temperature and poor-quality water. Tilapia production is presently the fastest-growing aquaculture enterprise in the USA, and the fish is favoured because of its firm yet moist meat, and because growth is rapid whilst fed a relatively inexpensive fish feed. In colder climates such as Canada, tilapia are farmed in intensive recirculation systems, using tanks kept in greenhouses or steel frame buildings.

Water quality

When ponds and natural reserves are stocked at low densities, they can usually maintain good water quality without the need for intervention. The water is reconditioned naturally by microbial action in the sediment removing ammonia and CO_2, and phytoplankton plus atmospheric gas exchange renewing the oxygen. If fish density rises, these processes may not keep up, and oxygen supply will be the first to fail. Supplementary aeration is needed, or water flow through the system has to be increased.

When a fish encounters water with a **low oxygen** content, it can show two responses. When there is an abrupt reduction in water oxygen tension, there are alarm behaviour and escape attempts (Erichsen Jones, 1952). This develops once the fish becomes hypoxic, as fish do not have a sensory system for directly monitoring oxygen tension in the water. The alternative more common reaction is reduced activity and shifting position vertically or horizontally to find better-oxygenated water. There are also periods of surface respiration to counteract the hypoxia.

Hypoxic conditions are a common hazard in tropical fresh waters, especially in permanent swamps and flood-plain pools. Excessive plant cover is undesirable. Floating large-leaved macrophytes on the pond surface slow down oxygen exchange at the surface, and shade from surface plants reduces oxygen generation from deeper photosynthesizing plants. Both factors contribute to hypoxic conditions. Decaying vegetation also consumes oxygen, as do suspended bacteria that live off leachates from rotting vegetation. Set against this, both live and dead plant material supports a community of invertebrates on which some fish species feed.

The plant community influences the day-to-night-time changes in oxygen content of the water. Normally, the water becomes progressively less oxygenated during hours of darkness, and the risk of hypoxia is greatest during early morning. This may become obvious in the fish if they are fed before it is fully light.

Overfeeding can lead to low oxygen concentrations in fresh water, as well as high levels of NH_3 and NO_2. The level of suspended solids may rise, and this is obvious from cloudiness or opacity of the water. Dense phytoplankton may also develop, with scum on the water surface. A short-term and simple remedy is to flush the pond with fresh water whilst reducing the feeding level.

Low dissolved oxygen concentrations can increase disease risk from opportunistic pathogens such as *Aeromonas hydrophila*, especially when combined with high water temperature or NH_3, CO_2 and NO_2 concentrations. Gill function is sensitive to water that has a low pH. The lower limit for water pH that is safe for freshwater fish is usually assumed to be 5.0.

One of the limitations to increasing stocking density is removal of nitrogenous wastes. **Ammonia** and nitrite produced during proteolysis and nitrification can be toxic to fish if they are not removed quickly enough. Fish do not seem to always react behaviourally to ammonia and sometimes they enter concentrations that lead to death (Erichsen Jones, 1952). At low concentrations, ammonia can promote growth, even in ammonia-sensitive species such as rainbow trout (Wood, 2004). At higher concentrations, chronic exposure can lead to gill epithelium damage, which will compromise the animal's ability to take up oxygen, and it may increase susceptibility to disease. It is sometimes recommended that withholding feed is the quickest way of correcting ammonia toxicity, especially in species that excrete ammonia as the main nitrogenous waste product. However, this can be counterproductive if it allows water pH to rise, as this favours the presence of higher concentrations of ammonia in its more toxic un-ionized form instead of NH_4^+.

Nitrates are not generally toxic to fish, but **nitrites** are highly toxic. NO_2 levels in water are highest during the cooler months, when ammonia assimilation by phytoplankton is decreased. NO_2 that is taken up by the fish binds to haemoglobin, forming methaemoglobin, and this reduces the oxygen carrying capacity of blood. The fish appear to be listless and may gasp for oxygen. The gills have a characteristic brown colour.

Suspended solids in the water can be a problem, especially in recirculating systems. It is the small particles (5 to 10 μm) that pose the greatest hazard as they can block water passage over the gills and damage gill epithelium, especially in juvenile fish. Larger particles settle out within the pond and can be filtered out more easily, but recirculating pumps tend to break these down to smaller particle sizes. The main sources of the smaller particles are fish faeces, uneaten feed and microfauna. Froth flotation systems can remove some of these finer particles, and the froth can also extract charged particles that attach to surfactants in the bubbles. Suspended solids have also been a problem where local deforestation has resulted in soil erosion and sedimentation in fish breeding grounds.

Common carp and other benthic species are sometimes considered a nuisance because they make the water turbid. They feed at the bottom of the ponds uprooting aquatic plants and disturbing the sediments. They can withstand suspended sediments whilst other species may be more sensitive, and where the pond serves as a source of drinking water they are regarded as a menace.

Some aquaculture operations have created their own water quality problems. Shrimp farming has become a sizeable industry in the mangrove swamps of India. Shrimps excrete ammonia through the gills as the main end product of protein catabolism. When shrimps are kept in stale water, death can occur from excess ammonia or from lack of oxygen. Under anaerobic conditions the shrimp may also experience H_2S exposure.

Some spectacular water quality catastrophes have occurred at fish farms affected by **algal blooms**. High-risk situations are when a fish farm adjoins an urban retention pond that is used for managing sewage and hazardous wastes. Low flow rates and low oxygenation in these ponds favour the growth of algal blooms, some of which can release toxic products. Overflow, seepage or leakage of relatively small water volumes carrying the toxins can wipe out a fish population downstream (Kempston *et al.*, 2002). In some cases the deaths have been linked to oxygen depletion of the water, in association with degradation of the algal bloom.

Eutrophication occurs when water is poorly oxygenated along with excessive amounts of nitrogenous compounds or phosphates. The N and P favour plant, algal and bacterial growth, which can upset the balance in the ecosystem that would otherwise favour fish culture. Some algal and bacterial decomposition products can be toxic to the fish.

Water **salinity** can be important as a stressor in aquaculture in three contexts. First, as unexploited freshwater sites become limiting, some people are looking at brackish water and estuaries as potential fish farms for freshwater species. Secondly, the effort and cost of running marine species hatcheries could be reduced if less saline water could be used. Thirdly, the stage at which salmonids are transferred from fresh water to seawater needs to be managed so that it does not needlessly cause stress for the fish.

The limits in osmotic adaptability need to be appreciated. In the case of the Atlantic salmon, before smolting they can be kept successfully in brackish as well as fresh water, and after smolting they grow well in seawater. If they are transferred to seawater too early, they grow slowly and may be more susceptible to disease. In the early days of rainbow trout farming, trout were grown only in fresh water. Since the 1980s, they have also been farmed in seawater. The right stage for transfer can be more difficult to recognize as there is no distinct smoltification process in trout. The risks from transferring to seawater too early include abnormally swollen bellies, which can be due to osmoregulatory disturbance. During cold conditions, osmoregulation can fail because the low temperatures suppress active ion transport by the gills and kidneys. In addition, low temperatures lower gut motility and drinking rate and this predisposes the fish to dehydration whilst they are osmotically challenged. Fat digestion may also be impaired, and this can be seen from fat droplets at the water surface, which have been expelled from the stomachs of the fish.

Pond fish and fish kept in brackish water have a reputation for **muddy flavours**. These flavours are due to 2-methylisoborneol and geosmin produced by microorganisms in the sediment and water. High concentrations of these compounds can be toxic to fish, and some people consider these off-flavoured products as 'unsafe' to eat. This is a problem for the US channel catfish (*Ictalurus punctatus*) industry, and on occasion as much as 50% of the catfish in a pond have been unacceptable for human consumption. This can be managed by **depuration**, where the fish are transferred to fresh uncontaminated water and held for up to 150 h without feed (Dionigi *et al.*, 2000). The off flavours often arise from overfeeding, as the culprit compounds are produced by bacteria from uneaten feed and faeces. Muddy flavour is also a problem in caviar produced from sturgeon farmed in polyculture systems (Cardinal *et al.*, 2002).

Live marketing is one of the high-risk periods for exposure to poor-quality water, and high temperatures exacerbate the situation. It is not always easy to control water temperature in fish displayed in shallow water at market stalls. High temperatures make the fish more active and they are more likely to foul the water. This places greater demand on oxygen in the water, and excessive activity can lead to the fish damaging themselves and each other, especially in species with sharp fins. Starving the fish before they are sent to market helps keep the water clean.

Water supply

Water shortages are limiting the future of aquaculture in some districts. For example, in Xinjiang province in China, where the annual rainfall is 15 cm, fish farms depend on bore water, and over-extraction has depleted some of the more accessible aquifers. This has introduced problems with salinity. In addition, pond hygiene has been deteriorating where organic matter has been allowed to build up on the bottom of the ponds. Stocking densities have had to be reduced to minimize the risk of fish diseases, and production has inevitably become less sustainable financially.

Competition for water has also affected fish stocks in natural reserve fisheries in some countries. For example, in the high-altitude lakes of Armenia, the Sevan trout (*Salmo ischchan*) has become an endangered species because of surges in demand for water for irrigation and by the hydroelectricity plants. The decline in trout population numbers has arisen from desiccation of the spawning grounds and eutrophication.

In future, rainbow trout may be farmed in cages, instead of flow-through systems, where water becomes more limiting. One system that is envisaged is a deep pond that can be easily drained, holding a trout cage. This would be too expensive presently, but if stocked at high densities it could be economically feasible in the future. Experience will tell what welfare issues this will raise, and how they can be managed.

Polyculture

Most pond systems in China are based on polyculture. These are multi-species systems, with each species occupying a different niche within the pond. The more species there are, the more complex are the controls. The key to success, and the key to avoiding welfare problems, is in understanding the feed requirements and the effluents produced by each species occupying the pond. The main species are silver carp (*Hypophthalmichthys molitrix*), bighead carp (*Aristichthys nobilis*), grass carp (*Ctenopharyngodon idellus*), common carp (*Cyprinus carpio*) and crucian carp (*Carassius carassius*). Herbivorous species such as grass carp, Nile tilapia and white amur bream (*Parambria pekinensis*) feed predominantly on grasses, benthivorous species such as common carp, crucian carp, mud carp (*Cirrhia molitorella*) and black carp (*Mylopharingdon piceus*) feed on organic detritus and bottom fauna, whilst bighead carp is zooplanktivorous and silver carp is phytoplanktivorous. Silver carp, bighead carp, grass carp and common carp have a short growth period and are easy to feed and harvest. They can be produced using a fertilizer application programme with little or no supplementary feed. On account of their versatility, they make up 79% of the inland aquaculture production in China. Grass carp can be sensitive to too much fertilizer application, and usually requires clearer water.

Overall fish growth in these systems depends on the growth of water grasses and algae. These in turn depend on water temperature. By the end of the winter, carp can be in a chronically underfed state. They have no adipose tissue reserve, and depend instead on fat deposits in liver and muscle for their survival during wintertime underfeeding. Those deposits are quite small. As a result, the fish are vulnerable to energy-demanding disorders and disease in spring, and this is accentuated by the low capacity of their immune system at low water temperatures. Strategic supplementation with feeds has obvious benefits.

In practice, the methods used for gauging when and how to manage the simpler polyculture ponds are not sophisticated. Manure application is usually adjusted according to the transparency of the water. For example, when visual depth is less than 25 cm, fertilizer may be withheld, but as soon as it exceeds 50 cm, the application rate can be increased. A normal application rate in China would be 30 to 50 kg dry matter (DM) pig manure/ha for an intensively managed polyculture pond.

In rice–fish polyculture, where fish are grown in paddy fields, there is an optimum stocking density for each species, and here again this is usually established from the quality of the water. If the stocking density is too low relative to the amount of pig manure or artificial fertilizer added to the rice field, the water becomes turbid and develops high levels of NH_4^+ and PO_4^{3-}. This may harm the resident fish population without diminishing rice yields. If stocking density is too high, rising fish mortalities can be expected.

Industrialized polyculture systems have moved towards year-round harvesting, in place of seasonal fish harvests during autumn and winter. The ponds are stocked and destocked on a regular basis, and there are separate ponds, with different depths of water, for the different growth stages. If unwanted fish species are occupying a pond, they are sometimes killed with bleach before a pond is restocked with the preferred species. The fish require regular supplementary feeding, which is often in the form of a dough made from rice bran and groundnut oilcake. In these systems, closer attention has to be given to maintaining water quality. Water pH is adjusted regularly with lime. At some farms the fish are graded for size, as this allows systematic harvesting and feeding according to stage of growth. Typically, live fish in polyculture systems are sorted and graded by hand, as the mechanical graders presently used for salmonids would not be suitable for large carp.

Dike ponds or bunded ponds are a common form of polyculture in delta swampland. The ponds are excavated from the swamp, and the spoil is banked up to form ridges, which support crops or, less commonly, livestock. In this way swampland has been brought into production in parts of China and South-east Asia.

Fish ranching

With the exception of tuna farming, fish ranching is not a major contributor in aquaculture production. However, new ranching systems are being developed and, if successful, they could have a significant role in the future. There are two types of ranching, and they are described using salmon, cod and tuna as examples. In the salmon industry **ocean ranching** is used for restoring numbers of wild salmon in three ways:

- restoring natural runs with hatchery fish
- mitigating habitat loss using hatchery-based populations
- conservation measures to save threatened stocks

Ocean-ranched salmon have a firmer meat than sea-cage or land-based tank salmon, and buyers claim they can often be recognized from the absence of fin erosion.

Cod ranching was first developed on the east coast of Canada and uses a similar approach to tuna farming. The fish are caught from the wild and transferred to cages, where they are maintained and fed until they are fit for slaughter or until the weather turns cold. Market prices are usually high in autumn and early winter, and growth rates in the cage environment up to this time can be impressive. Large fish can double their weight in a 4- to 6-month period when fed capelin, herring and mackerel. The wild cod used for stocking the cages are caught in traps as they move inshore during early summer, and final stocking density in the sea cages is typically 12 kg/m³. They are inactive much of the time, resting at the bottom of the cages.

There are two types of tuna farming. In North America, Europe and Australia the wild-caught fish vary between 15 and 200 kg and they are kept at the farms for a short period (2 to 6 months). In Japan, the farms are stocked with small seed stock (0.2 to 1 kg) and the period lasts longer (2 to 5 years). Trauma and mortality are substantial hazards in tuna breeding farms, and so catching wild stock is generally preferred.

Stocking density

The hazards associated with overstocking fish farms include:

- poor water quality
- disease
- cannibalism
- competition for feed and uneven growth
- social stress

Fish density inevitable rises as the fish grow, and by the end of the grow-out period in salmon it can be as high as 60 kg/m³. The hazards associated with such densities depend, in part, on the rate of water flow through the cage, and some authorities think that it is more appropriate to set standards in terms of water quality rather than spatial needs.

The trout and salmon industries have been criticized for overstocking during the grow-out period. High densities can depress growth rate when they lead to greater competition for feed, and long-term overcrowding can increase the risk of disease. Whether this happens depends on other management interventions. For example, diseases such as furunculosis can spread more rapidly in heavily stocked units, but this particular disease can now be controlled quite effectively by vaccination (Ogut and Reno, 2004). High stocking densities can also result in chronic stress and reduced immune competence. In addition, if overcrowding increases the risk of scale abrasion, for example from striking the cage wall or from aggression, the risk of skin infections and mortality can be higher.

Overstocking has been a common problem with tilapia in natural reserves. Tilapia are precocious and prolific breeders, and when a pond becomes overpopulated, the fingerlings outnumber and outcompete the older fish, which fail to reach a marketable size. One approach used for correcting this is to introduce a predatory fish such as African catfish. Another approach, used in India, was to replace *O. mossambicus*, which is prone to stunting from over-competition, with *O. niloticus*. Overstocking also affects growth performance and survival under more intensive tilapia farming conditions (Fig. 8.2).

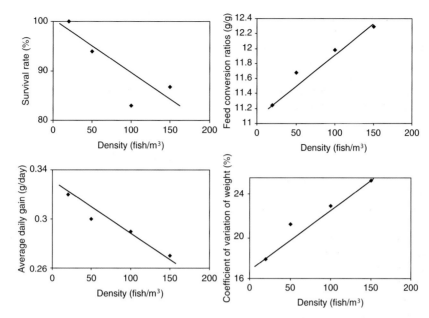

Fig. 8.2.   Relationships between stocking density and survival or growth performance in tilapia. This study was conducted on caged black-chinned tilapia (*Sarotherodon melanotheron*) in a man-made lake in Côte d'Ivoire. It showed that as stocking density increased, growth rate in the individual fish became more variable and average growth rate declined. Fish at higher stocking densities made better use of feed in supporting growth, but overall, the feed conversion efficiency was poor for this particular species of tilapia. Survival rate declined with increasing stocking density, and this was thought to be largely due to greater aggression. (Ouattara *et al.*, 2003).

Overstocking can have opposite effects on cannibalism and aggression. It can increase the risk of cannibalism if the fish are underfed, whereas it can reduce antagonistic behaviour if it breaks down social dominance or disrupts territorial behaviour. This is not inevitable, and in fact overstocking has sometimes increased fighting intensity and dominance aggression between individuals, whilst aggression involving mobs was milder.

Species such as Arctic char (*Salvelinus alpinus*) show stress-related behaviours when understocked as well as when they are overstocked. In some situations increasing the stocking density has been associated with reduced aggression, increased feed intake and improvement in overall growth rate.

The type of swimming behaviour can be an indicator of overstocking. For example, when salmon are stocked at high densities they swim as a school, whereas this behaviour would be unusual at low densities and in wild salmon. In Atlantic salmon, schooling has been noted when the density in the cage increases to about one fish per cubic metre. **Schooling behaviour** occurs during daylight hours and it reduces the risk of collisions and confrontations. In this respect it is a beneficial response to high stocking density, and it has been claimed that it is linked to reduced aggression and improved growth efficiency. However, the fish will stop schooling when they compete for feed at feeding time, and continuous feeding disrupts schooling behaviour. Schooling stops at dusk and the fish disperse.

Salmon that are heavily stocked are more prone to injury from seals attempting to catch the fish through the net wall, and this causes substantial economic loss in north-west Europe. Trout reared at high densities show more fin erosion, and this may reduce their market value. There is no effect on the fat or protein content of the meat.

Stocking fish at high densities has additional complications for species that are naturally solitary during the growth period. This applies to eels, which would normally have their own territory along a river or creek, whereas in eel farms they are crowded together. At high densities, there is no obvious territorial aggression, and instead this seems to be displaced or replaced by dominance aggression. Territoriality is more likely to be expressed in confined individuals. Similar concerns have been raised about holding wild-caught male breeder sturgeon (*Acipenser gueldenstaedtii*) at high densities in hatcheries, but these fish are normally polyandrous and polygynous, so this may not be a severe social imposition.

There are few unequivocal standards for the maximum stocking density for fish. For example, in grow-out cages for tuna, the recommended maximum ranges from 4 to 12 kg/m^3.

Feeding

Feeding time is a good opportunity for gauging the general well-being of fish. Pelleted feeds are designed to have a low sink rate. This helps reduce feed wastage as many farmed species do not scavenge feed from the bottom. It allows the grower to assess hunger, competition and vitality in the fish. Bottom-feeding species, such as channel catfish, are also fed floating feeds because of the benefits from observing their behaviour. Behaviour at feeding time also allows the grower to assess whether more or less feed should be provided according to their appetite.

In salmon farms, the fish are apt to chase, nip, bite and physically damage each other, especially when they are hungry or they are defending a feed source. The way the feed is delivered is important. If it is delivered at a single point from a dispenser alongside the cage, there can be guarding behaviour at this position. Dominant fish try to monopolize the feed, especially if feed is restricted. Restricted feeding carries greater risks of aggressive behaviour and variation in growth rate. Continuous delivery at a single site from an automatic dispenser leads to localized overcrowding. Instead, the feed should be scattered over a wide area to help separate the fish and reduce competition. It is usually the larger fish that win the fighting bouts and guard a feed source, and so they grow faster. This leads to greater divergence and unevenness in size, which is usually corrected by grading for size. Grading introduces stress and physical damage, but this is justified by the reduction in stress and injury through competition and aggression. Under natural conditions, the smaller members in

a cohort can emigrate and fend for themselves, whereas this is not possible in aquaculture.

Where there is a strong predation threat, the smaller fish in a group often obtain more feed. It is as if they are more willing to feed under an apparent threat. In addition, there is less aggression within the group when there is a predation threat, and this favours the smaller individuals. There is greater uniformity in growth.

It is usually recommended that feed particle width should be about 2.6% of a fish's length. This is only a rough guide as it depends on the gape size of the mouth, which varies between species. Feeding a range of particle sizes can be an advantage, if it means that the fish do not need to be size-graded so frequently (Juell, 1995).

Cannibalism and aggression

The predisposing factors for cannibalism include:

- growth stage
- stocking density
- underfeeding or restricted feeding
- genetics
- size disparity
- water temperature
- light

Cannibalism is most common at the larval stage, and in some situations it has been a limiting factor in production. It is one of the contributory reasons why northern bluefin tuna (*Thunnus thynnus*) are not farmed successfully from larvae. Larval cannibalism can get out of control in species such as sea bass (*Dicentrarchus labrax*) when they are weaned off live feed (*Brachionus* and *Artemia* species) on to artificial feed. Cannibalism may be worse at high stocking densities at this time (Hatziathanasiou *et al.*, 2002). Young fish in general have to learn to recognize formulated feeds as food as early as possible. Otherwise there is a higher risk of so-called 'non-starters' (fish that never eat any feed).

The relationship between stocking density and cannibalism is complex. Overcrowding can intensify cannibalism (Hseu, 2002). However, some species start shoaling once a certain threshold density is attained and this will reduce the risk of aggression and subsequent cannibalism. Shoaling helps limit the need for defence of territory, establishment of social hierarchies and other forms of aggression.

Farmed fish are usually fed at below their *ad libitum* intake. This helps to limit feed wastage and water pollution. However, it can result in some species directing their hunger at other fish in the pond or tank, and it may facilitate cannibalism through differential growth and size heterogeneity.

In rainbow trout fry (about 270 mg), it has been found that there was less mortality from cannibalism and more uniform growth when they had a self-feeding system compared with an automatic feed dispenser that delivered a set amount of feed per day (Shima *et al.*, 2001). Presumably the fry in the self-feeding system had fewer periods of hunger. There are cases where cannibalism has been reduced by providing more feed, and supplementing the feed with tryptophan has produced small reductions in cannibalism in grouper juveniles.

Most aquaculture species display cannibalism when the conditions are right. Exceptions are some of the cichlids that show a high degree of parental care, and species that have a small gape (mouth) size. In some species, such as grouper (*Epinephelus coioides*), there can be mass cannibalism amongst juveniles with overall mortalities exceeding 50%, and in *Brycon moorei* it has been as high as 40% per day.

There is conflicting evidence as to whether **domestication** is increasing or decreasing aggression in fish. The outcome probably depends on species. In brook trout (*Salvelinus fintinalis*), coho salmon and cut-throat trout (*Oncorhynchus clarki*), domestication seems to be resulting in increased aggression, whereas, in medaka (*Oryzias latipes*), Atlantic salmon and tilapia hybrids (*O. mossambicus* × *O. hornorum*), there has been reduced aggression. Increased aggression has been linked to competition for feed or some other resource. In principle, genetic selection for increased growth rate will favour those individuals that can compete best for feed and it favours aggressive behaviour.

It has been argued that some of the aquaculture species are unsuitable for farming. For example, eels are by nature a solitary animal for much of their lives. Under farm conditions, the dominant individuals are 'feeders' and continue to grow, whereas subordinates become 'non-feeders' and show signs of stress, including gastric ulcers. There are frequent confrontations between individuals and this can lead to aggression and cannibalism. Losses of about 30% from cannibalism and failure to thrive have been known.

There are two views about the importance of genetics in cannibalism. One theory proposes that within a population there is a group of natural-born killers with a propensity or ability to become cannibalistic. The other view is that at the outset all fish are equal in this respect, but individuals become cannibalistic as the winners of an intense initial competition. Cannibalism can be heritable, suggesting that in some species there could be a subpopulation of natural-born killers. It has been suggested that domestication might be favouring this type of fish, through selection for fast growth, but, as yet, there is no evidence that has tested this.

High temperatures are likely to increase the risk of cannibalism by increasing the fishes' hunger and feed requirement. Low light intensity, or water turbidity, can reduce the likelihood of cannibalism in species with good vision, but cannibalism can still occur at night in larvae that bunch together when it is dark. Providing shelter could improve survival in some species. However, there have been situations where fish provided shelter were shorter in length, presumably because they were reluctant to leave the shelter to feed.

Cannibalism is often associated with enhanced growth in the cannibals, and higher overall growth in the production unit. It is both a cause and an effect of size heterogeneity.

It is not easy to generalize about the ways cannibalism causes suffering, as this depends on whether the capacity for suffering has developed at the time it sets in and on the way a fish is eaten. There are two eating methods in fish larvae. In type 1 cannibalism part or all of the victim is eaten tail first. The prey can be similar in size to the predator. It is sucked in, up to its head which may be broken off and discarded. In type 2 cannibalism there is size heterogeneity between the two fish. The whole of the prey is consumed head first. Swallowing the prey in this orientation is more efficient as there are fewer escapees, but there is a greater hazard of oesophageal abrasion.

Type 1 is the preferred method amongst small larvae, and as the size difference between predator and prey increases they progress to type 2. Some characid species (*Bryconinae* and *Serrasalminae* species) have sharp teeth, and adults take bites out of each other, reducing a victim to small pieces.

Cage or pond conditions

In rough weather, cage **noise** can be stressful to fish. The noise is generated at joints in the cage frame and from impacts by waves. Normally fish move to deeper water and cluster together when they experience sudden or disturbing noises, but they have limited opportunity to express this behaviour in cage conditions. They also show less directed swimming behaviour and, following the flight response, they show more milling. The precise signs of **fear** vary between species (Pfeiffer, 1962). Some freeze, others flee to the surface where they crowd together, swimming hastily and frequently jumping out of the water, whilst others hide or swim excitedly whilst disturbing debris on the bed to conceal themselves in turbid water. The common carp, for example, is a sediment disturber and hider.

Salmon usually avoid the bottom of the cage, presumably to keep away from predators and to avoid contact with the net. The distance they establish from the bottom depends on **lighting** conditions, feed availability and hunger level. Under good lighting they descend at dawn and ascend at dusk, and presumably this is an adaptation to predation risk by birds. When hungry, they congregate near the surface, and so there are times when there is crowding within the cage. It is thought that providing shade promotes a more even distribution, but this is not provided in marine facilities. The shape of the cage influences the way the fish swim. In square cages, they have to adjust their course at corners, and this can disrupt the group structure. Circular cages encourage schooling.

Social stress is probably greater during daylight hours, especially in high-density situations. When it is dark, fish do not seem to be aggressive to each other, and they show less evasion generally. They are often easier to catch with a dip net when it is dark. Subordinate fish modify their behaviour in the presence of dominant fish, and this includes their feeding behaviour as well as more general activity.

Male brood fish are often held upstream of the females. Otherwise pheromones from the females provoke excessive activity in the males and in some species this incites male–male aggression.

Parasitism

Ectoparasites are a common cause of irritation for farmed fish and production loss in aquaculture. The irritation is apparent from the fishes' behaviour. For example, flashing occurs when fish infested with sea lice rub against the net, and when repeated frequently this leads to skin wear and ulceration. *Costia* infestation of the gills can cause respiratory distress, gnathids cause erratic behaviour as the fish react to the unwelcome penetration of the skin, and *Argulus* species can also lead to frenzied rubbing of the skin. *Neobenedinia* infestation has been a cause of losses from blindness and failure to feed and thrive in the cobia (*Rachycentron canadum*) industry in Taiwan. Other reactions to ectoparasites include breaking the water surface or rolling at the surface.

Sea lice (*Lepeophtheirus salmonis*) are presently one of the most important ectoparasites in aquaculture. They are blamed for the collapse of sea trout (*Salmo trutta*) enterprises in several countries, and they are a continuous threat to the Atlantic salmon industry. Their main effect is reduced growth and feed conversion efficiency, but in severe infestations they lead to downgrading from surface haemorrhages and open wounds at the back of the head (Mustafa *et al.*, 2001). They feed on the mucus, skin and blood of the fish, and cause skin ulcers, dehydration, immunosuppression and secondary infections. Ulceration allows the lice to burrow deeper. They reproduce rapidly once populations become established and so regular surveillance and prompt reaction are needed. Treatment often involves crowding the fish in an enclosure within a pen and adding an appropriate anthelmintic. The fish need to be immersed in the solution for an appropriate period to ensure exposure, and there is inevitable scale loss and damage to the eyes during crowding and struggling, which adds to losses from downgrading.

Disease control methods

As the aquaculture industry has become more intensive, the risk of losses due to bacterial and viral pathogens has increased. Rising stocking densities and the close proximity of farms have contributed to this risk. Poor water quality can contribute to disease outbreaks, especially when the deteriorating conditions favour survival of the pathogens. In intensive systems good hygiene is a necessary precautionary measure, and it is sometimes ignored when farmers are trying to save money and manpower. Common **disinfection** procedures that are used include treating the pond with quicklime or bleach, decontaminating feeding equipment with bleach, and disinfecting incoming juveniles and fingerlings with a hypertonic salt/bleach/copper sulphate solution or a solution of $KMnO_4$. These chemical disinfectants are no doubt irritating for the fish, but in the absence of a better alternative their use is justified.

There are about 200 known diseases that affect aquaculture, but only about 20 are sufficiently serious to halt production. Some are controlled by **vaccination**. Some of the oil-based vaccines that are injected intraperitoneally have caused problems with carcass downgrading because of an inflammatory response and adhesions (Poppe and Breck, 1997). They can also cause melanin staining of the peritoneum, which looks superficially like bruising. In addition, the vaccination procedure itself can cause bruising from handling and struggling. Several approaches have been tried to overcome these problems. They include improving the design of vaccination tables to control the level of bruising from handling injuries, reducing the handling by using immersion vaccination methods, and eliminating handling by using oral vaccines. The immersion vaccines can cause other problems. With hyperosmotic immersion vaccines, the fish are first placed in a strong saline solution for about 2 min and then netted into the vaccine solution (Huising et al., 2003). The soak in the hyperosmotic solution allows greater uptake of the vaccine, but is inevitably stressful. One major advantage of immersion vaccination is that it can be used for very small fish, which would be too small to inject.

Trauma and mortality

Fish are usually crowded before slaughter to enable transfer or catching. They may be handled in large numbers at a time, and, at trout farms, thousands are sometimes killed at a time. **Crowding** is one of the main causes of stress and injury at slaughter. The fish become excited, and in this situation many species tend to burrow. The increased activity magnifies the oxygen deficit created by the high density of fish. It also damages the eyes, snout and skin and leads to bruising and poorer water-holding capacity in the meat. In Atlantic salmon

crowding stress results in quicker onset of rigor, and this can make prompt processing more challenging (Skjervold et al., 1999). In Arctic char (Salvelinus alpinus), crowding for as short a time as 5 min increased subsequent weight loss from brine-treated fillets (Jittinandana et al., 2003).

Fish in ponds are often caught with a net that may span the entire width of the pond and is pulled along the length of the pond. In large ponds, the net is pulled from the deepest part towards an accessible point where the fish can be removed by elevator, pump or hand net. Seine nets are used in ponds that cannot be drained in a controlled manner. Some carp species are difficult to catch by this method (e.g. silver carp). They manage to escape and form large shoals behind the net. One way of controlling this, which is used in China, is to have an aerial net above the water attached to the floats of the capture net. A less common **catching method** is to position a net on the bed of a carp pond at a feed station. The net is lifted suddenly at feeding time, and the fish are held in a detachable bag at a cod end. The bag is taken to the bank where the fish are sorted. Small fish are returned to the pond and the larger ones are placed in a holding facility for live distribution and marketing.

The usual method in small carp ponds is to drain away part of the water to crowd the fish, which can then be brailed (removed with a hand net, basket or bucket) or scooped out. In some systems the fish are confined and removed as they swim upstream of the outflow sluice or monk (modified sluice pipe), whereas in others there is a cage trap or tank downstream that collects the fish as they leave the pond with the water. A clever alternative is to empty some of the water from the pond, whilst retaining the fish, and then turn on the water inflow. The fish normally move to the inflow pipe and swim against the current. If unimpeded they continue swimming into and through the inflow pipe. A system has been developed that allows automatic grading into separate tanks through different sized grills as the fish swim along the inflow pipe network. This is known as the countercurrent system for catching and grading fish.

Transfer to the killing area from a cage or tank is either with a cage, landing nets moved by a hydraulic arm, vacuum pump, Archimedes screw or an airlift pump. Modern screw-type **pumps** cause limited damage to scales and skin, but there can be injuries to the eyes and fins, especially if the fish are too large for the pump (Helfrich *et al.*, 2001). Lifting the fish into the air with a cage or net is very stressful because they are piled on top of each other, they have limited ability to breathe and they are prone to drying out. With all the methods, live fish should not be dropped on to hard surfaces, and there should be no rigid or other hard surfaces that inflict a wound. Needless to say, the size of the inlet funnels and pipes used in the pumps and airlifts must accommodate the fish without any wedging.

The physical injuries acquired during handling can lead to fatal infections. **Abrasion and scale loss** are common problems amongst salmon smolts when they are transferred from freshwater to seawater net pens. The injuries allow fungal infections (e.g. *Saprolegnia* spp.), which can cause mortality. A polyvinylpyrrolidone water additive has been developed (Polyaqua®) that temporarily bonds to exposed tissue, sealing it against infection. The adhesive is gradually sloughed off as healing takes place and the normal mucus layer re-forms. This is used to reduce mortality in juvenile salmonids following transport.

Tuna are prone to injury because they panic and collide with tank or cage walls. This has been one of the factors that has prevented bluefin tuna breeder farming (*Thunnus thynnus*). From about 30 days after hatching the juveniles and young adults developed **broken jaws** and dislocated vertebral columns from impacts with tank walls. On one occasion, daily mortality reached a peak of 9%, and mortality was not immediate when there were broken jaws (Miyashita *et al.*, 2000). Stimuli provoking panic behaviour and collisions with the tank included flashing lights at night, loud noises around the tank and vibrations in the tank wall. All-night lighting helped reduce collisions.

Storm damage is a hazard for fish kept in sea pens or cages in exposed regions. At times it has been catastrophic. For example, in 1996 a storm hit the coast of South Australia and wiped out 70% of the farmed tuna. One way this can be managed is to tow the cage to deeper water or a protected site when a storm threatens, but this may only be realistic for small cages. In other cases, exposed sites have been abandoned as unsuitable for fish farming. Another approach, developed by the cobia (*Rachycentron canadum*) industry in Taiwan, is to use submersible cages, which can be lowered to a depth of 10 m during typhoons.

Halibut farming gained popularity when salmon prices fell because the market was oversupplied. The halibut (*Hippoglossus hippoglossus*) was seen as a way of diversifying into a higher-value fish. However, halibut farming has its challenges. The young are prone to injuring each other when they are weaned off live prey, which is used as the starter feed, on to a formulated crumb diet. When hungry they nip each other, causing damage to the eyes, tails and pectoral fins. Badly damaged fins and tails can become infected and **eye damage** that leads to blindness reduces a fish's ability to compete for feed. It has been estimated that 3 to 5% of halibut juveniles (150 g weight) have eye injuries, with one or both eyes removed or damaged, and up to 30% have tail or pectoral fin damage (Greaves and Tuene, 2001). Aggression can persist after weaning, and attacks are commonly directed at fish that won a feed pellet. This behaviour would not be so common under natural conditions, because the halibut is normally a solitary animal for much of its life.

In Atlantic salmon, **erosion of the dorsal fin** is often a sign of aggression. The dorsal fin is a preferred region for attack, and when fish are heavily stocked the frequency of attack and prevalence of damage can be high. Splitting of the fin is a sign of recent damage because it usually heals rapidly, whereas regrowth from erosion takes longer. Within a pen, the smaller salmon are often the recipients of both forms of injury.

Juvenile rainbow trout can detect skin damage in other trout in the same pond. This may be due to detection of compounds released from the wound into the water, and to stress-associated pheromones. When stressed by a predator, crayfish (*Procambarus clarkia*) release compounds in their urine, which alert other crayfish, which then move away from the area (Zulandt Schneider and Moore, 2000).

Injuries associated with **fish pumps** have declined with the introduction of the screw-type pump with a centrifugal impeller (Fig. 8.3). Fish pumps were originally developed for removing dead fish from the holds of marine trawlers. When they were used for handling live striped bass (*Morone saxatilis*) on fish farms, there were cases of damage to the head, eyes, skin and fins. The extent of the injuries depends on species, fish size, fish density, pump speed and the distance or time that the fish are pumped (Helfrich *et al.*, 2004). Eels are relatively resilient, whereas yellow

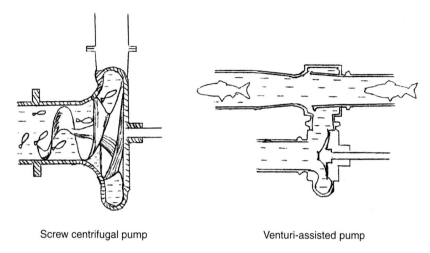

Screw centrifugal pump   Venturi-assisted pump

Fig. 8.3.   Two types of fish pump.

perch and channel catfish are easily damaged at fast pump speeds. There is an optimum pump speed for each situation, size and species of fish. It may not be easy to identify bruising before the fish are filleted. However, a method using near-infrared (NIR) spectroscopy is being developed which can detect bruises to a depth of 1 cm through the skin (Lin *et al.*, 2003). With further development, this technique could find application as an in-line method for grading table fish. Seining a pond with a net is a common alternative to pumping fish and it can cause skin damage (Grizzle and Lovshin, 1996).

Grading

Grading fish according to their size is a necessary procedure as it reduces the opportunity for type 2 cannibalism. Grading, before stocking a pond or cage, also reduces size variation when the fish are harvested. This allows easier marketing. Grading allows fairer competition for feed and provision of the right-size feed for all the fish, as well as reducing victimization and cannibalism of smaller individuals.

The disadvantage with grading is the bruising, surface damage and stress it imposes during handling. Fish pumps are used to harvest the fish for grading and this procedure is stressful and potentially damaging. In the salmon industry it is common to grade up to five times during early development (between 1 and 70 g). In the grow-

out pens they are usually graded by crowding them in one corner of a pen, and then pumping them out with an Archimedes screw-type pump, which moves the fish in pulses of water. The grading machine then sorts the fish into two or three pens according to size. Modern pumps and graders inflict little damage on the fish, and they are less stressful than older designs.

Small fish (e.g. fingerlings) can be graded with in-pond sock or box graders. Sock grading is a passive process and can only be done under warm conditions when fish are active. Box grading is an active grading method, where the smaller fish fall through spaced bars. Some mechanized graders allow simultaneous vaccination, whilst recording a sample weight of the fish, as well as counting the fish in each grade. Some species (e.g. channel catfish) extend and lock their pectoral spines when handled or removed from water, and this makes mechanical out-of-water grading more difficult. Mechanical methods for grading flatfish have not been perfected yet.

The stress that occurs at grading is obvious from the subsequent reduced feeding and the growth check. At worst it can increase the risk of disease outbreak. Some species (e.g. bass) are particularly prone to this check, and to stress from crowding and noise, and they are often sedated or anaesthetized when they have to be handled. It also helps to perform grading under cold conditions, when the fish are less active.

Uniform hatching can reduce the need for grading and is recognized as a part of good hatchery practice. If hatching is staggered, late-hatched, undersized larvae can provide an incentive for the start of cannibalism. Early-hatching fish may have a growth advantage, and they should be screened out at the hatchery as potential cannibals.

Transport

Farmed fish are transported in the following situations:

- stocking farms or natural reserves
- transfer from sea cages or inshore farms to nearby slaughter premises
- transfer to urban markets or restaurants where they are kept alive until needed for consumption
- live export to overseas markets

Some transport methods are costly: salmon smolts are sometimes transferred from freshwater hatcheries to the more remote seawater net pens by helicopter. Other methods are very simple, such as a plastic bag containing water and fish secured to the handlebars of a bicycle.

The size of the international trade in live fish is shown in Table 8.4. The main importing countries are Hong Kong, the Korean Republic, Japan and Malaysia.

Two important stress factors influencing survival when transporting fish are maintaining the **supply of oxygen** and avoiding high temperatures. The fish's oxygen requirement depends on water temperature, its size and the quality of the water (pH, CO_2 and build-up of waste). In sealed containers, the CO_2 will build up at a rate of 0.9 ml CO_2 for every ml O_2 consumed. In practice, the O_2 concentration should not be allowed to fall below 5 mg/l when transporting carp.

Table 8.4. World trade in live fish during 2002.

Species	Tonnes
Carp	44,738
Eel	34,327
Trout	13,499
Other fish	114,447
Total	207,011

Ideally, the fish should be conditioned to the transport container and water. Sick and dead fish must be removed, and the fish need to be fasted to reduce fouling and the build-up of NH_3. Reducing the temperature helps to subdue physical activity, reducing the risk of oxygen depletion and injury. Filtering the water and adding oxygen through a recycling system are needed on long journeys or where water temperature is likely to rise. The exact requirements depend on species and length of journey. Some fish can be kept alive during transport by simply changing the water periodically, whereas others have more exacting demands. The lungfish (*Protopterus* spp.) is an unusual case as it can be transported live out of water provided the skin is kept moist.

Fish are sometimes sedated when they are transported, if they are not due to be slaughtered immediately. This reduces their activity, but it is still important to maintain water quality, because CO_2 expired through the gills can build up quite rapidly. For example, when benzocaine-sedated matrinxã (*Brycon cephalus*) were transported at 150 kg/m^3, water pH declined from 7.0 to 6.2 in 1 h due to the build-up of CO_2 (Urbinati and Carneiro, 2001). Ammonia accumulation in the water was, in comparison, relatively slow. MS222 is also used to tranquillize fish during transport, but under US Food and Drug Administration (FDA) regulations there has to be at least a 21-day withholding period before human consumption. Where this is impractical, fish are sometimes exposed to bicarbonate-treated water (150 to 650 p.p.m.) for 5 min to sedate them before transport.

Mortality during transport has been a significant problem in the cobia industry when young stock were transferred from nursery tanks to inshore cages. This species has a particularly high oxygen requirement because of its high metabolic rate. In addition, young stock are usually transported in large numbers and at high densities, and mishaps can be costly. Transport mortality in other species has been controlled to some extent by conditioning and removing diseased or weak fish before transport. During conditioning the fish are kept in clear running water in separate tanks, without feed. On arrival at their destination, the fish should be transferred gradually to the new water by blending it with the transport water; otherwise there is a risk of mortality.

The Southern bluefin tuna (*Thunnus maccoyii*) and Atlantic bluefin tuna (*Thunnus thynnus*) are caught and transferred to the fattening pens in **transport cages**. The fish are caught in the western Mediterranean or South Pacific by purse-seining and held in the transport cages, which are towed to the farms by tugboats. Towing speeds are extremely slow (about 1 knot), so as not to excite or damage the fish. Towing distances between the capture site and the farm in the Mediterranean can be over 480 km, and in the South Pacific the journey can last up to 14 days.

Well-boats that are used for stocking salmon cages with smolts can hold up to 30,000 fish. The boats are also used as a holding facility during grading and delousing, and for transporting finished fish from the sea cages to the processing facility. First the fish are netted or pumped from the cages into a tank on the well-boat. Fish density in the tank is high (about 125 kg/m^3), and the risk of oxygen depletion is usually controlled by continuously irrigating the fish with fresh seawater. On arrival at the quay, the water level in the tank is reduced, and the fish are netted, transferred in air to a platform, where they are stunned electrically or with CO_2 in a water tank, stuck and passed on to a water tank to bleed out. The stress of loading, unloading and killing is sufficient to accelerate muscle phosphocreatine and adenosine triphosphate (ATP) depletion and the accumulation of IMP (5-inosinic acid) (Berg *et al.*, 1997). Time to rigor is reduced, but it can be increased by resting the fish overnight in the well-boat before unloading and slaughter. Towing the salmon in a small cage is an alternative to transporting the fish in well-boats, but it is not common because of potential problems with winds and tides. There are also concerns about damage from crowding at the back of the towed cage.

In hot climates fish are transported live to metropolitan and other **urban markets**. The fish may be held in floating cages, wells, tanks, fish yards or basins alongside a river before dispatch to the markets. Some systems are equipped with O_2 control, water filtration and temperature control. In China, street vendors sell catfish from metal drums on their tricycles. Other species sold live at markets include carp, eel, tilapia, lobster, crab, prawn, shrimp, mussel, oyster and cockles. The primary aim in choosing live transport is to avoid spoilage and loss of quality.

Farmed Crustacea

Water quality problems are the main welfare concern in farmed crustaceans. The problems are the same as those described for farmed fish. Large market-size **prawns** are particularly prone to hypoxic water, and this has probably been the most common cause of losses in *Macrobrachium rosenbergii* at farms in hot climates. This has happened when staff failed to recognize the signs connected with the onset of low oxygen conditions. When dissolved oxygen approaches the critical minimum, prawns display characteristic stress behaviour and gather along the edge of the pond. This reaction concentrates the prawn biomass, which makes the local oxygen deficit worse. Localized build-up of ammonia exacerbates the situation further by increasing the animal's oxygen consumption rate. Prompt reaction by staff is essential.

Some forms of **shrimp and prawn farming** (*Penaeus monodon* and *Marsupenaeus japonicus*) are highly intensive. Japanese kuruma prawns can achieve a feed conversion efficiency that approaches 1.0 when fed 55% protein four times daily. Control of water quality is inevitably critical in these systems, especially with the inshore *P. monodon* farms. In parts of Asia, hypersaline water is taken to these farms in road tankers, where it is shandied with fresh water when the ponds are filled. Thereafter, no water exchange is used except for topping up the water level with fresh water to replace evaporative loss. Disposal of the salt water at the end of a cropping cycle can impact on vegetation in the neighbourhood.

The success of a shrimp grow-out business also depends on the vitality and strength of the post-larvae. The post-larvae that are used for stocking the ponds must be robust. This is often assessed from the post-larvae's ability to swim against a current. A sample of the post-larvae in a delivery is decanted into a white basin, which is then swirled, and the alignment of the post-larvae relative to water flow is assessed. Another test that is being promoted is to take a sample of post-larvae and assess their survival to either a solution of formalin or saline. This is known as the **post-larvae stress test**, and a high mortality rate justifies rejection of that batch.

The giant river prawn (*Macrobrachium rosenbergii*) is the most popular farmed freshwater prawn. In the tropics, brood stock are netted in natural reserves and the egg-carrying females transferred to farms in buckets, double plastic bags or tanks. When bags are used, the rostrum has to be cut off: otherwise the bag is punctured. The chelipeds are also tied with rubber bands and covered with plastic tubing to prevent bag rupture. Another transport method is to wrap them in a cloth or enclose them in a polyvinylchloride (PVC) pipe, which is then placed in a double plastic bag. On arrival at the hatchery, they are disinfected with a solution of copper sulphate or formalin for about 30 min.

When grower prawns are ready for consumption they are often harvested at the farms with a seine net, which is pulled through the pond. The risk with this method is that the net becomes overloaded and the weight of the catch kills the smaller prawns as it is lifted out of the water. This risk can be managed by dividing the pond with more than one net. The other methods used for harvesting the prawns are to drain the pond, but this carries the risk of killing them from oxygen deficiency, or using electric shocker nets.

In warm climates, the majority of cultivated prawns are sold live in a chilled state. Some authorities consider that placing prawns live into ice is not a suitable live chilling method because the decline in body temperature is too slow. When kuruma prawns (*M. japonicus*) are exported live from Australia to the lucrative sashimi market in Japan, they are usually packed in chilled sawdust, and survival expectancy is between 10 and 30 h. During live marketing in street markets, the display needs to be kept in the shade and the water changed regularly to eliminate ammonia. Heavily stocked tanks should also be aerated. Those prawns that are sold dead are usually killed at the farm of origin by removing them from water, and then they are chilled.

Eyestalk ablation is common in the prawn industry and at some shrimp farms (*P. monodon*). It is used to induce precocious reproduction. One of three amputation methods is used in unsedated animals. The most common method is unilateral ablation where one eyestalk is removed with a pair of scissors. In enucleation, the eyestalk is grasped between index finger and thumb and the eyestalk is squeezed as it is rolled outwards. The contents of the eye pop out under pressure. With some species (e.g. kuruma prawn), the eyeball is incised with a razor blade before pressure is applied. In cauterization, a pair of heated artery forceps is clamped across the eyestalk of a restrained prawn and the eyestalk is then clipped off with a pair of scissors. The aim with all three methods is to remove the sinus gland, which releases a hormone that inhibits reproduction. In South-east Asia, gravid wild-caught female shrimps are transferred to hatcheries and, after an initial spawning, four or five additional spawns are induced by ablation.

Electroejaculation has been used to collect spermatophores from prawns and shrimps, but dissection or manually squeezing the ampullae at the base of the pereopods is more common.

Several attempts have been made at farming **lobsters, crayfish and yabbies**. A range of lobster species are farmed in China, India and Taiwan (*Panulirus* spp., *Thenus orientalis* and *Puerulis sewill*). They are wild-caught in bottom trawls, bottom-set gill nets, trammel nets and traps, and then transferred to farms. There is a large export trade of live lobsters from India to Taiwan and China, and they are shipped out in tanks or boxes with wet straw or paper plus ice bottles. The live trade is more lucrative than the frozen or cooked meat export trade, which has now been almost completely displaced.

In clawed lobster farming, the main constraint has been fighting and cannibalism when they are held together in a communal tank or pond. Periodic claw amputation has been considered, but it is not used on a commercial scale. In general, most of the cannibalism in crustacea farming occurs when the victim is at the soft-shelled post-moult stage. The consumed parts range from small sections of the tail, legs or antennae to the whole carcass.

Fighting in lobsters often takes the following form. Before a fight, both lobsters face and direct their antennae towards each other. A fight is unlikely if one of them backs away at this stage, but a fight will develop when one advances and pushes into the other. There is mutual clasping, pushing and chasing, which ends when the loser breaks off and flees. Most contests last less than 10 s.

Crayfish are farmed in natural ponds and semi-intensively in ponds and raceways. When they are overstocked there is less territorial behaviour but greater risk of cannibalism and chelae injury. High stocking densities can reduce growth performance and create greater size variability if there is competition for feed. In this situation there can also be higher mortality rates and cheliped loss from fighting (Savolainen *et al.*, 2004).

In South Australia enforcement of quota systems requires that the fantail is removed from freshly caught marine crayfish destined for private consumption. This prevents private catch being sold on the open market. Removal is performed with a knife, and the crayfish struggle when the fantail is cut.

Crab megalopae are prone to fighting when they are boxed together during transfer from hatcheries. This can escalate to cannibalism if they are not unpacked in time. One method used to limit this behaviour is to pack them in seaweed. Adult crabs are often transported with their claws bound, and more recently people have been gluing the claws. They can survive up to 17 days when held in chilled, humid conditions, but this depends on ventilation and the substrate within the box. Live lobsters are often packed in chilled wood wool in polystyrene or cardboard boxes, and journey distances can be long. For example, live lobsters are exported between Canada and France, and between Australia and Japan.

By-catch

Young dolphins have been an important by-catch in the tuna aquaculture and fisheries industries. This is now controlled by a 'back-down procedure'. When a dolphin is spotted as the seine net is closed, it is allowed to escape before the net is hauled in. Pingers and other sonic devices are used for scaring dolphins from gill nets, but they would have little use in seine capture for aquaculture.

Shark are also a by-catch of the tuna industry. They are caught by tuna longlines and drift nets, and sometimes form over 25% of the total number of caught fish. There is a good market for the fins, which are cut off the conscious animal and the body discarded back into the sea. In Fijian tuna fisheries, shark fins are a bonus for the crew. They sell the fins to dealers, who dry and store them in warehouses that have to be closely guarded because of their high value in the traditional soup market.

Only a proportion of the shark by-catch is sold for its meat. Shark meat has a specialized market. It is best chilled and processed promptly because it is prone to forming ammonia through breakdown of urea and reduction of trimethylamine oxide post mortem. Shark carcasses normally contain high levels of urea, which is used in the live animal to combat osmotic stress. In some countries the meat is refused because of local taboos, but there are good markets for particular species in France, northern Italy, Australia, the Caribbean, Japan and parts of the USA. Blue shark (*Prionace glauca*) is one of the least favoured species for its meat, but it is the most highly valued species for its fins. The most valuable fins are the first dorsal fins, the pair of pectoral fins and the lower part of the tail. Traditionally, the fins are traded as sets of these parts from individual animals. They are valued according to the percentage yield of fin needles and needle size, texture and appearance. Other species popular for making shark fin soup are the dusky (*Carcharhinus obscurus*), sandbar (*Carcharhinus plumbeus*), hammerhead (*Sphyrna lewini*) and mako (*Isurus oxyrinchus*) sharks.

9

Other Species

This chapter examines welfare issues in the following species when they are destined for meat consumption:

- goats
- water buffalo and yaks
- deer
- camels
- horses and donkeys
- ostriches and emus
- pigeons
- guinea pigs
- dogs
- bushmeat species

These are often considered minority species, but each has its niche in particular parts of the world. Goats are important in Africa and the Indian subcontinent, and their numbers have been growing, especially in poorer communities. Dog meat, bushmeat and whale meat consumption have an uncertain future, because of concerns about eating companion animals and about conservation of dwindling species. Each species has its own welfare problems, and these need to be examined within the context of the aims of their production systems.

Goats

There are about 770 million goats. Over 60% are in Asia and about 30% are in Africa. They are the single most important source of meat in many of the world's poorest communities. They are also raised in pastoral regions for selling instead of

home consumption, and the money is used for buying grain.

Owing to their link with poverty, goats are unfairly regarded as a symbol of old and primitive farming systems, and they are sometimes called the poor man's cow. In harsh environments, keeping several goats is better security than a single cow. If one goat dies, it has less impact on family wealth than the loss of the cow. Goats are hardy animals in arid and semi-arid environments. They tolerate trekking long distances between feed and water sources. They are preferred to sheep as a source of milk and where there is a high proportion of woody species in the range. Goats are opportunistic feeders, selecting browse and sparse grazing, and they are often the last species to survive in regions where rangeland grazing has been spoilt by overgrazing. They eat browse even during the green season, and when available browse constitutes at least 40% of the goat's diet over the whole year (Landau *et al.*, 2000). These attributes mean that goats are often the last animals to be sold or slaughtered during adversity, such as severe droughts.

Goats are often blamed for **environmental degradation**, largely because they are the last species to be found in a spoilt environment. Before that stage is reached there is usually a history of land overuse, and other species besides goats usually contribute to the degradation. Goats cause less pasture and soil trampling than cattle, and their preference for browse helps to even out grazing pressure in mixed-vegetation areas. Sheep tend to pull grass with roots attached when grazing on sandy soils, whereas goats help relieve grazing

©N. Gregory 2007. *Animal Welfare and Meat Production*
(N. Gregory)

pressure but make regeneration of trees and bushes more difficult.

The welfare problems associated with goat keeping are similar to those of cattle and sheep, even though they show better survival in extreme conditions. The welfare hazards are mainly connected with **under-watering** and underfeeding. Typically, in parts of Africa, goat graziers stop at a waterhole for half a day to allow the stock to fill up with water. Then they head off for 2 or 3 days without any prospect of coming across water. The goats feed on pasture, browse and seeds pulled down from trees by the herdsman. When the animals become dehydrated they feed less, and the herdsman knows he has taken them too far. He tries to get the goats back to a waterhole before this depression in appetite occurs, and whether this happens depends on his skill in judging how the animals are coping. Under these marginal herding conditions, when a drought sets in it becomes more difficult to allow the goats to eat and drink adequately. The goats become emaciated and kid mortality increases.

Infrequent drinking can be hazardous to health for some animals. On reaching a water source, intake of water can exceed 40% of the animal's body weight. For many species this would present a risk of systemic haemolysis in the bloodstream. The goat avoids this through its ability to store the water in its rumen and draw on that reserve as needed to maintain osmotic homeostasis. In contrast, sheep show wider fluctuations in blood osmolality. Taking in large volumes of water can alter rumen flora in goats. Microorganisms with pliable cell walls, such as the protozoa, are prone to osmotic bursting, whereas bacteria have a more rigid cell wall and are less susceptible. Goats can withstand long periods of water shortage. For example, there used to be a herd of Angora goats near Lake Eyre in central Australia where the average annual rainfall at the time was only 7.5 cm. The breeds that become adapted to hot arid conditions usually have a slim body and slender legs, which aid convective heat loss, and the structure of their sinuses enhances heat dissipation during panting.

Overgrazing is a significant problem that has been caused by goats in the Chihuahuan desert of Mexico. It has resulted in pronounced changes in rangeland flora, away from productive species and towards more coarse and less palatable species such as Spanish dagger (*Yucca carnerosana*) and creosote bush (*Larrea tridentata*). In this situation, keeping goats becomes less feasible as they lose body weight and condition (Mellado *et al.*, 2003). Goats that lose condition become less selective in what they eat. They eat more of the fibrous and resinous species, whereas the goats in better condition choose the tender-leaved, more nutritious shrubs (Mellado *et al.*, 2004). This adaptation exaggerates the divergence in body condition within a herd.

Until recently there were large herds of goats in the south of the former Soviet Union, but with conversion to the CIS these herds were allocated to smallholdings, usually comprising less than a dozen animals. Seasonal migration, according to feed availability, has stopped and problems have developed with localized overgrazing and poor upkeep. Declining body condition in does is leading to poorer kidding rates, and in extreme cases there have been **abortions**. Abortion is a common response to neglect and underfeeding in goats. For example, during droughts in Mexico, abortion rates increased from 5% to 36% because of declining body condition during previous breeding cycles (Mellado *et al.*, 1996). Whilst goats are decreasing in numbers in the CIS, they are increasing in other countries, such as those bordering the Mediterranean. They are being reintroduced to control woody scrub, because of the bush-fire risk.

Besides underfeeding and under-watering, other **welfare issues** associated with goat keeping include:

- Diseases partly arising from underfeeding.
- Absence of veterinary care.
- Castration methods. A traditional method in some countries is to use two pieces of wood plus a hammer or stone.
- In the wet tropics of Asia and Africa, goats are kept in hutches or raised pens made of bamboo or picket wood for most of their lives. They rely on the owner to cut and carry the feed. This work can be harder than anticipated especially when the feed has a high moisture content and is heavy to handle.
- In raised pen systems, the feet and limbs of young animals sometimes get trapped in gaps between the slats and in the fencing wire that keeps them in (Nfi and Ndamukong, 1997). There is a risk of nasal botfly (*Oestrus ovis*) during the dry season if manure is infrequently removed.

- Inbreeding is present in some systems that impose continuous confinement, and in some of the dairy goat breeds (e.g. Shami breed).
- Kid predation.
- Kid mortality, especially from inadequate milk supply or poor mothering ability. Failure in milk supply can occur during the monsoon season in the humid tropics as well as the dry season in semi-arid regions.
- Some of the blood collection methods are inappropriate. For example, in East Africa some goat keepers value drinking the animal's freshly drawn blood. First the animal is smothered by placing a hand over the mouth and nostrils, while an assistant holds the animal down. When struggling ceases, the jugular vein is opened with a longitudinal slit, and the blood is drunk straight from the wound. The animal is then allowed to regain sensibility.
- In some urban areas of Africa, hobbled goats are sold in car boot sales. They are taken to and from the market in the boot (trunk) of the car.
- Sun exposure, especially in short-haired breeds.
- Fighting and bullying amongst goats kept in enclosures. This is usually controlled by tethering them separately.
- Attacks on goats by members of the community. This has become an increasing problem in recent times. It either occurs when a goat is found eating someone's crops, or when a crop grower retaliates for crop damage; or it may be due to outright opposition to goat keeping and the animals are attacked and injured out of prejudice.
- Traffic accidents.
- Aerial culling of feral goats from helicopters in Australia.
- Goats raised in semi-arid regions can show a strong fear of open water. This needs to be recognized when it is relevant.

The advantages of keeping goats in poorer semi-arid regions in place of cattle are:

- Does can be more easily managed to produce an even supply of milk through the year in comparison with one or a limited number of cows.
- Does routinely produce twins with a kidding interval of less than 12 months, whilst under the same conditions a cow would typically calve only once every 16 months.
- Under favourable conditions, the annual offtake of young goats for sale or slaughter from three to five does may be six per year, compared with 0.5 calves per cow per year.

These advantages disappear if the goats are not adequately cared for. In particular:

- The does need adequate feed to support lactation for one or two kids.
- Attention to feeding and watering may slip if a farmer is unwilling to divert labour towards goat management.
- Sufficient milk needs to be left for the kids where it is also taken for domestic consumption.
- Breeding bucks need to be available to ensure a short kidding interval. This can be a problem because of reluctance to keep bucks where they are considered unproductive. Community antagonisms sometimes frustrate the loan of a buck.
- Production declines when morbidity or mortality due to disease or parasites gets out of control.

African goats are more prolific than sheep, have lower pre-weaning **mortality** and are more suitable for milk production. They are said to do better than sheep when underfed. This is partly due to their wider choice of vegetation, and also to their higher digestive efficiency of low-grade roughage, plus adaptability to water shortage (Karim and Rawat, 1996). Their ability to tolerate water deprivation enables them to exploit the sparse desert pasture and range more widely over water-depleted areas. However, following a period of severe feed shortage, they are inevitably prone to emaciation and at this extreme they show less pronounced compensatory growth compared with sheep. Controlled feeding trials during the dry season have shown that goats are more likely to lose a higher proportion of their body weight than sheep when they are unable to scavenge (Table 9.1). Initially, during the period of underfeeding, there is mobilization of body fat, and under extreme conditions this progresses to muscle loss. The level of body fatness at the start of the dry season helps determine the individual's ability to regain weight during re-feeding in the wet season. Fatter breeds of sheep regain their lost weight earlier than lean breeds. Similarly, goats regain weight relatively slowly compared with sheep that are fatter. The fatter animals probably maintain fat mobilization for longer than lean animals, and once an animal starts losing body

Table 9.1. Weight loss during winter and summer in goats in India at increasing levels of feed restriction.

| Level of feed restriction (% of *ad libitum* level) | Weight loss (%) | | | |
| | Sheep | | Goats | |
	Winter	Summer	Winter	Summer
80	4	4	6	8
60	8	6	12	9
40	12	10	17	13

protein it is weakened and its ability to recover is compromised.

Where feed is adequate, pre-weaning growth rates of 40 to 70 g per day and post-weaning daily gains of 20 to 50 g should be realistic targets. About 60 kg of milk is needed to reach a weaning weight of 10 kg and, for does rearing twins, this does not leave much surplus for the family.

In crop-growing regions of Central Africa, goats are kept tethered or confined by a picket fence. Production figures for this system in Cameroon are shown in Table 9.2. These values would be typical for other parts of Central Africa, except for the age at first kidding, which is early. Kid mortality rates are high, but, on large goat stations in South Africa, the average kid mortality between birth and weaning is about 12%. On small goat farms in the same region it is about 47% (Sebei *et al.*, 2004). Frequently reported causes of kid deaths in West Africa are starvation (35%), pneumonia (18%) and helminthiasis (25%) (Manjeli *et al.*, 1996). Kid mortality rate varies, especially where there is predation by jackals or risk of pneumonia during the wet season. In one report from Ghana, kid mortality up to 3 months of age was 55%. Mortality is higher among twins (54%) compared with singles (11%). Worm burdens impose an additional burden for the undernourished animal, and goats in these parts of Africa have been called a 'walking helminth zoo'. Worming is considered unnecessary or an extravagant cost. Some of the methods used for treating other goat diseases lack refinement. For example, in India gid cysts are removed by craniotomy and aspiration using only a subcutaneous local anaesthetic (Kalita, 1997).

Although a short kidding interval might as a first impression seem to be a sign of efficient production, it can indicate the opposite. In year-round kidding systems a short kidding interval is often linked to high kid mortality, with the doe returning to breeding early. In practice, it is difficult to achieve twice-a-year kidding in the absence of high kid mortality.

In South-east Asia, many goats are raised on a small scale as a secondary enterprise to crop production or aquaculture, but some are also raised on a large scale intensively. About half the goat keepers raise them for consumption during religious ceremonies. The conditions under which they are kept are simple (see Fig. 1.1). **Shelter** may be provided in high-rainfall districts, but, in one study in Thailand, 12% of farmers considered that shelter was not necessary (Pattamarakha *et al.*, 1997). Three-quarters of the farmers considered that slatted floors were unnecessary, whilst others recommend slats for night-time accommodation if the goats are to be kept clean. When goats are kept indoors or on raised pens, it is helpful to provide additional trough space if some of them are horned. In addition, it is important to ensure that the feed remains in front of them, rather than being pushed out of reach as they feed. Raising the height of the trough by between 25 and 50 cm above foot level improves the animals' reach and allows better access to the feed.

In Australia, the feral goat population can sustain an offtake rate of over 15% per year without showing signs of a decrease in numbers (Southwell and Pickles, 1999). Aerial **culling** of goats from helicopters is less common than it used to be, because of the cost of hiring helicopters and because there is a now a good market for live

Table 9.2. Production figures for West African dwarf goats in the western highlands of Cameroon.

	Mean	± SD
Age at first kidding (days)	304	62
Kidding interval (days)	278	83
Litter size	1.64	0.58
Kids born per doe per year	1.70	0.54
Kidding (%)	79	1
Kid mortality to 5 months age (%)	37	1

goats. Instead, the goats are trapped in pens at watering points and trucked to either an abattoir or a port for live export. These removals are part of a feral goat control programme, but they seem to be having less impact on overall numbers than droughts.

In India, **goat meat** (chevon) is one of the more expensive meats even though goat keeping is associated with the poorer classes. The goat population has been increasing rapidly, as this animal fits well with the declining size of landholdings in India. Over 40% of the goat population is slaughtered every year, and the meat is accepted amongst most castes. Between 30 and 40% of goat meat is from animals over 5 years old and is potentially tough (Kumar and Pant, 2003). In East Asia there is a strong preference for carcasses with a high lean content. Buyers look for animals that are well muscled in the forequarters, rib and hind areas, plus a wide back, particularly at the loin. However, there is also a preference for medium weights, because they are easier to sell. The custom is to consume the whole goat on the day it is slaughtered. If there is too much meat, it has to be cut into strips and air-dried to preserve as pemmican or biltong, or it may be deep-fried until crisp and dry and then stored in sealed containers. In Mediterranean regions and parts of West Asia there is a good market for kid meat (*capretto*). The preference is for carcasses from 15-day-old kids as this meat is paler in colour, milder in flavour and more tender (Argüello *et al.*, 2005). In many Islamic communities in Africa, sheep are valued more highly than goats. Sheep are preferred for both social and religious ceremonies, and so they command a higher price. Goat carcasses have a large shoulder joint, less leg meat, a smaller plate joint and a higher dressing percentage than sheep (Gaili, 1978). Sheep have more fat in the pelvic cavity compared with goats,

but other non-carcass depots are similar. Within the carcass, goat meat has less intramuscular fat but this is compensated by its higher moisture content. Overall juiciness of the cooked meat is often the same (Sen *et al.*, 2004). Goat meat is darker than sheep meat, and it can be tougher when grown under comparable stall-feeding conditions.

Water buffaloes and yaks

There are two types of water buffalo (*Bubalus bubalus*). The **swamp buffalo** is a semi-aquatic animal kept largely for draught power and cultivating rice paddies. The mature female weighs about 500 kg, and spends the hotter part of the day in swamps or wallows. The **river buffalo** is slightly larger and is kept mainly for milk production. It produces between 1000 and 2000 kg per lactation. Both types have thick black skin with well-developed sebaceous glands, which improve water repellancy.

Buffaloes wallow for two reasons: first, to keep cool and, secondly, to avoid insect attack. They do not rely on sweating as a cooling mechanism and, in the absence of a swamp or other wallow, they will seek shade.

Post-partum complications can occur in water buffalo because of unhygienic calving conditions. Dystocia can add to this, and some individuals seem prone to exhaustion during calving.

Buffaloes are seldom dehorned, but, when they are, anaesthesia is not usually used. Practical experience shows that the animals soon (within 20 min) start grazing after the horns have been lopped off. Tail docking is occasionally done in cows that are milked.

In parts of the Philippines there are feral herds of buffalo that are recruited periodically to assist with preparing the rice paddies. Selected animals are split off and herded to the paddies, where they puddle the beds. Once they have completed the work, they are returned to their herd. In some parts, feral swamp buffalo are tamed for lifetime use on farms, and according to most accounts from South-east Asia taming is a relatively innocuous procedure, but it is less so in the case of feral river buffalo in the Middle East.

Over 70% of the world's buffalo population is in India and Pakistan, and male buffaloes are an important source of meat. About 10 million male buffalo calves are born in India every year. Those that survive are usually slaughtered as yearlings at about 85 kg live-weight, producing a 36 kg carcass, but some are grown on for slaughter at 200 to 300 kg live-weight. Buffalo meat is also an end product from draught animals. It is from very old, unproductive animals and, because of their age, it can be outstandingly tough and dark in colour. It has to be finely cut or ground to overcome the toughness.

In Mongolia, the **yak** (*Poephagus grunniens*) provides about half the country's meat, butter and milk. In other Central Asian countries, the number of yaks has been declining, partly because members of the younger generation avoid farm work and prefer to be daily wage earners, and partly because the need for yaks as transport animals has declined with tighter border controls in the region. The yak can tolerate very low temperatures and it is the only large mammal that can graze at 6000 m above sea level, even at –40°C. It has a broad chest, large lung capacity and a high red blood cell count. It can carry loads of 50 to 60 kg on very rough terrain and survive on scanty supplies of mountain feed.

Deer

There are about 5 million farmed deer, half of which are in Australasia and over 20% in East Asia. The main species are shown in Table 9.3.

Wapiti are sometimes used as terminal sires on red deer in New Zealand. The cross-bred slaughter generation are heavier at weaning and have higher post-weaning growth rates than pure-bred red deer, and they produce heavier carcasses when slaughtered as yearlings. Specialist sire lines have been developed with three-quarters or seven-eighths wapiti parentage, and there is a

Table 9.3.   Main species of farmed deer.

Red deer (*Cervus elaphus*)
Sika or spotted deer (*Cervus nippon*)
Wapiti or elk (*Cervus elaphus nelsoni*)
Fallow deer (*Dama dama*)
Reindeer (*Rangifer tarandus tarandus*)
Rusa deer (*Cervus timorensis*)
Sambar deer (*Cervus unicolor*)

corresponding increase in mature body size with the proportion of wapiti genes. These sire lines are also called 'wapiti'. Concerns have been raised that wapiti sire lines will increase the prevalence of dystocia when used on red deer hinds. However, in practice the overall prevalence of dystocia in New Zealand red deer is low (0 to 3.6%), and disturbance at calving time and over-fatness are more important causes.

Red deer farming is a relatively new industry. So far, they have been genetically selected for growth rate, for antler velvet production and for suitable temperament, especially ease of handling. Genetic selection for large body size is thought to be associated with more consistent early breeding in the females, and selection for antler size has caused complications with survival in sedated animals. In New Zealand, sales of antler velvet represent about 30% of the deer farmers' income, but the market varies substantially from year to year. Through selecting for large antlers with corresponding high yield of **antler velvet**, stags now carry formidable weights on their heads. There are occasional reports of stags having problems with weak neck muscles. This becomes obvious when a stag is sedated for a routine veterinary procedure and on recovery from the tranquillizer the animal is unable to hold its head up. This can interfere with breathing and has been known to result in death from asphyxia. The problem is said to be less obvious when the rut approaches and stag's testosterone improves the strength of its neck muscles.

There are four systems of farming:

- extensive ecotourism and trophy hunting
- rangeland
- intensive hill and lowland breeding farms
- intensive finishing units

The recreational and hunting farms are mainly in mountainous regions. They are enclosed, and full antler stags are released on to the range at the start of the hunting season. In New Zealand, rangeland deer farming is in hill and high country. During the summer, these properties are stocked at low rates and they use limited labour. In winter, the deer are held on lower ground and may be provided brassica crops. These animals experience infrequent handling, and animals with temperament problems have to be culled. Overstocking in winter is sometimes a problem.

Most of the world's deer population is raised on **intensively managed** hill and lowland breeding units. These farms are subdivided, with fencing allowing grazing control, and all the young stock are cropped, except for replacements. An important management aim is to achieve a condensed calving pattern. This allows easier management and it is achieved by close attention to weaning date and culling for breeding failure. However, there can be a burst of reproductive failure in a year that follows a dry summer, and managing this situation depends on early adjustment of stocking rate (Pearse and Drew, 1998).

At weaning, the 3- to 4-month-old weaners are transferred to intensive grazing units in low-lying ground, where they are finished. Finishing weaners or replacement hinds in feedlot pens or sheds is practised in some regions to improve growth rate and limit pasture damage. Indoor **housing** can lead to coat damage from deer biting each other, and it is often the smaller animals that are victimized in this way (Pollard and Littlejohn, 1998). Fighting has usually been managed by removing the bullies or the bullied individuals and housing them separately. The isolated calves grow more slowly, even though their feed intake is the same as that of group-penned calves. In addition, they spend more time lying and less time self-grooming and feeding. Keeping deer indoors on grain or a pelleted feed has also caused problems with grain overloading and with abnormal chewing behaviour, often directed at the enclosure. Destruction of the wooden pens can be controlled by providing some hay or a tree trunk to chew on.

One of the temptations when overwintering calves indoors is to save on feed costs by maintaining them at constant body weight and then relying on compensatory growth in the next grazing season. When this was tested in New Zealand, it was found that compensatory growth was slow, and it was only by the end of the summer that they had caught up with calves fed more generously during the winter period (Webster *et al.*, 1997).

Deer are good jumpers, and so high fences, high close-boarded yard walls and enclosed transport crates are needed to prevent escape. Excitable deer are difficult to handle. Mechanical restraint is used for routine procedures such as drenching, tuberculosis testing, blood sampling, pregnancy scanning and antler velvet removal. Specialized handling and restraining systems have been developed (Fig. 9.1).

In a survey of health problems on deer farms in Korea, 60% of farmers reported that they had experienced problems with bone fractures in their animals, and 57% reported dystocia (Kwak *et al.*, 1994). Other common problems

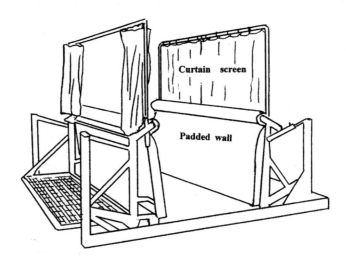

Fig. 9.1. Hydraulically operated deer crush.

were enteritis, bloat, dermatitis, osteomyelitis and anaemia. Overall **mortality** rate was 7% per year. In New Zealand, stress-related problems encountered during the early days of the industry, such as abomasal and duodenal ulcers and yersiniosis, are now less common, but death from stress-related injuries is still a problem (Audigé *et al.*, 2001). Part of the reduction in yersiniosis is due to the introduction of a vaccine.

A study of diseases and defects in red deer carcasses at New Zealand abattoirs during the late 1980s showed that most chronic pathological lesions were trauma-related (Selwyn and Hathaway, 1990). Major **injuries** included fractured ribs (recent and long-standing), broken and fractured long bones, extensive bruising of the forelegs and rump, and pleurisy, peritonitis, pericarditis and arthritis arising from earlier injuries. Often there was atrophy of muscles surrounding a wound. The prevalence of these problems has been reduced, but the reports from 20 years ago show how difficult this species can be to manage without the right facilities and experience. One estimate of the prevalence of bruising in deer carcasses put it at about 24%, and this was higher when there were large numbers of deer in a truck, presumably because they were more crowded and more likely to climb on top of each other (Pollard *et al.*, 1999).

Mismothering can occur when hinds are stocked too heavily in a calving paddock. It creates problems when the mother accepts an alien calf in preference to her own progeny, which then becomes motherless. However, under intensive deer farming systems there is substantial allosucking, where calves take milk from non-maternal hinds, and so orphaned calves can and do thrive. Allosucking takes two forms. If a hind accepts an alien calf, the calf drinks from the hind whilst standing in the usual anti-parallel position. Alternatively, the calf may be able to suck from between the hind's rear legs whilst the accepted calf is drinking in the normal anti-parallel position. Attempts at allosucking increase as the supply of milk diminishes, and it tends to be more common in fallow than in red deer.

Red deer calves are often weaned 3 to 5 weeks before the rut. This break from lactation allows the hinds to regain some condition before mating, and it helps to synchronize oestrus. **Weaning** is stressful for the calves, and five systems are used to try to reduce its impact:

- Keeping the calves in the same paddock. Hinds are moved to a fresh paddock that is at least one field away from the previous paddock where the calves remain.
- Fence-line weaning. The calves and hinds are in adjacent paddocks.
- Transferring the calves to indoor housing. Indoor confinement with good feed is said to settle the weaners and allow them to develop new social bonds.
- Providing surrogate adults. Introducing the calves to a stag, or a quiet group of dry hinds, or calves that have already been weaned.
- Soft weaning. About 10% of the hinds are removed each day for 10 days. This allows a progressive change in social structure of the herd, and gradually reduces the comforting presence of the adults.

There is no consensus as to which is the least stressful method, and there may be merits in each system. Joining a group of dry hinds encourages the calves to settle (Pollard *et al.*, 1992). They sit down sooner and remain sitting for longer. They also show greater curiosity and less timidity when an adult is present in the group, but some of this could be because they copy the adult hind. Some hinds are more appropriate as matriarchs than others, and they may need to be selected specifically for this role.

The most common problem encountered in early-weaned calves is injuries due to poor handling (Bao *et al.*, 2004). The risk situations are mustering in the paddock, handling groups that are too large in the yards, and fence or gate charging during the immediate period following separation of the calves from the hinds. High wire fences need to be clearly visible to reduce the risk of fence charging, and this is achieved either with natural vegetation, with scrim or by installing rails as visual barriers. In weaners, broken necks are the most common form of lethal injury, followed by broken legs. These risks are not inevitable, and they can be managed. The stress of weaning can reduce the immune response to vaccination against yersiniosis, if the two coincide. Late weaning at 6 to 7 months of age is practised mainly on rangeland farming systems, where it is inconvenient to muster on a regular basis. Disadvantages of late weaning are failure to drench the calves at the right time against lungworm, and weaning can coincide with deteriorating weather conditions, which presumably add to the stress effect.

The stress of weaning, bad weather or under-feeding can precipitate yersiniosis. Typically, there is an enteritis and lymphadenitis with a smelly diarrhoea, which may become bloody. In the past this has been a significant cause of calf mortality.

Wapiti are more susceptible to gastrointestinal nematodes than red deer (Mackintosh, 1998). *Ostertagia* species can cause considerable damage to the abomasal wall, resulting in a rise in pH and poor protein digestion and absorption. It is thought that stressful events including insufficient feeding render wapiti more prone to these parasites.

Winter survival is a welfare issue for deer and reindeer kept in **cold** climates. It is the juveniles that are most susceptible, and if they are under-sized or have insufficient body reserves at the start of winter they may not survive. Their condition depends on the availability of feed during the summer, so a dry summer with limited feed, followed by a harsh winter, can result in high losses. An incompletely characterized condition known as 'wet belly' can develop before death from cold plus emaciation (Ahman *et al.*, 2002).

In New Zealand, calf mortality is mainly associated with dystocia. Hypothermia and mismothering have been important in colder climates, and, in cold conditions in Scotland, calf mortality has been greatest in small female calves born to small hinds. Hinds need to be sufficiently well developed and have sufficient body condition when mated to control this risk (Blaxter and Hamilton, 1980). In the wild, a hind will normally separate from the herd for up to 3 weeks when she calves. The calf spends much of the time hidden whilst the hind is feeding. On intensive deer farms, if there is insufficient cover to allow the calf to hide it may attempt to go through fencing in search of a hide.

'Winter death' has been a problem in breeding stags that lost condition during the rut and had insufficient body reserves during a severe winter. During the rut, fallow deer stags may lose up to 30% of their body weight, at a rate of 450 g/day within 3 to 4 weeks. This is not usually regained until the spring.

Mixing unfamiliar deer can lead to aggression, reduced self-grooming and immunosuppression (Hanlon *et al.*, 1995). Red deer have a preorbital gland, which is used in scent marking and can serve as a behavioural indicator of excitement. Stags often open their preorbital gland when they are about to fight, and the gland is also open during feeding in calves. Social mixing is best done in the handling yards, rather than introducing unfamiliar deer to the main herd in a paddock. Aggression can take the form of biting, chasing, head butting and flailing with the forelegs. In all-male herds, small animals may be repeatedly chased and mounted during the breeding season. In hind herds, victimized individuals are bitten, often on the back. The victims are often unhealthy or injured animals and there is a risk that they will become debilitated even further. Mature breeding stags should not be kept together in small mating paddocks during the rut. It has been known for a subordinate stag to be hounded to death by a mature stag when they were closely confined. Fighting through a fence can be avoided by having at least one intervening paddock between breeding stags.

Fence pacing is a problem on intensive deer farms that have light soils and moderate to high rainfall. It is thought to be a sign of anticipation, frustration or social discomfort, and it is influenced by siting of the fence line, hunger, incompatibility between individuals in a group, an urge to roam more widely, a need to find shelter and a drive to join another group of deer that can be seen in the distance. However, none of these factors have been proven causes except for the drive to join another group. Fence pacing can cause substantial erosion in the paddock, and it has been known for the fence line to collapse into a gulley created by the deer and rain.

Deer can be trained to come to call, and this is helpful when giving them a change of paddock. Some deer appreciate wallowing and create their own wallows at trough sites. It may provide some protection against ticks.

Abnormal life-threatening behaviours can be linked to social isolation or excessive exercise. When isolated, deer show increased activity and may attempt to break through fencing (Pollard *et al.*, 1993). This can lead to trauma, which in turn calls for destruction of the animal. Occasionally there are cases of stress-related **exhaustion**, such as the following incident in New Zealand:

> Strong winds that blew all day across a large Otago deer farm caused the deaths of a large number of deer. Approximately 1000 deer were in two adjoining fields; one held a large number of spikers and the other hinds and their fawns. All day the two mobs were seen running in circles around their fields, continuing all day and into the night. Next morning approximately 128 were found dead among some low scrub – the

only shelter. Unusually, most of the dead were hinds and only eight fawns and six spikers died. Necropsies of two showed few gross abnormalities apart from emphysematous lungs. Histopathological examination of a variety of tissues, including skeletal muscle, showed minimal changes and no evidence for a post-capture myopathy in the muscle sections. The cause of death was assumed to be a combination of stress and exhaustion.

In some parts of the world deer have to withstand high summer temperatures. Normally, feral reindeer move towards higher altitudes when it gets hot and wind speeds are low. This allows them to escape harassment by insects as well as hot temperatures. Under farmed conditions, ensuring that deer can benefit from shade is a helpful alternative, but it can make insect nuisance greater.

Biting can occur during **transport** to the abattoir, and it is the smaller animals that are usually the victims. Sudden impacts with the walls of the livestock compartment and with other animals often occur at the start of a journey and when travelling on winding, steep roads. When the journey lasts for 3 or more hours, some deer are likely to try and sit down. However, they usually immediately stand up again because they get trodden on. In New Zealand, carcass bruising from these plus other causes is usually seen in 1 to 10% of the carcasses, and it has been found to be greater following longer journeys.

In the yards at the abattoir, red deer are usually moved by staff working amongst the animals. Wapiti are larger and may need to be worked from outside the race or pen if their temperament is uncertain. Wapiti tend to stand their ground when yarded and do not move as freely as red deer. Signs of imminent **aggression** towards people include the following. The deer stands squarely in front of the person, staring directly at him or her, and may start grinding its teeth. The upper lip may be raised, the tongue extended and air hissed out through the nostrils. If the antler is hard instead of being in velvet, the stag may lower its head before butting or attacking. Otherwise, the animal is more likely to raise its chin and strike with its forelegs. People routinely handling unfamiliar deer in abattoirs need to recognize these signs. They may equip themselves with plywood shields and protective boots, and a sensible policy is to arrange for the haulier to alert lairage staff of any potential problem animals when they arrive at the abattoir. Non-fractious deer that are used to being handled generally respond well to close contact with stockpeople, and talking quietly plus placing a hand on the back can have a calming effect. Deer are relatively sensitive to sudden noises in handling facilities (Price *et al.*, 1993). This is a common problem with unbaffled metal gates and fittings. Fallow deer are particularly flighty and show remarkable ability at escaping over walls and partitions. They are apt to rear up in the lead-in race, and are also prone to producing high ultimate pH (pH$_{ult}$) meat.

Reproductive performance and **longevity** are closely linked to feeding and body condition in the hinds. In red deer this can be managed through target weights at first and subsequent matings (Moore *et al.*, 1985). Hinds typically produce six calves during their lifetime, and they usually stop breeding at 17 years of age. One study showed that there was no difference in meat tenderness in hinds between 1 and 21 years of age.

In China and Korea, venison is a co-product to antler velvet production. In this situation, stags are kept for antler velvet harvesting, and the blood from the antler wound is added to alcoholic beverages. Chemical restraint with succinylcholine was widely used in Korea for **antler velvet harvesting**, but now that this drug is only available on veterinary prescription its use has declined (Kwak *et al.*, 1994). Succinylcholine provides immobilization through neuromuscular blockade, with collapse in 5 to 8 min of injection and recovery after 25 to 30 min. The hazards, besides the stress of paralysis and difficulty in breathing, include heart failure and injuries between administration and immobilization, as well as death from asphyxia when given an overdose. In other countries, antler velvet is harvested using a local anaesthetic and sedative.

Camels

There are over 19 million camels in the world at present, half of which are in Somalia and Sudan. Recent wars in Somalia, Ethiopia, Sudan and Chad have reduced camel numbers through malnutrition, dehydration, disease and neglect, but when peace stabilizes the population will probably rise. Other countries with modest camel populations are Mauritania, India, Kenya and Pakistan. Nigeria imports camels for slaughter from seven countries between Mali in the west and Kenya in the east, and this wide-ranging recruitment raises problems with long-distant movement and cross-border disease control (Baba *et al.*, 1994). The demand for camel meat in Nigeria has been growing because of rising beef prices.

Camels are put to a wide range of purposes, including carrying freight, such as grain, transporting people, racing, tournament fighting, cultivation, bride price, sacrifice and producing meat, milk and hair. Camels are sometimes preferred to a pair of oxen for ploughing because they are faster and can be turned more easily at the ends of rows. In some cultures camel ownership is a sign of wealth and a status symbol. Those communities differ in their attitude towards using camels for meat. Some appreciate camel meat whilst others think it is inedible. In the past camels were often slaughtered in back streets, but now there are dedicated camel abattoirs that supply a recognized trade in quality camel meat, as well as slaughtering animals that no longer serve a useful purpose. In India, dromedary camels are used more for investment, milk, dung and hair production, than for meat. The hair is used in clothing and rope making, but the hide is also valued for making shoes, saddles and water containers. Bactrian camels in India are valued more for their meat and hair, instead of milk.

In herds where camels are used for transport the gender ratio is usually equal, but only the males are used as the pack animals. A normal load on a pack animal would be about 300 to 400 kg. In milk- and meat-producing herds the males are slaughtered early and so females predominate. In East Africa, calving occurs between February and May. A newborn calf weighs about 35 kg, and calving percentage for a breeding herd is typically 70%. In parts of Asia, lactation is sometimes made to last up to 16 months.

The welfare issues involving camels include:

- inadequate protection from the sun
- inadequate cover from rain, in wet districts
- inadequate protection from cold at night
- unregulated water intake when rehydrating
- underfeeding during droughts
- colic
- cutting the canine teeth
- nose-puncturing methods (for fitting nose pegs)
- restraining and hobbling methods

- poor drenching techniques
- inappropriate introduction to busy road traffic
- spinal trauma
- saddle and rope sores
- whip injuries
- cross-fostering methods
- branding
- slaughter methods

In addition, camels are used in competition wrestling and racing, but the prevalence of injuries from these sports is not known.

Camels have their own niche within pastoral farming. They have taken over from cattle in parts of Asia where the landscape has been degraded through either natural desiccation or human activities. Camels feed on trees and shrubs that cattle and sheep would find inedible, and they do not compete strongly with goats because they browse at different heights. They are more successful in hot arid regions, and they are also salt-tolerant. Nevertheless, they can suffer from sun exposure and this is a particular hazard for animals in a weak or emaciated condition. The signs of sunstroke are an excited behaviour, followed by collapse. The eyes are closed and the conjunctiva congested. Rectal temperature may be 43°C or more, whereas in the cooler parts of the day normal rectal temperature would be about 38°C (Rathore, 1986). Eventually the animal goes into coma and then dies.

Cold exposure can be a problem in seasons when there are low night-time temperatures. In the feral state, camels manage this through seasonal migrations. Feral camels in the Gobi desert concentrate along the riverbeds during the winter season, and move to the desert as soon as the snows melt and pasture becomes available. In some rangeland systems in Asia, calf mortality up to 3 months of age is about 12%, with most of the mortality attributable to cold, and drownings being a secondary cause. Hypothermia can also occur when hot exercised camels overdrink before cooling down, and it can occur when sweating camels have their saddles suddenly removed. The signs include muscle rigidity, especially in the neck and limbs. This leads to a crouching action in the hindquarters when walking.

During the drought of 1979/80 in Kenya, the offtake of milk for human consumption was high, because of the food shortage. Most of the calves born during this year were malnourished and 55% died, mainly from the high parasite burdens, which the weakened calves could not withstand.

The frequency of drinking in working camels in Ethiopia is usually every 2 to 5 days. During the rainy season, when bushes and shrubs are green, camels may go without drinking for up to 20 days, and during the dry season the maximum interval without water is 10 days (Mekonnen, 2004).

The camel survives in regions supplied by moderately saline aquifers because of the efficient way they excrete sodium. This allows the consumption of salt water without the need for adjusting plasma volume. In contrast, sheep increase their fluid intake when confronted with a salty water supply, at the expense of marked falls in plasma volume, weakness and dehydration (Assad et al., 1997). In addition to being more salt-tolerant than other livestock, the camel has a higher salt requirement. Salt deficiency leads to cramps and skin necrosis, whist excessive consumption of salt water can lead to gout if there is kidney failure.

During colic the camel stops cud chewing, appears distressed and lies down. Whilst on its right side, the left flank appears distended, there may be dyspnoea and the animal kicks in the air whilst moaning. It often refuses to get up and walk. Sand ingestion is one cause of colic. It occurs at the beginning of the rainy season when green plants emerge and are pulled up and eaten with roots and sand attached.

One of the hazards of drenching camels amongst inexperienced stockpeople is cutting off the airway through the nose whilst the head is being held. This causes the animal to breathe through its mouth and increases the risk of drenching fluid passing into the trachea and lungs, leading to pneumonia.

Vicious males may have their canine teeth shortened, and this should be done under anaesthesia.

Spinal trauma has been a problem in fence-line injuries, and saddle sores develop in pack animals that are bony and in poor body condition (Manefield and Tinson, 1996). Restraint is often achieved by tying a rope around the mandible between the canine and premolar teeth. The two ends pass back to the rider as reins. This can lead to damage to the mucous membrane, and to fracture of the mandible if the animal is startled and throws its head whilst tethered. Fracture of the mandible also occurs during falls and fighting, and it is one of the more common bone fractures

in this species. It results in difficulties in selecting feed and chewing.

One of the Bedouin methods used for fostering calves is unpleasant for the dam. It involves blocking the foster mother's nostrils with cloth plugs and its rectum with a large ball of its own faeces wrapped in a cloth, which is held in place with a suture across the anus. The foster mother is then walked for a couple of kilometres in the desert, and on return the calf is given access to the cow, which presents its udder with a well-arched back from straining on the rectal blockage. If she allows the calf to drink, the obstructions to the nostrils and rectum are removed and the animals are kept together.

Fire-branding the skin is used in Africa and Asia as a traditional treatment for a range of ailments, including abdominal hernias, but in particular for dermatophilosis. Chronic wounds such as abscesses and inflammations are also treated by firing with a hot rod, and other cauterization techniques are used for treating bleeding, fractures, sprains, dislocations and nervous system disorders. Hot-iron branding is common for identification purposes. The prevalence of hot-iron branding in camels presented for slaughter at two abattoirs in Ethiopia during 1998/99 was 28% (Woldemeskel and Gumi, 2001).

In Kazakhstan, where two-humped camels are raised for slaughter, the main killing months are October and November, when pasture becomes scarce. In the Sahel, single-humped camels are mainly slaughtered as cull animals when they are over 10 years old. Slaughter is either by a chest stick without stunning in the couched animal followed immediately by head removal, or by a low-neck stick whilst the animal is standing (Wilson, 1978). Following the low-neck stick the animal is prone to running off, and this is prevented by tapping a cane in its face as a distraction. The skin is often removed by the 'bush-dressing method', where it is peeled from a mid-line incision along the back, and the dropped hide serves as a ground apron around the couched carcass during dressing. Live-weight in males is about 445 kg and in females 440 kg. The dressing percentage is 54 and 48%, respectively, and the heaviest wholesale cut is the forequarter. Castrated two-humped camels can produce carcasses with considerable fat, which has a white colour. The meat is eaten fresh or dried, and the fresh meat is said to be sweet because of a high glycogen content.

Mange (*Scarcoptes scabiei*) is commonplace amongst camels in Sudan. Principal disorders seen in slaughtered camels in Africa are liver abscesses, hydatidosis, septicaemia and haemorrhagic diseases. The prevalence of bruising in camel carcasses was 15% at two abattoirs in Ethiopia during 1998/99 (Woldemeskel and Gumi, 2001).

Horses and donkeys

Countries with the largest numbers of horses and asses are China, Mexico, Brazil, the USA and Ethiopia. Many of these animals are destined for human consumption or inclusion in pet food once they are no longer needed. About 580,000 t of horse meat is consumed each year, and it is a favoured meat in China, Mexico, Kazakhstan and Italy. Smaller amounts are consumed in Belgium, the Netherlands and France. Poland has several live horse and horse-meat centres that serve a significant proportion of the European trade. Over the past 10 years, horse-meat consumption has increased in Europe because of the BSE food scare. In Italy, local production cannot meet demand and 80% of the horse meat is imported.

In most countries horses, asses and mules are kept for work or recreation, but a smaller number are reared in France and Italy with meat consumption as an important goal. The local breeds raised for meat consumption include the Haflinger, Avelignese and Franches Montagnes. Many horse abattoirs provide a salvage or disposal service for animals that are no longer useful or valued. The most common disorders in horses presented for slaughter within Poland are tying-up syndrome, limb fractures, pneumonia, liver disease, joint and tendon disorders, tetanus and trauma. Routine meat inspection records at one abattoir in Poland showed that, between 1988 and 1992, 21% of the carcasses were unfit for human consumption and 81% were casualties. There is a sizeable live horse export trade from the USA to Asia, leaving from Washington State.

In a survey of over 1000 horses sent for slaughter at two abattoirs in Texas, 3% of the animals were emaciated and 78 had a condition that was rated as a severe welfare problem. Nineteen per cent of those were considered unfit for transport when they arrived at the abattoirs, and two died shortly after arrival (Grandin *et al.*, 1999). Twenty-five per cent of the carcasses had bruises.

Kicks and bites caused over 51% of these bruises. Falls during the journey are one of the causes of these injuries. In heavily stocked trucks, when a horse goes down, it is crowded by the other animals, making it difficult for it to get up again (Collins *et al.*, 2000). If it stays down, it gets trodden on. Horses held at high stocking densities also tread on each other's hooves and pasterns, whilst repositioning their feet in an attempt to maintain balance.

Horses are a prestige animal in parts of Africa. They are only slaughtered as sacrifice animals during important funerals. The horses are transported in trailers to specialized markets that supply this trade.

Donkeys are not a conventional source of meat. Donkey meat comes mainly from old, sick or exhausted work animals. Islam prohibits the consumption of donkey meat and people in several other cultures refuse to eat it. Many Muslims also consider that it is forbidden to sell donkeys for meat. This has driven the donkey meat trade underground and the meat is sometimes passed off as beef.

Traditionally, in West Africa, feral donkeys were caught, trained and used for draught purposes as and when needed. When a donkey was no longer needed, it was returned to the feral herd. As the taste and demand for donkey meat developed, the feral herds became depleted. Originally, donkey meat was dried and smoked. This allowed easier transport and distribution, but this has now been replaced by transporting live feral and domesticated animals to the destination, where they are slaughtered and consumed. Feral donkey numbers have now declined and they are no longer a free resource for farmers.

Ostriches and emus

Most of the value in an ostrich is in its skin, and for an emu it is the fat depot at the base of the tail on the back. Ostrich hides are used for making handbags, belts, footwear, small leather goods, office desk furnishing, upholstery and clothing. The quill pattern gives these products character and high value in the luxury boot market, whereas the glove market prefers a soft, thin leather that is more waterproof. Leather made from the skin on the bird's shank is sought after by handbag manufacturers for its similar appearance to reptile skin.

Above all, the skin must be free of graze marks, scratches, pecking marks and skin burns from pour-on insecticides used against ticks.

Top-quality hides have a well-developed quill pattern and fall into the 130 decimetres size class. Traditionally this has been produced from birds slaughtered at 14 months. However, in recent times the leather market has been oversupplied and prices have been depressed. This has meant that farmers have had to maintain margins by reducing feed costs and slaughtering the birds earlier. Slaughter age is presently 11 months, at a live-weight of 90 to 95 kg.

An unusual feature of the ostrich meat market is that it has come to expect a certain number of portions from each of the high-value cuts. A 95 kg bird is acceptable, but there is market resistance to reducing slaughter weight below 90 kg live-weight because it limits portion size.

The fashion feather market is presently depressed, and a lot of the feathers are being used for making feather dusters. Ostrich meat is a co-product to leather goods, and is sold mainly in the restaurant trade.

The welfare problems in ostrich and emu farming include:

- incompatibility and aggression between breeder birds
- fence charging and entanglement
- leg injuries
- stress during restraint and transport
- exposure to cold and wet conditions
- proventriculus impaction
- declawing emus
- feather-plucking ostriches
- ostrich riding at tourist farms

Matching breeding stock for **social compatibility** is important for the harmony of an ostrich farm. Incompatibility usually arises when small paddocks are used and a hen cannot escape from an over-aggressive male. The usual way this is managed is to keep the birds in small groups or in large paddocks. About 15 males can be kept successfully with 20 females, and this group composition ensures good coverage of the females and higher egg fertility. The exact number of males that can be managed in this way depends on the level of aggressive rivalry. In small breeder paddocks there may be one compatible male to one or two females, but sex ratios of one to eight have been successful on some properties.

Aggression amongst ostriches can take the following forms: gaping, open-mouth hissing, threat displays with the tail erect and wings open, chasing, chest ramming, kicking, mounting or trampling. Aggression starts to increase at the time of puberty, which usually occurs between 11 and 13 months of age. At the start of puberty the males show more pacing, and this is a common territorial behaviour in older breeding males.

Experience in Namibia has shown that kantling (cloacal intercourse) can be common in male-only groups. This was thought to be a form of aggression rather than a way of releasing sexual tension, as it did not involve normal courtship behaviour.

Ostriches can develop a number of **abnormal behaviour patterns** that cannot be put down to any specific cause. They include sand-snapping, fence-snapping, air-snapping and plucking the bird's own feathers (Sambraus, 1998). Pecking at the feathers of another bird can also become a problem, and the target area is usually on the back above the hips. Normal comfort behaviours include care of plumage and sand-bathing. Sand-bathing is greatest in the afternoon. Fence pacing is often a form of territorial behaviour, but in some circumstances it may instead be an indication of anxiety or stress.

Chicks are prone to impaction of the proventriculus. It can occur from hunger and eating excessive amounts of fibre whilst there is a lack of grit, or eating sand, bedding or novel objects. The impacted chick shows lethargy, loss of appetite and constipation. There are three ways of reducing this risk: first, by limiting the availability of coarse feed that can give rise to impaction whilst giving adequate amounts of feed to satisfy hunger; secondly, by rearing the chicks alongside adults – the chicks seem to copy and learn from the adults what to take as feed; and thirdly, by removing debris from the birds' pens or paddocks and providing the feed in a trough, preventing inadvertent consumption of harmful materials on the ground. Stress can precipitate impaction, through inhibition of upper intestinal tract motility and through exaggeration of the stone-eating habit whilst stressed.

Chicks reared in the company of adult birds are also more active than those reared with only their hatch-mates. Inactivity can lead to leg disorders. The chicks have strong filial imprinting when reared with adults or surrogate parents. If the parent figure is a person, the chicks may show distress when that person leaves them, and they may develop inappropriate sexual imprinting when mature.

Emu hens sometimes attack their own chicks when they hatch. This is usually managed by removing the hen just before hatching. Emu chicks reared away from their parents can be slow at learning how to feed and drink. It is usually recommended that feed and water are placed in well-lit areas as this helps the birds find them.

Although ostriches can withstand short periods of snow, their feathers are not waxed naturally and so they lack good waterproofing. This makes them susceptible to chilling when it gets very wet. Protection from rain with a **shelter** is necessary except in warm regions with low rainfall.

When alarmed, ostriches will run, and if the paddock is small they may charge a fence. Fencing needs to be clearly visible to deter this, and oblong paddocks may be preferred as they allow a larger flight distance. Some producers favour a radial **paddock arrangement**, where up to 16 paddocks radiate from a central partitioned shed. The shed acts as a shelter and collecting/handling area, and may also serve as a feed store. Fencing between the paddocks is with $1\frac{2}{3}$ to 2 m wire mesh. Double fencing between adjoining breeding paddocks helps prevent males reaching each other when trying to fight.

Trauma is the single greatest cause of mortality in birds over 3 months of age. The injuries occur during handling, restraint and fence charging and from dog attacks.

Manual restraint can provoke vigorous reaction. The bird jumps and flails with its legs and this can injure the handler. This reaction can be controlled either by hooding the bird, catching the bird in subdued lighting or holding the bird's head horizontally and below the level of its back. Whilst in this position, the bird is less able to kick out with its legs, but, even so, cuts to the neck, including the oesophagus, have occurred when birds have been restrained in this position.

Some farmers use a shepherd's crook to catch a bird. The width of the narrowest part of the crook is sufficient to place around the neck of the bird. The crook slides up, behind the bird's head, and is used to pull the head down. There must be an assistant to control the bird once the crook is engaged. This is usually done by grasping the lower beak, but holding the body may be sufficient in smaller birds. It is held around the girth from

behind or above its back. When under control, the crook should be disengaged. The crook method can be rough on the bird if it panics.

Male ostriches can be dangerous, especially during the breeding season. In Africa, people working with ostriches sometimes arm themselves with a long thorn branch with a good stock of thorns. This can lead to hide damage. Sticks are often used, or a set of bicycle handlebars attached to a pole and arranged like an extended pair of cattle horns. The handlebar device or a long-handled Y-shaped stick can be helpful in turning the bird's head away when in a dangerous situation.

More than one person may be needed when catching and handling individual birds. Ostriches are not always easy to steer in a particular direction, and so two or more people should be available. If there is plenty of assistance, an unhooded bird can be moved with a person alongside each rump guiding the bird from behind, whilst a third person grasps the lower beak with the thumb in the mouth and draws the head down. It can be led forwards whilst in this position.

On some farms emus are caught whilst on the run. The catcher is positioned close to a fence line and the birds are driven by another person between the catcher and the fence. When the required bird is recognized, the catcher runs alongside, grasps it around the body and slows down. Emus have long sharp claws and can use them to strike out when restrained. It is best to approach and to hold a penned bird from the rear, and to restrain the bird from across the back with an arm over the far-side wing.

Emus do not usually respond well to being hooded, and this procedure is not generally recommended. It can be a useful practice in ostriches, as it helps to quieten the bird. It is important to maintain hand contact to placate the bird whilst it is hooded, otherwise it 'gets lost' and walks into fences, which it attempts to climb. Hooding can be used in birds from 6 months of age.

Scratches and cuts acquired early in life can affect the quality of ostrich skin by the time the bird is ready for slaughter (Meyer et al., 2003). Much of this skin damage is due to birds contacting each other with their claws either during handling, through aggression or when they are cold and huddle close to each other. Where **declawing** has been tried in ostriches there was a substantial reduction in skin damage, but they are not usually declawed (Meyer et al., 2002).

In Australia, emus are sometimes declawed to avoid injury to staff and other birds. In other countries radical declawing is disallowed because of the likely pain associated with the procedure, and because neuromas can develop as secondary complications. Radical declawing is normally done in the chicks, and involves partial amputation and cauterization of the distal phalange to remove the toenail and its growth point (Fig. 9.2). A hot-blade chick debeaker is used and all three toes, or just the middle toe, are amputated. The cut is usually sloped through the centre of the second scale behind the claw to minimize the removal of foot pad, and this leaves the sole longer than the dorsal toe surface. In modified (non-radical) declawing, only the claw is removed.

Following radical declawing, the chicks show depressed appetite and reduced live-weight gain for about 2 weeks. Fighting is reduced, and the final quality of the skins is greatly improved. The birds are also easier to handle.

Most of the handling and **transport** injuries that occur in ostriches and emus are inflicted on the legs. The birds have the habit of putting their legs through gaps in pen sides. Partitions in the trucks must be solid, and they are often padded, to help reduce injuries. Leg damage also occurs from other birds' claws.

The birds are caught and loaded individually into the truck. Because of their size and shape, ostriches do not always have good control of their balance. One producer commented, 'Imagine yourself standing in a truck for 3 hours with your hands in your pockets and trying to keep your balance.'

In Africa, ostriches are transported to the abattoirs in open-top trucks, and handlers are distributed

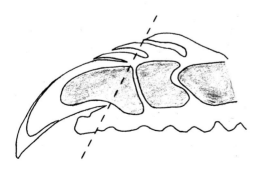

Fig. 9.2. Position of radical declawing cut in an emu chick's foot.

amongst them inside the livestock compartment. The handlers' job is to quickly lift any birds that sit down or go down during the journey. This helps reduce skin damage. The handlers are included at a rate of one man to 12 birds. Head hoods can be used during journeys, but this is not common.

Signs of stress during transport include fluffing out the wings, and hyperventilating with the beak permanently open. When this occurs, the birds are sometimes sprayed with water to cool them down.

When the birds have been unloaded at the abattoir the white wing feathers (white plumes) and the by-locks (darker feathers at the ends of the rows of white plume feathers) are usually removed whilst the birds are in the kraals. This helps avoid damage and contamination of the feathers during stunning and slaughter. The larger feathers are clipped with sheep shears and the smaller feathers are sometimes plucked by hand. In some countries **plucking** all the feathers from the live bird is prohibited by law, while in others live plucking is limited to the hard body feathers, floss and blood (unripe) feathers. When hand-plucking these smaller feathers, it is considered kinder to remove a limited number with each pull. When feathers are removed by cutting, they should be cut in the blood line of the shaft.

Slipping whilst moving birds to the **slaughter** point at the abattoir yards can be a problem. It is not necessary to use hard-restraining methods when stunning birds that are familiar with being handled. For example, weighing can be done single-handed by standing the hooded bird on a platform weigher in a shed that is holding other birds. Similarly, birds can be manoeuvred to the stunning point quite easily without upsetting them unduly. When the birds are presented for stunning, great care is taken in controlling the convulsions because of the risk of skin damage. Captive bolt stunning causes quite severe convulsions and is considered impractical where the skin is destined for the quality leather trade. Various systems are used for holding the bird at stunning. A recently developed method uses a semi-automatic, vertically mounted, walk-in, swivel crush fitted with a leg clamp. The bird's head is positioned inside a box, the lid is closed and dry electrodes within the box deliver a 400 to 800 mA current through the head for 8 to 10 s using 105 V. The recommended stunning current is not less than 500 mA, with a stun-stick interval of 10 s or less (Lambooij et al., 1999). A more common electrical stunning method is to restrain the bird in an upright wedge restrainer and deliver a current through the head using calliper electrodes fitted with metal tumbler electrodes. A leg clamp is often used to control movement and stop the bird falling over backwards once it is stunned. The clamp is applied either before the stunning current or whilst the current is being delivered. Yet another method is to handle the bird into the stunning pen using two men at the bird's rear and one holding the lower beak with the neck horizontal. When the bird is in the stunning pen, a loose chain is looped around the base of the wings. The chain is secured to an overhead hoist and the slack in the chain is taken up. When the bird is under control, a person with calliper stunning tongs approaches from the front and in synchrony with the person holding the beak, electrically stuns the head of the animal. The beak holder has to let go at the appropriate moment, and with experience this works well. As the bird is stunned, its legs go into flexion and it is suspended by the chain without falling to the ground. Pre-slaughter shackling around the wings helps reduce skin damage that would otherwise occur if the bird were convulsing whilst on the floor. Whilst the bird is still rigid, the beak holder cuts its neck at the jaw so that the carotid arteries are severed. A chest stick is usually performed to relieve blood from the carcass once the convulsions have subsided and the carcass has been hoisted by a leg shackle. Ostriches do not have a crop, and so there is little risk of spilling ingesta when doing a chest stick.

Ostrich meat has a dark colour and a springy texture. Darkness is due to its high myoglobin content, and it may be enhanced by its naturally high pH_{ult} (≥ 6.0). When pre-slaughter stress provokes an even higher pH_{ult} value, there is poorer colour stability (Naudé et al., 1979).

Pigeons

In the USA and Australia, young pigeons raised exclusively for the meat market are known as squabs. This differentiates them from older birds, which are known as pigeons and have a reputation for being tough. Pigeon meat is often marinaded for 2 days when attempting to overcome its toughness.

Specialized lines are used in **intensive squab production** (e.g. Canadian King × Carneaux). They have been selected for white skin, growth rate and breast meat yield. The flocks are usually managed as small enterprises with between 500 and 1500 breeding pairs. Ideally, there are pens of 20 to 30 breeding pairs equipped with perches, a bathing facility, nest boxes and nesting material. The pens used for raising replacements and 'weaners' are not usually equipped with nest boxes. Partitions between nests, or pairs of nests, can help control fighting. Water must be in open-top drinkers, but not allow head access if bathing is to be avoided. Nipple drinkers are not suitable as pigeons normally immerse their beaks in water and suck, when they drink. A breeding pair can be expected to produce between 10 and 16 marketable squabs a year.

Some people argue that pigeons are not suited to intensive production. They are monogamous, the young are altricial and they have considerable space requirement for flight. Instead, **semi-feral production** systems are preferred, and, in parts of Asia and Mediterranean countries, they are the main method. An example of a semi-feral system is as follows. The birds are provided with a sheltered night-time roost and a morning feed. During the day they scavenge food from the surrounding areas, including communal rubbish tips. There is no rigid ownership of the birds, and there is some interchange between flocks. Sometimes birds are lured between flocks, and there are a range of feeding and watering strategies that try to counteract this. The birds are cropped at night-time by catching them from their perches, or the squabs are cropped at about 4 weeks of age just before they are able to fly.

The housing systems include wood and wire cages, clay pots suspended from rafters, dovecotes and attics with perches for semi-feral flocks. Predation can be a problem in some housing systems, but chick mortality is lower than for other avian species kept under subsistence conditions. In the Sahel, pigeons can produce 15 eggs a year, with two eggs per clutch. Hatchability is typically about 68%, and daily growth rate in the chicks to 3 weeks of age is about 14 g.

Guinea pigs

Guinea pigs were originally domesticated in the Andes, and this is where they are used most for producing meat. They are also grown for meat consumption in some urban areas of sub-Saharan

Africa, where they are known as 'the white man's rat' and the 'rat of Egypt'.

Guinea pigs are kept in a variety of household situations. Sometimes they are in a spare room, or under the bed, in purpose-built sheds or a corner of the yard. Concrete floors are advisable as this prevents burrowing and makes cleaning easier. They scavenge for feed and are supplemented with household scraps, vegetables, harvest waste and forages. Partly because of the poor nature of their diet, feed conversion efficiency is not high (about 10 : 1). In wet regions in West Africa they are left to find their own water. They do not always grow well in close confinement, and pneumonia can be a problem during the rainy season.

Litter size is between one and four pups. Still-births usually account for more than half the pre-weaning mortality. The young are prone to hypothermia in wet conditions, but mortality is unusual once they start finding solid feed. They can be weaned from 2 weeks of age. First kidding may be between 6 and 22 weeks of age. Adult males and females can be kept together continuously, and mating is often uncontrolled and opportunistic except where first parturition is intentionally delayed. Early breeding is associated with small subsequent litter size and stunting in the doe. Normally the adults weigh up to 1 kg, and offspring are typically slaughtered when they reach about 450 g. These features allow low-effort production in urban areas.

In West Africa, a typical household keeps about seven animals, but in Peru some specialist producers who supply restaurants maintain as many as 200. In Cameroon, the offtake of guinea pigs for home consumption, for sale and as gifts are 40%, 55% and 5%, respectively (Manjeli *et al.*, 1998).

Dogs

Dog meat is eaten in southern China, Korea, the northern Philippines, Cambodia, Laos and Nigeria. The meat is often considered a delicacy, it is expensive and it is a prestigious meat. For example, it may be served in special dishes at weddings, christenings and other celebrations or commemorative feasts. In other communities eating dogs is regarded with horror because the animals have pet status. Open criticism of dog consumption has led to slaughter being performed in secrecy.

Dogs are not usually grown specifically for slaughter. More often they are consumed as road-kill or they are sold for meat at the end of their useful life. In Nigeria, where there is a large dog meat trade, it is eaten on a regular basis, and at festivals, and it is mainly old, sick, fractious or stolen animals that are sold for slaughter. Semi-feral dogs are periodically rounded up as a community service and then used for meat, and sometimes surplus puppies are passed on to dog meat traders (Anon., 1992). Feral dogs come from most parts of the country (Fig. 9.3). They are sold to traders, who walk them to the major dog markets at Bedi, Kagoro, Dawaki and Zing. They are then transported in trucks to the cities and larger towns. In some parts this trade is conducted discreetly so as not to offend the community, and the dogs are transported in confined and unhygienic conditions. In the Ogun region, a small proportion of dogs are used for sacrificial purposes and are not used for meat consumption. Ogun is the god of iron and worshippers appease the god by killing a dog with iron, which is usually in the form of a knife, but, more recently, running a dog over with a car has been gaining popularity.

The dog meat trade in the Philippines is somewhat different. In some rural areas dogs are raised solely for human consumption. Dog meat consumption is accepted as a cultural and religious right, and it is consumed as a fried snack food (*pulutan*) while drinking beer in bars, as are the testes. In some rural areas it is customary to beat the animal just before it is killed. One approach is to hit the dog with a cane whilst it is tied or suspended. The purpose is to inflict bruising and allow the blood to coagulate. This results in blood retention after the carcass is dressed, and is considered desirable. It also pulverizes the muscle, making the meat more tender. Another approach is to beat the dog to death with a heavier stick. These practices are defended on the basis that it is part of the culture of an ethnic group, and no one has the right to deny those people their right even though others may find the practices unacceptable. Chickens are sometimes killed by beating, in the same way, and similar practices were used in the past in many other countries. For example, in the UK some pigs destined for bacon production were whipped before slaughter to improve tenderness, and in Italy cattle were chased through the streets by mounted drovers equipped with goads for the same purpose (Anon., 1910).

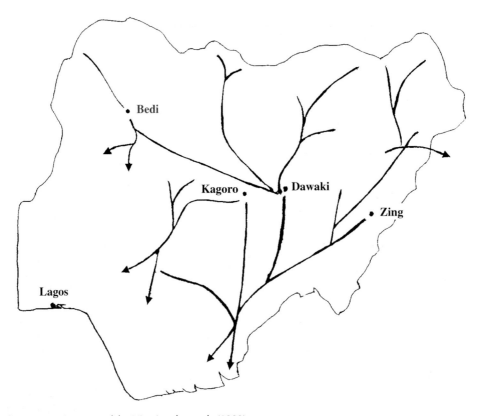

Fig. 9.3.   Main routes of the Nigerian dog trade (1992).

Bushmeat

In some communities bushmeat is preferred to meats from domesticated animals. In the Congo basin, bushmeat contributes between 30 and 80% of the animal protein consumed by forest-dwellers. It is hunted to provide meat for the family as well as income from selling the carcasses. The removal of forest game for meat consumption is now recognized as a conservation and animal welfare issue. It is of particular concern in Central Africa and this summary focuses on that region.

Bushmeat costs nothing, other than the price of a cartridge and the time taken to find and shoot an animal. Large-bodied species are usually preferred, so, for example, red duiker are depleted before blue duiker. The main factor influencing the offtake rate is availability and supply of the preferred species. The two most important Central African bushmeat animals are the duikers (*Cephalophus* species) and porcupine (*Atherurus africanus*), and the depletion of guenon (*Cercopithecus*

solatus) is the one causing most concern presently from a conservation perspective. However, many other species are caught or shot. In a recent study of animals sold through Mundoasi market in Equatorial Guinea, 4328 wild animal carcasses from 48 species were examined during a 2-month period (Table 9.4; Puit *et al.*, 2004).

The majority of these animals had been shot or snared. Snares are set for duiker and porcupine, whilst the hunter stalks arboreal game with a gun. In this situation the snares are checked about once a week (Fa *et al.*, 1995). Snares can cause limb injuries in duikers that manage to escape, and they are wasteful because snared animals are often eaten by scavengers or the carcasses spoil before the snares are inspected. According to one estimate, up to 39% of trapped or snared animals either rot or are lost to scavengers. The frequency of checking traps and snares depends on their distance from the villages, and on the season. Snares set near villages are checked quite frequently and wastage is correspondingly lower. Distant traps may be left

Table 9.4. Main species sold through a bushmeat market in Equatorial Guinea in 2003.

	Number	%
Cephalophus monticola	1273	29.4
Atherurus africanus	717	16.6
Cercopithecus cephus	461	10.7
Cercopithecus nictitans	348	8.0
Cephalophus dorsalis	255	5.9
Cercopithecus pogonias	160	3.7
Colobus satanas anthacinus	117	2.7
Kinyxis erosa	114	2.6

uninspected for 3 months during the dry season, when hunters become distracted by work on field crops. In the forests of Kenya, duiker, baboons and monkeys are hunted with dogs and shot with bows. The animal welfare hazards in bow hunting are much greater than for rifle hunting (Gregory, 2005b).

Net hunting is used in the Central African Republic. It requires the cooperation of a large number of people, and they are deployed in a number of ways. In one system, the nets, which can be up to 100 m long, are erected in a semicircle, and the animals are driven into them from up to 1 km away (Noss, 1997). Alternatively, a net circle may be erected around a patch of woodland or brush. The animals are flushed into the nets, seized, killed and then butchered where they are caught. In most systems the nets are set up many times through the day, whilst moving over an area. There may be up to 30 casts in a day. Net hunting is favoured for small ungulates such as blue duiker (*Cephalophus monticola*), which are not strong enough to break the net.

Logging has led to an increase in bushmeat offtake in Central Africa. New logging concessions open up remote forest regions when roadways are constructed. The roads allow easier access to the game and rapid removal to urban markets. In addition, the logging staff place a demand for local bushmeat.

The guenon is the most recently discovered primate species, and it has the smallest distribution range, being restricted to the forests of central Gabon. It is an occasional by-catch in snares set for duikers. Dead monkeys are found occasionally bearing signs of snares on wrists and ankles.

It is not known whether the demand for bushmeat is price-elastic. This is important as it will determine whether hunting will escalate or contract in future. There are two counteracting issues. First, we do not know whether forest communities would revert to meats produced on farms if they were available. The attraction of farmed meats is that they are easier to procure. Secondly, if the price of bushmeat rose, either through scarcity or offtake constraints, the higher individual value of a catch might provide a stronger incentive for hunters to intensify their hunting. Experience has shown that, where bushmeat has been depleted to unsustainable levels, protein deficiency can occur where the community relies on groundnuts and squash seed as the main alternative sources of protein. Other communities, which reverted to pig, rabbit and cane rat production as the main sources of animal protein, have fared better.

10

Livestock Presentation and Welfare before Slaughter

This chapter considers the transfer of animals from farms to the point where they are stunned at the abattoir. It focuses on:

- meat safety issues
- microorganisms of public health importance
- stock cleanliness
- injuries and bruising
- skin damage
- animal health issues
- drug residues
- transport stress
- animals dead on arrival (DOAs)
- holding animals before slaughter
- casualty animals
- ease of handling
- noise in abattoirs

Meat Safety Issues

In the past, the main effort in controlling meat safety hazards has been directed at improving dressing and meat handling standards. More recently, the focus has shifted towards reducing cross-contamination between live animals and improving live animal cleanliness. When animals are transferred from farm to abattoir some will inevitably acquire faecal soiling. A balance has to be struck between controlling this contamination and not compromising animal welfare. In practice, there are three ways in which managing the meat safety hazards can compromise welfare. They are:

- hunger from fasting the animal to empty its gut before slaughter
- stress involved in cleaning the animal
- minimizing the resting period at the abattoir to control the risk of contamination and infection

Pre-slaughter fasting reduces gastrointestinal tract contents and the likelihood of spillage from the crop, reticulum and rectum during dressing. It also helps minimize defecation during transport and yarding, and this reduces the risk of faecal soiling of other animals. The fasting period that provides these benefits varies between species. In ruminants and pigs it is generally assumed that a 12- to 16-h fast before slaughter should balance the meat safety and animal welfare hazards, but this estimate is not supported by precise evidence. In broilers, it takes about 8 h feed deprivation before defecation frequency is reduced to that of the completely fasted state. Only 4 h is needed if water is withheld as well, and in practice this depends on the stage at which the drinkers are raised before the catching gang moves into the shed (Warriss *et al.*, 2004).

It is usually recommended that cattle, sheep and pigs should be rested after they arrive at an abattoir and before they are put up for slaughter. The aim is to allow excited animals to settle and then they will be easier to handle. It introduces risks of faecal contamination by prolonging the exposure period. The resting period is not necessary in animals that are not aroused or excited.

©N. Gregory 2007. *Animal Welfare and Meat Production*
(N. Gregory)

The optimum fasting and resting periods from the meat safety perspective depend on the particular hazard that needs to be controlled. This depends on the microorganism that poses the greatest threat to public health.

Microorganisms of Public Health Importance

The meat-borne microorganisms that are of greatest concern in public health are shown in Table 10.1. This table describes the prevalence of contamination of the final product. It does not, however, equate to the number of cases of food poisoning. In the USA, the laboratory-confirmed cases of *E. coli* O157 infection, salmonellosis and campylobacteriosis are in the ratio of 1 : 9 : 7, respectively. Taken together this indicates that the risks associated with beef are low, but risks associated with *E. coli* contamination are the primary concern in cattle. The greatest overall risks are with campylobacteriosis and salmonellosis from eating chicken.

In cattle and sheep, the holding pens at farms, markets and abattoirs are some of the main

sources of cross-contamination. Not surprisingly, pen floors have been shown to be the single most important source of *E. coli* O157 in cattle lairages, whereas in sheep lairages it is the unloading ramps (Table 10.2). *Campylobacter* are more widely distributed in sheep lairages.

Campylobacter

Undercooked poultry meat is one of the main sources of *Campylobacter jejuni*, but the size of the risk varies considerably between flocks and countries. For example, in European countries the prevalence of *Campylobacter*-colonized flocks varies between 18 and 90%. Some vertically integrated broiler companies are monitoring their flocks for *Campylobacter* through faecal swabs taken at farms, in an attempt to understand the causes of the variation and the routes of contamination. The findings so far are as follows.

It is suspected that a poultry shed becomes contaminated with *Campylobacter* through the catching gangs and equipment used when the shed is depopulated. This may not be an important risk

Table 10.1. Prevalence of *Campylobacter*, *E. coli* and *Salmonella* in retail raw meat in the USA.

Meat	Number of samples	% of samples positive for		
		Campylobacter	*E. coli*	*Salmonella*
Chicken	212	71	39	4
Turkey	194	15	12	3
Pork	209	2	16	3
Beef	210	1	19	2

Table 10.2. Severity of contamination with *E. coli* O157, *Salmonella* and *Campylobacter* in UK cattle and sheep lairages.

	E. coli O157		*Salmonella*		*Campylobacter*	
	Cattle	Sheep	Cattle	Sheep	Cattle	Sheep
Unloading ramp		++				+
Corridor						+
Holding pen floor	+++					++
Water trough			+	+		+
Stunning pen entry gate	+				+	
Stunning pen floor			+			

if the shed is emptied and then sanitized. However, this is not the case when a flock is thinned. Instead, the retained birds can become infected from bacteria left behind following an earlier thinning, and so the pathogen becomes established in previously uninfected birds (Newell and Fearnley, 2003).

When a shed is contaminated with *Campylobacter*, chicks acquire the bacteria during the first 7 days following placement in the brooders. By day 14, most birds are infected (Wallace *et al.*, 1998). On acquiring the infection, some birds have a short bout of diarrhoea, but otherwise the infection is uneventful. In broilers the counts of *Campylobacter* in faeces are often highest at the time they reach slaughter size, whereas in slaughter-weight turkeys this peak has passed and the prevalence of excreters is relatively low.

Campylobacter thrive under moist conditions, and are commonly found in water sources, including drinking equipment. There are two views about how this should be managed. Some take the view that troughs and drinkers should be cleaned regularly, and the water should have no visible growths or contaminants. Others take the view that allowing protozoa to survive in tanks or trough water helps control *Campylobacter* and *E. coli*. The approach is to remove visible contaminants, especially manure and sediment, which can harbour bacteria, as and when they are noticed, but to otherwise empty and clean the water trough or tanks infrequently. The risk when encouraging a protozoa culture is that species such as *Tetrahymena pyriformis* engulf live *Campylobacter* and act as a reservoir for the live bacteria in the broiler shed water system (Newell and Fearnley, 2003). In addition, co-cultivation of *Campylobacter jejuni* with such protozoa appears to reduce the susceptibility of the bacteria to chlorine and some disinfectants. This increases the viability of the *Campylobacter*, and it indicates that allowing protozoa to accumulate is a risky strategy.

When birds are fasted before they are collected for slaughter, they become hungry and start eating the litter. This increases the risk of acquiring *Campylobacter* in their crops (Byrd *et al.*, 1998). Crops are often ruptured during processing, because of their attachment to the neck flap, and this allows crop fluid to spill on to the carcass. In this way, withholding feed for 12 h before slaughter has been associated with greater contamination of broiler carcasses with *Campylobacter* (Northcutt *et al.*, 2003).

When feed has been withheld for as little as 6 h before slaughter, there has been an increase in the proportion of broilers carrying *Campylobacter jejuni* in caecal plus rectal contents (Willis *et al.*, 1996). In turkeys and broilers, transport to the processing plant can increase the prevalence of *Campylobacter*-excreting birds, and providing a rest period before slaughter does not bring the counts of *Campylobacter* down (Whyte *et al.*, 2001; Wesley *et al.*, 2005).

If there is large variation in bird size when a shed is depopulated, there is a greater risk of carcass contamination with *Campylobacter*. This is because size variation increases the risk of rupture of the gastrointestinal tract and release of gut contents on to the carcass during processing. This has been seen, for example, in flocks affected with air-sacculitis (Russell, 2003). The diseased birds were smaller, and their carcasses were more prone to machine damage.

In free-range or organic systems where broilers have access to soil, there is a higher risk of contamination and infection with *Campylobacter* (Rivoal *et al.*, 2005). Some processors are reluctant to slaughter organic broilers because of this risk.

A population of *Campylobacter* can become quickly established in poultry and pigs. When a pig ingests *Campylobacter*, the caecum is the first region of the gut to be colonized with large numbers of these bacteria. Fluid containing the *Campylobacter* reaches the caecum in as little as 2 h. Finishing pigs and sows can also become infected with *Salmonella* within this period when held in a pen at an abattoir or market.

Salmonella

In developed countries, farm animals that produce meat and eggs are the main reservoir of zoonotic *Salmonella*, whereas, in developing countries, contaminated vegetables, drinking water and human-to-human transmission are more important.

In Europe, pigs and poultry are common sources of *Salmonella*. A recent study in the Netherlands showed that 47% of pigs carry *Salmonella* when they are slaughtered. Infection can be tested by screening the tonsils, lymph nodes and rectal contents for this microorganism, and swabbing the surface of the liver and carcass gives information about contamination during carcass dressing. When pigs inhale or ingest *Salmonella*, the gastrointestinal tract and lymph nodes are quickly colonized.

Salmonella are taken up by the tonsils, and they have been recovered from faeces within 2 h (Hurd *et al.*, 2001). This implies that it is relatively easy for pigs to become infected with *Salmonella* during the normal course of events between the farm and the slaughter point. It has also been reported that the number of pigs excreting *Salmonella* can double within a 6 h period of leaving the farm, and it was suggested that half of these cases were new infections (Berends *et al.*, 1996). Sources of cross-contamination include the transport vehicles, lairage pens and direct contact with other pigs. Cross-contamination in lairage is probably a particularly important route. Drinking equipment can mediate cross-contamination (see Table 10.2), and, in a study of pig abattoirs, one-third of the troughs were found to be contaminated with *Salmonella* (Rostagno *et al.*, 2003). Once a pig is infected, the bacteria can become quickly established in lymph nodes. Mandibular lymph nodes can be infected within 3 h and ileocaecal lymph nodes within 6 h of ingesting the *Salmonella*.

The aim should be to get the pigs to the point of slaughter as quickly as possible, to minimize the opportunity for colonization in otherwise uninfected pigs. This is not consistent with recommendations given for resting pigs on arrival at the abattoir in order to make handling easier and to minimize the risk of PSE meat. In addition, holding pigs overnight at the abattoir inevitably increases the risk of *Salmonella* infection. Stress can increase the shedding of *Salmonella*, and so minimizing pre-slaughter stress should theoretically reduce contamination of the transport vehicle and lairage. Simple hygiene and sanitizing measures cannot be relied on as effective ways of eliminating this risk once contamination has occurred (Swanenburg *et al.*, 2001).

Most of the meat from breeding sows is used for making manufactured products such as sausages, and so there is less emphasis on managing the holding period at the abattoir to optimize meat quality. Instead, in the USA, sows are often held overnight at the abattoir. This increases the risk of dark, firm, dry (DFD) meat and presumably increases the risk of infection with *Salmonella*. It is suspected that it contributes to the higher prevalence of *Salmonella enterica* seen in sows (Larsen *et al.*, 2004).

The Danish pig industry has put substantial effort into minimizing cross-contamination with *Salmonella*. Farms are classified according to their prevalence of *Salmonella*, into low-, medium- and high-prevalence categories. Pigs from high-prevalence farms are managed separately. They have separate transport, lairage and slaughter from those in the other categories. *Salmonella* status is assessed from the presence of *Salmonella* antibodies in drip collected from pork after freezing and thawing. The medium- and high-risk category farms receive 2 and 4% less money for their pigs. This scheme has decreased acute infections in pig herds, presumably by reducing the exposure to below the minimum dose needed to cause acute *Salmonella* infection, but there are residual medium- and high-risk category farms where it is proving difficult to improve their status. A similar scheme has recently been introduced for broiler farms (Wegener *et al.*, 2003).

The optimum pH for the growth of *Salmonella* is between 6.5 and 7.5. In the broiler's crop, the pH is usually about 3.6, and a resident population of lactobacilli helps control the *Salmonella*. When the birds are about to be slaughtered the pH rises during the first 8 h of fasting, and *Salmonella* can become established. The crop is contaminated with *Salmonella* from eating litter in the same way as for *Campylobacter*. Contamination of broiler caeca is, however, less common (Corrier *et al.*, 1999), but pre-slaughter stress increases the susceptibility to caecal colonization with *Salmonella*. This might be due to disturbance of the normal flora of the intestinal tract. Including a yeast (*Saccharomyces boulardii*) in the finisher ration has reduced the number of caecal carriers.

In a study in Australia, the prevalence of cattle coming from feedlots that were excreting *Salmonella* was similar to that of cattle coming directly from pasture. The overall prevalence was 7% (Fegan *et al.*, 2004). As with monogastric species, transport stress promotes the shedding of *Salmonella*, leading to contamination of the trucks and cross-contamination of animals (Barham *et al.*, 2002).

E. coli O157

The importance of *E. coli* O157:H7 to the image of the meat industry is highlighted by the following story of a 10-year-old girl, Brianne Kiner, who contracted haemolytic uraemic syndrome (HUS).

One day Brianne visited a fast-food restaurant and she asked for a hamburger, which turned out to be undercooked but she nevertheless ate it. Within a few days, she developed a fever and abdominal cramps, turned pale and started to void blood from her bowels. Her urine was heavily stained with blood, and for the next 80 hours she experienced severe pain every 10 to 12 minutes from abdominal cramping. She fell into a coma and had to be placed on a ventilator. At one stage her heart stopped beating, and the doctors were obliged to open her abdomen from sternum to pelvis to allow her swollen organs to expand. Her intestines were swollen to three times their normal size. Ongoing monitoring showed that there was damage to liver and pancreatic function, and an insulin pump was instated. Blood clots developed in her eyes, and her electroencephalogram (EEG) revealed episodes of epileptiform activity. She remained in a coma for 40 days, and was hospitalized for 167 days. She is now disabled by a heart disorder, brain damage and diabetes, plus lung and liver damage.

This is the type of story that makes fast-food companies want to impose strict health and hygiene standards on their meat suppliers. Let us consider the following fictitious problem, to decide how the risks could be managed.

Suppose that *E. coli* O157:H7 is identified within country X in some beef originating from country Y. It is identified without causing any health problems to consumers but the incident is now threatening the future of importations from country Y. Using a trace-back system, the property where the cattle originated is identified. The cattle were raised on pasture and finished in a small feedlot on adjoining ground, and as far as we know the bacteria are limited to that region. The next stage, which is more difficult, is to decide how the microorganism should be managed or controlled. Clearly it would be unrealistic to decontaminate the property other than by destocking. Not surprisingly, the farmer is not prepared to destock, and no organization is stepping forward with an offer to buy the farmer out with a view to destocking. To help maintain market access, the abattoir company is considering introducing a zero tolerance policy for all *E. coli*, which would be monitored on each slaughter batch of beef carcasses. In addition, the company wants to introduce a standard operating procedure (SOP) that reduces soiling and the burden of surface and enteric *E. coli* in all cattle entering its abattoirs. Implementation of the SOP would be audited by livestock agents. The challenge is in knowing which standards should be in the SOP, and how to manage the repercussions on cattle welfare.

E. coli O157 is now considered ubiquitous in many beef-producing regions of developed countries. Release of the organism in faeces from animals is episodic. The mechanisms leading to release are not understood, but, when animals are slaughtered during one of these bouts, there is greater risk of cross-contamination of other animals. There are conflicting reports on the scale of this risk. One study showed that there was no increase in the number of cattle shedding *E. coli* O157 in their faeces during transport and lairage, whereas another study showed an increase from 18% to 43% during transport (Barham *et al.*, 2002; Minihan *et al.*, 2003). In the study where there was an increase, the proportion of beef heifers carrying *E. coli* O157 on their hides rose from 6% to 87%. More often, the number of animals carrying *E. coli* O157 on their hides is lower than this. A survey of 12 beef plants in the USA showed that 3.6% of the hides and 0.4% of carcasses were carrying *E. coli* O157:H7, whilst a study in the UK found that 33% of cattle hides were contaminated with *E. coli* O157 (Avery *et al.*, 2002). The US plant with the highest level of contamination had 19% of its hides identified as O157-positive. Much of the carcass contamination with *E. coli* O157:H7 occurred before the carcass was eviscerated (Barkocy-Gallagher *et al.*, 2001). Taken together, these findings indicate that a primary focus for reducing contamination should be on improving skin cleanliness and hide removal. Skin contamination occurs when cattle sit on the ground or inside the truck, and especially in heavily contaminated manure.

One abattoir company reacted to the *E. coli* O157 threat by introducing a one-in-ten tolerance level for carcasses carrying generic *E. coli*. This changed the emphasis away from controlling *E. coli* O157 specifically towards minimizing all contamination. Part of the focus was directed at reducing the entry of bacteria into the abattoir by manipulating the feed during the final days before dispatch from the farm. The feeding regime affects *E. coli* numbers in faeces in two ways. First, it influences rumen pH, and, if this falls outside the favoured range for *E. coli*, their proliferation will decline. Secondly, if there is undigested carbohydrate in large-intestine contents, *E. coli* numbers can increase (Gilbert *et al.*, 2005). One of the worst things, in terms of allowing *E. coli* (and *Salmonella*) proliferation, is to take the animal off feed (Gregory *et al.*, 2000). *E. coli* in faeces increased by 1.6 $\log_{10}$/g and the total burden in the alimentary tract increased by 1.7 $\log_{10}$/g when cattle were fasted for 24 h before transport to an abattoir.

Finishing feeds have been used to manage the flora and number of *E. coli* in faeces before slaughter. Feeding cattle hay for 48 h before transport reduced *E. coli* numbers by about 3 $\log_{10}$/g faeces in comparison with withdrawing feed, which was a sizeable reduction. Similarly, counts of *E. coli* in rumen contents were reduced by 3.8 $\log_{10}$/g. The type of hay that is provided could be important. For example, lucerne hay has a high buffering capacity, producing a high rumen pH, which favours *E. coli* growth (Gregory, 2002).

It is unrealistic to attempt to eradicate *E. coli* O157 from farms and feedlots. In regions where there are high levels of hide contamination with *E. coli* O157, the temptation is to clean and disinfect the body surface of the animals shortly before slaughter to minimize the risk of transfer to the carcass. The welfare issues associated with pre-slaughter washing are discussed in the next section. There may also be some benefit in finishing beef on an all-in–all-out basis, with resting or cleaning of the premises between batches. There should, of course, be effective cleaning of the transport vehicles between batches. It has been suggested that, because calves are more likely to carry *E. coli* O157, they should be last up for slaughter during the day. This would not be consistent with welfare requirements for bobby calves that are held overnight in the lairage, as they would be off feed for an extended period.

In Dutch veal calves, *E. coli* O157 is more common in pink veal than white veal herds (Schouten *et al.*, 2005). Pink veal calves are grown to heavier weights, and they receive solid feed, which can encourage *E. coli* O157. There is no difference in the use of oxytetracycline in feed between the two types of veal, and so antibiotic use has not been implicated.

Like *Campylobacter*, *E. coli* O157 can survive in sediments in water troughs, and dirty water is thought to aid survival during cold conditions. Replication of the microorganism in the water inevitably narrows the difference between the infective dose and the prevailing concentration in the water.

Dags and Cleanliness

In the early 1990s in the USA, it was estimated that 38% of cattle had obvious dirt on the hide where the opening cuts were due to be made (Boleman *et al.*, 1995; Table 10.3). The *E. coli* O157 food scare was a major incentive for introducing clean stock policies. These policies penalize farmers or dealers for presenting dirty animals and this has reduced the prevalence of dirty animals in many countries.

Cattle that are penned in abattoir yards at high stocking densities are likely to be particularly dirty, as are cattle that perform a lot of mounting behaviour. Dirtiness is also greater when stock have been grazing lush spring pasture. At this time of year the faeces are loose, have a low dry-matter content (sometimes as low as 7%), and poor colloidal gelation. Emotional stress and physical activity during transport increase the dispersal of this type of **runny faeces**. This stress response is mediated by the release of corticotrophin-releasing factor in the brain (Gue *et al.*, 1991).

There is a good understanding of the features that influence stock dirtiness. Cattle fed silage are prone to getting dirty during transport because of their softer faeces and long-haired animals accumulate dirt more readily than cattle with short coats. In some countries there is a requirement to provide bedding during transport, for welfare reasons, but there is little evidence that this influences cleanliness.

In some high-rainfall countries, when transport vehicles are not bedded, runny faeces are a public nuisance on roadways. When cattle are transported to markets and abattoirs in the rain, a spray of slurry is blown from the floor of the stock compartment on to the road and other motorists. This is being managed by installing sumps below the stock compartment in the vehicle to collect and hold the slurry. The sumps are unloaded at abattoirs or central depots.

Where cattle are sold on a live-weight basis, surface soiling can introduce sizeable errors in estimating carcass weight. In a study in the UK, hides classed as dirty carried on average 3.7 kg of dung.

Table 10.3. Prevalence of dirty cattle at 27 US plants in the early 1990s.

Visible dirt	Prevalence (%)
None	62
On legs	19
On legs, belly and side	14
On legs, belly and back	5

On the dirtiest hide, there was 20 kg of dung (Auer *et al.*, 1999).

Genetic selection against dags could be used to reduce dirtiness problems in sheep. The h^2 for dagginess in sheep is 0.13 to 0.60. We do not know the mechanisms responsible for this inheritance, but they may be linked to one of the following features. When faeces are shed, they usually roll off breech wool without leaving a dag residue. However, caecal contents and faeces with a high moisture content are more likely to leave a residue attached to the wool, especially if the faeces and wool are pressed together, as would happen when a sheep sits in a patch of dung. Once the faeces starts to stick to wool, the process accelerates because faeces cohere well. Surprisingly, the serum in faeces from lambs with heavy worm burdens does not increase the adherence of faeces to wool (Waghorn *et al.*, 1999). Instead, it is probably the plastic properties of high-moisture-content faeces that allow them to flow around and encase wool fibres and allow a dag to form. Lambs that experienced a period of watery diarrhoea neonatally are more likely to be daggy later in life, even though they overcame their episode of diarrhoea (French and Morgan, 1996). This could be linked to lamb size, as low-birthweight lambs are more likely to be dag-bearers, perhaps because they have reduced colostrum intake during the critical early period.

The evidence that **pre-slaughter shearing** or crutching improves the microbial quality of lamb carcasses is conflicting. Some studies have shown a benefit, whilst others have reported no effect (Vipond *et al.*, 1992; Biss and Hathaway, 1995). Similarly, there have been conflicting reports on the effects of live animal cleanliness and dag removal on microbial quality in cattle carcasses (Ridell and Korkeala, 1993; Van Donkersgoed *et al.*, 1997; Rowland *et al.*, 2000). During dag removal, contaminated wool and hair are clipped from the breech and/or belly with shears. This is an additional stress for the sheep, and it is not a preferred pastime amongst yard staff. Abattoirs levy a charge for this service, which acts as a disincentive for presenting stock in a dirty condition. Sheep handshears have been used for removing matted hair from cattle and this causes skin injuries.

Washing procedures are used mainly in large abattoirs in warm climates. The aim is to remove readily dispersible manure, mud and dust. The methods include:

- hosing cattle and pigs
- immersion washing
  - plunge washing using swim wash tanks for sheep
  - soak tanks where cattle stand in water to soften leg and belly dirt
- spray-washing sheep and cattle
  - walk-through spray races for cattle and sheep
  - spray-washing pens for pigs, sheep and cattle

Repeated or prolonged washing is usually needed to soften and disperse the dirt, and this adds to water and effluent disposal costs. These costs are reduced by reusing dirty water for the initial washes, and giving a final rinse with potable water. In the case of cattle this is done with a hose as they move up the race to the stunning pen. Depending on how it is applied, this rinse can either interfere with stock movement, for example if there is erratic escape behaviour, or it can act as a convenient forcing method in getting them to the stunning pen.

Some washing methods are overzealous. For example, some spray-washing races have been poorly designed, and the animals have to be bullied to get them through. When there is repeated washing, poorly designed systems are evident from the increasing reluctance of the animals to enter the race. Some spray washes have included quaternary ammonium compounds, which are irritating to the eyes. Warm-water showers are more effective at removing dirt than scrubbing with stiff brushes, and the latter can scratch the skin to the extent that it devalues the hide (Auer *et al.*, 1999).

The method used for deciding whether a sheep needs repeated washing is to squeeze the wet fleece and examine the colour of the water in the palm of the hand. Washing is repeated until this test gives a clean sample. Repeated washing with a shower can be sufficiently stressful to affect meat quality. At one abattoir it was found that pH_{ult} in lambs that were not washed was 5.71 (± 0.01 SE). Following a single shower wash it was 5.78 (± 0.03), and after two shower washes it was significantly higher (5.96 ± 0.03). After washing, it is helpful to stock the sheep at low densities in the holding pens to allow air circulation and encourage drying.

It has been claimed that some washing procedures are counterproductive for carcass hygiene. Giving cattle a final rinse as they pass up the race to the slaughter point means that the suspended carcass is dripping and this allows transfer of water to the carcass during leg and hide removal. When pigs are spray- or hose-washed in abattoir pens, they often suck on or drink puddles on the floor. This allows ingestion of microorganisms of public health importance.

Stock cleanliness is equally important in developing countries, but the causes of dirtiness may be different. In these countries, stock that walk to abattoirs become laden with dust in the dry season and are mud-spattered when it is wet. In the rainy season some abattoir yards turn to mud and this creates problems for vehicles as well as trying to keep stock clean.

Transmissible Spongiform Encephalopathies (TSEs)

At one stage, the threat of TSEs from eating beef created concerns about the future of the captive bolt-stunning method. This was because the brain is damaged by the bolt and brain matter can be irrigated by venous circulation from the head to the heart and lungs (Gregory, 2005a). Nevertheless, the proportion of shot cattle with brain matter entering the venous system is quite low (4% following captive bolt and 2% following percussion bolt stunning). The actual risk to meat consumers is difficult to assess, but because of the rapidly falling prevalence of BSE in the cattle population it is now thought to be low. TSE prion-contaminated meat has also been a concern in sheep and goats in the EU, and in deer where chronic wasting disease is prevalent in North America.

Aquaculture

Some of the microorganisms that are of public health importance originate in the animal's feed used at the farm. A particularly high-risk situation is using human sewage as a fish feed in aquaculture. This is practised in several Asian countries, but there is a withholding period before the sewage is fed to the fish, and this should reduce the risk. In India the municipal effluent is usually held for 1 day in an anaerobic pond, followed by 5 days in a facultative pond, before discharge into the fish ponds (Jana and Jana, 2003). A fish farm can produce over 3550 kg fish/ha/year in this way, without the need for any supplementary feed or fertilizer.

The risk of people contracting poliomyelitis either at these farms or from handling the product has not been reported, but in view of the prevalence of polio in India this could be a concern.

Injuries and Bruising

Bruised tissue has to be trimmed from a carcass. This disfigures the carcass, making it less marketable. Sometimes it is difficult to remove all the bruising because it is diffuse, whereas in other cases it is localized and more easily removed (Fig. 10.1).

Bruising is an obvious sign of suffering. Pain comes not only from the blow but also from subsequent longer-term pain at the injured site. In some countries the prevalence of carcass bruising in cattle has been very high. For example, in Nigeria it has been as high as 75% (Antia and Alonge, 1982), and in the USA during the 1990s it varied between 40 and 50% (Lorenzen et al., 1993; Boleman et al., 1995).

Some of the injuries that cause bruising are:

- Collisions with gates, head-bails and entranceways in crushes and raceways used for grading cattle at markets.
- Collisions with gateways leading from sale rings in markets.
- Excessive use of sticks at markets when the rate at which cattle are put up to the sale ring falls behind the selling rate in the ring.
- Impacts during loading on to vehicles. In pigs, shoulder bruising has been reduced by

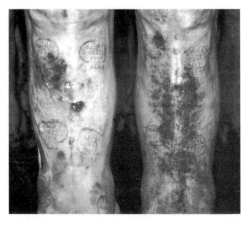

Fig. 10.1.   Two types of bruising in lamb carcasses.

loading and unloading with hydraulic lifts instead of walking them up and down ramps (Faucitano, 1998).

• Impacts in crushes used for reading ear tag numbers before slaughter at abattoirs.

Putting fat cattle through markets leads to bruising and stick marks and some of these injuries could be avoided by selling directly to the abattoir (Weeks *et al.*, 2002; Table 10.4). The greater bruising in market cattle is due partly to extra handling and partly to the longer transport journeys. In the USA, the prevalence and severity of bruising increases with transport distance beyond about 325 km.

Horns are also responsible for additional bruising (Fig. 10.2). Under Australian conditions, it has been estimated that dehorning could prevent a loss of 1.5 kg of trimmed bruised tissue per carcass. Although horned animals are less likely to show head butting, an animal standing close to a wide-horned animal shows signs of discomfort by raising and lowering its head in an attempt to avoid the horns. Withholding feed has been associated with increased levels of carcass bruising in horned steers (Dodt *et al.*, 1979). Fatigued animals

Table 10.4.   Prevalence of commercially significant bruising at an abattoir in the UK.

Source	Number of cattle	% bruised	% stick-marked
Farm	21,638	3.4[a]	0.06[a]
Dealer	20,252	4.7[b]	0.12[b]
Market	7,036	4.6[b]	0.21[c]

Means in the same column without a common superscript letter were significantly different at $P < 0.05$.

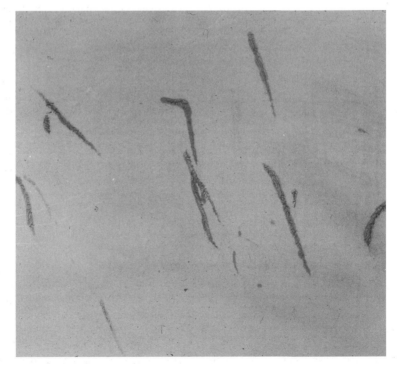

Fig. 10.2.   Horn damage in cured hide.

are more reluctant to move and avoid other cattle, and this probably explains their greater risk of carcass bruising.

In the USA, the Brucellosis Eradication Program has contributed some bruising, and no doubt this also applies to tuberculosis testing and checking ear tag numbers against passport numbers in EU countries (Hoffman *et al.*, 1998). In the US scheme, cattle are tested at their first point of concentration in the marketing chain. These points include livestock markets, buying stations, registered stockyards, assembly points and feedlots. The prevalence and severity of carcass bruising in beef cows subjected to brucellosis testing and passing through an auction market were double those in beef cows sold directly from ranches to abattoirs.

Mature breeding stock are easier to handle and are less likely to panic, but they often acquire more bruising. In a study of 344 cull cow carcasses at three abattoirs in Texas, it was reported that 86% showed bruising (Lorenzen *et al.*, 1994). This was over double the prevalence seen in steers and heifers. Beef cows were more likely to be carrying horns than steers and heifers, and this could explain some of the additional bruising, as well as their lighter finish and greater boniness, which can make bruising appear worse. Their larger size may also be a contributing factor. Bruising was greatest in the loin, followed by the round, ribs and then the chuck. Bruising on the back can be due to animals hitting their backs on low truck ceilings, guillotine gaits descending on the back, trampling or stick injuries at sale yards. Bruising at the hips occurs when the animal strikes gates, gateways and corner posts. Cushions are sometimes used as padding on high-risk protrusions. Drovers are sometimes instructed to confine their hitting with sticks to regions of the body that correspond to lower-priced cuts of meat, but this did not seem to be the case in the study in Texas. In the UK, young bulls are less bruised by the time they are slaughtered than steers and heifers, presumably because less pressure is used when they are being moved (Weeks *et al.*, 2002).

In some countries, dogs are used for herding sheep in abattoir holding areas. At one plant in New Zealand, the prevalence of bruising and bite wounds fell dramatically when dogs were no longer allowed (Table 10.5).

Fasting for 24 and 28 h before long-distance transport has been associated with a higher prevalence of bruising in cattle compared with no fasting.

Table 10.5. Prevalence of wounds and bruises in a sheep abattoir according to dog use in yards.

	Prevalence of wounds and bruises (%)	Dogs used in yards
1994	2.64	Yes
1995	2.36	Yes
1996	1.75	No
1997	1.69	No
1998	1.29	No
1999	1.02	No

Presumably the fasted animals were weaker and less adept at maintaining balance (Dodt *et al.*, 1979). High stocking densities during the journey (1 m² per head) have been associated with a twofold or greater increase in carcass bruising when compared with 2 and 3 m² per head (Tarrant, 1990).

Some injuries arise from **slippery floors**, and the risks depend on floor type. There are four types of flooring used in markets and abattoir yards. In low-cost situations, the yards are earth and compacted stone. This is relatively comfortable to lie on, but it is unhygienic. Slatted metal and slatted concrete floors are used in sheep and cattle abattoirs, respectively. They are more hygienic and provide good grip for animals passing across the slats but poorer grip when turning. There are two types of solid non-slatted concrete floor. Flat solid concrete floors are the most common type, and sliding is prevented by shallow profiling in the surface, achieved either when the concrete is laid or by cutting grooves by machine after it has worn smooth. The profiling is often in one direction only and greatest foot purchase is perpendicular to that direction. The other type is a deeply contoured surface as a diamond pattern (see Fig. 13.5). The intention is to break a foot slide in any direction by providing purchase in a groove. It does not eliminate sliding. It simply arrests a foot slide.

Concrete floors become slippery when wet or covered with manure. For example, a floor with a skid resistance value (SRV) of 85 when dry can be 45 when wet, and for pigs an SRV of less than 60 is the critical safety limit (Applegate *et al.*, 1988). Various materials have been considered for improving the grip of the flooring. Epoxy screeds are being installed at critical points in some abattoirs and are proving to be very effective in improving grip. Crumb rubber or basic slag can increase skid resistance when added to asphalt or concrete mixes. In fresh asphalt–stone aggregate mixes, high proportions of asphalt can reduce skid resistance, but exposure of aggregate as the surface wears will improve grip, until the aggregate becomes polished through further wear. Mats are sometimes used to reduce the risk of slips in cattle and sheep in high-risk areas (Gjestang, 1979).

Some slipping is almost inevitable when animals move over unfamiliar ground within abattoirs. An important issue, however, is what is a tolerable level of slipping. When Temple Grandin inspects an abattoir, she usually scores slipping and falling in 50 animals in the following way:

- excellent – no slipping or falling
- acceptable – slipping in fewer than 3% of the animals
- not acceptable – 1% falling down (body touches the floor)
- serious problem – 5% falling down or 15% or more slipping

In a survey of 29 Canadian abattoirs, it was reported that 27% had excellent non-slip floor surfaces, 52% were acceptable, and 21% had unacceptable slippery floors. In the beef plants, slippery floors were worst in the stunning pens or in high-traffic areas where the rough finish had worn off the concrete. A study of 11 beef abattoirs in the USA showed that the prevalence of falls at two of the plants was 8 and 12%. Here again, the raceway leading to the stunning pen and the stunning pen itself were the high-risk areas.

Old markets tend to have greater problems with slippery floors, and, once an animal experiences a slip, it is more reluctant to move and more likely to be hit with a stick. Sliding gates are sometimes used inappropriately when regulating the flow of animals and may be responsible for some bruising. Casting in the raceway and weighbridge leading up to the sale ring is also an occasional problem. A common design problem with raceways is the gap between the floor and the bottom rail of the race. When an animal goes down, its leg is prone to getting lodged under the rail, and it is difficult to release the leg to allow the animal to stand up. Some injuries are inevitable. Others can be managed to some degree by improving the facilities or encouraging more appropriate handling methods.

Skin damage

Skin damage from an injury is inevitably painful for the animal. It can lead to infections and parasite infestations, which can also be irritating and painful. Some common forms of skin injury and skin disease that are important from a welfare perspective are listed in Table 10.6. All these conditions contribute to hide or pelt downgrading. In Africa, ticks are the main cause of skin damage in cattle, but in one study, scratches explained 28% of skin damage and brands a further 24% (Mersie and Bekele, 1994). Bruising was also evident in 27% of the hides.

No one system of livestock production produces superior hides or skins. Skins from organic beef are no better or worse than those from feedlot beef. Organically grown beef may be older at slaughter and the hide stronger, but it can also have scratch marks if the animals were raised in bush country. Free-range pigs sometimes have more skin damage than indoor pigs (Lambooij et al., 2004). Pigs reared in an unenriched environment are more likely to be aggressive when confronted with an unfamiliar pig in the lairage and so fighting injuries may be greater (de Jong et al., 2000).

In a survey of five EU countries during the 1990s, the prevalence of moderate to severe skin blemishes in slaughter pigs was 10% (Warriss et al., 1998a). The main causes of skin injury were fighting damage and impacts during transport and handling. Pigs are often mixed as they are loaded on to a vehicle. The intention is that the confusion created by loading plus the early part of the journey will distract the pigs from fighting. Feeding management may influence the level of damage as fasted pigs fight more aggressively and for longer than fed pigs. Some trucks are fitted with dividing gates, which allow pigs from different pens in the finishing shed to be kept separately in the truck and this should reduce fighting damage. Pigs transported on the lower deck can have more skin damage because they tend to spend more time standing and presumably acquire more injuries when they lose balance (Faucitano, 2001). Mixing at the abattoir, because of limited pen or lairage capacity, is less easily managed. Showering or hosing the pigs can help, as can reducing the lighting. In one case study, skin blemishes fell substantially when the pig abattoir stopped using electric goads (Faucitano et al., 1998). In another study, holding pigs overnight at the abattoir almost trebled the prevalence of skin damage (Costa et al., 2002).

Table 10.6.   Injuries affecting hides and skins and which are of welfare importance.

Type	Cause
Skin damage	
Brands	Hot-iron, freeze branding
Scratches	Thorn bushes, rough fencing
Horn rakes	Fighting (see Fig. 10.2)
Fighting scars	Fighting, biting
Abscesses	Injections
Yoke and harness scars	Improper designs or fitting
Goad damage	Excessive use of sharp or heavy sticks
Dung irritations	General dirt and filth
Vegetation damage	Weed seed penetration
Shearing scars	Improper technique or undue haste
Skin diseases and parasites	
Dermatomycoses	Fungus (*Trichophyton verucosum*)
Contagious pustular dermatitis	Bacteria (*Corynebacterium pseudotuberculosis*)
Lumpy skin disease	Herpes virus
Hyperkeratosis	Allergic response
Streptotricosis	Bacteria (*Dermatophilus congolensis*)
Demodicosis	Mite (*Demodex bovis*)
Warble flies	Fly (*Hypodermis bovis*, *Hypodermis lineatum*)
Ticks	*Boophilus micropilus* and others
Lice	*Lignognathus* and *Dalmalinia* spp.

Animal Health and Meat Inspection

In ruminants, disease states that warrant exclusion from the slaughter hall and can be recognized from the animals' behaviour during ante-mortem inspection include rabies, listeriosis, tetanus and foot-and-mouth disease. Equally serious skin and mucosal lesions that can be identified ante-mortem include tuberculosis, anthrax and contagious ecthyma. Signs of diarrhoea are, of course, less specific but can suggest salmonellosis, campylo-bacteriosis, colibacillosis or yersiniosis. In most of these conditions there is suffering from illness in the animals.

An important role of meat inspection is to identify and reject carcass defects such as abscesses. In the USA about 14% of beef livers are con-demned because of abscesses and 63% of liver con-demnations are due to **abscesses** (Lorenzen *et al.*, 1993; Boleman *et al.*, 1995). Pus-filled nodes are common in the necks of CLA-infected adult sheep presented for slaughter at abattoirs in the Middle East. The pus is usually released during the halal cut, and the neck wound is inevitably contaminated.

Some abscesses have a non-infectious origin but are equally objectionable. For example, *Pasteurella haemolytica* and *Actinobacillus* bacterin vaccines containing oil adjuvants can provoke a sterile inflammatory response, resulting in abscesses with granuloma formation at the site of injection.

Meat inspectors also examine carcasses for **tumours**, especially in cull breeding stock. At one abattoir in New Zealand, 41% of all cattle carcass condemnations were due to neoplasia and 63% involved cancer eye.

Carcasses from emergency-slaughtered heat-stressed animals have few distinguishing features. In one case, the carcass showed slight peripheral congestion and a small amount of fibrinous exudates on the surface of the spleen, but other-wise appeared normal and was passed for human consumption (McQueen, 1972).

Standards used for condemning carcasses can vary according to the sensitivity of the market. This works two ways. Sometimes, a condition becomes common and condemnation standards have to be lowered. In North America poultry carcasses with **cellulitis** used to be totally condemned. Cellulitis is now so common that the objectionable parts are trimmed and rejected, whilst the remainder of the carcass goes to the portion meat trade. Alternatively, processing companies that have been aiming for high standards and have been putting pressure on growers to control a problem are in a good position to increase condemnation rates through the control they have over market access.

The main cause of cellulitis in broilers is skin scratches, which allow a subcutaneous infection. The risk of scratches depends on stocking density, feeder and drinker space allowance, and activity of the birds, which in turn is influenced by the lighting programme and whether feed restriction is used early in life to control leg disorders (Table 10.7; Dozier *et al.*, 2005).

Some infectious diseases that cause lameness in broilers also cause losses from carcass or portion downgrading (Dobson and Glisson, 1992). For example, **osteomyelitis** of the femur can lead to more bone breakage during automatic deboning of the thigh, and this leaves unwanted bone fragments in thigh portions. Not all cases of femoral head necrosis are infectious in origin but commonly it is due to a *Staphylococcus aureus* infection through the navel of the chick (Iordanidis *et al.*, 1998). Similarly, recent outbreaks of *Enterococcus hirae* have caused rejection and downgrading from septicaemia, endocarditis and visceromegaly (Chadfield *et al.*, 2005). Antibiotic resistance is presently at a low level in this pathogen.

Drug Residues in Carcasses and Antibiotic Resistance in Microorganisms

There are minimum drug-withholding periods before an animal is fit for slaughter. Otherwise, drug residues will be present in the meat or offal that consumers eat. Compliance with drug residue regulations becomes a welfare problem when treatment is withheld from a sick animal, for example if it has been unsuccessfully treated and it is decided to send it for slaughter.

Antibiotic residues have been a concern in US cull dairy cows and young calves. It has been difficult to give reliable estimates of the scale of the problem because the residues are only confirmed in follow-up investigations after initial detection of a suspect case (Guest and Paige, 1991). In one study, involving 23 dairy herds, 33 out of 877 cows that were sent for slaughter violated drug-withholding requirements (Marteniuk *et al.*, 1988). Some antibiotic residue cases in slaughter calves occur when the calves have been given milk from cows being treated for metritis, mastitis or lameness. The tissues in calves most likely to carry the residues include kidney and fat. In the past it was recognized that the single most important reason for failure to meet the withholding period regulations was lack of awareness of the risks and ignorance about the withholding periods (Van Dresser and Wilcke, 1989). This has strengthened the role of veterinary practitioners in dispensing antibiotics and other drugs to farmers, as they are in a good position to emphasize the withholding period requirements.

The use of antimicrobials as growth promoters in pig and poultry feeds is being phased out in the EU and it is now limited to only a few compounds in most other countries. The policy is now towards prudent use of antimicrobials before the development of resistance amongst microorganisms severely reduces their usefulness. It is considered imprudent to use an antimicrobial as a growth promoter if it still has value for therapy. Nevertheless, the arguments for including antimicrobials in

Table 10.7. Effect of space allowances on the prevalence of skin scratches in broilers.

	Stocking density (kg/m^2)			
	30	35	40	45
Feeder space (cm per bird)	2.95	2.54	2.18	1.95
Birds per nipple drinker	4.1	4.8	5.6	6.2
Skin scratches (%)	41.9	51.4	58.7	53.5
Mortality (%)	3.6	7.8	7.3	7.5

feeds as growth promoters are better growth performance and feed conversion efficiency, better nitrogen and phosphorus utilization and hence less waste nutrient loading, improved health and welfare of the animals, reduced pathogen load and hence safer meat products (Cummings, 2006). Including antibiotics in broiler grower rations has been effective in reducing mortality when mortality rates would have been high, but the benefits may be small compared with those achieved with probiotics (Sun *et al.*, 2005).

In Denmark, antibiotic resistance to antimicrobials is being monitored, with enterococci and *E. coli* as indicator bacteria (Bager *et al.*, 2000). Identifying antibiotic-resistant strains of zoonotic pathogens is difficult, but it is suspected that fluoroquinolone resistance has developed in *Campylobacter jejuni* from the use of enrofloxacin in poultry, avilamycin resistance has developed in *Enterococcus faecium* from using this antibiotic as a growth promoter in broilers, vancomycin resistance in *E. faecium* has emerged from the use of avoparcin and possibly tylosin in pigs, flavomycin–erythromycin–tetracycline– virginiamycin-resistant strains of *E. faecium* have also come from the pig industry, and similar multi-resistant strains of *Salmonella typhimurium* DT104 are beginning to emerge in poultry. These routes of development are, however, difficult to prove.

Enrofloxacin has been promoted as an antimicrobial that can be included in feed, but it is intended as a medication rather than a growth promoter. Enrofloxacin residues in turkey meat are associated with suppression of antioxidant enzymes (Carreras *et al.*, 2005). A new concern is that their overuse could compromise defence against free-radical formation in animals before they are slaughtered, as well as increasing the opportunity for resistance in *Campylobacter*.

Drug residue problems also exist in developing countries. In these countries it is often the animals that are sick or are not thriving that are submitted first for slaughter. There is a higher risk that these animals will be carrying drug residues, and some may even be in a poor condition to withstand the journey to the abattoir. The scale of the drug residue problem is not appreciated, but in a study in Nigeria it was found that 7% of cattle submitted for slaughter were excreting antimicrobial substances in their urine. Some of these could, however, have been false positives (Kabir *et al.*, 2002).

Transport

The types of stress that can occur during transport include:

- water deprivation
- feed deprivation
- physical fatigue
- sleep deprivation
- social disruption
- injuries
- unpleasant motion
- noise
- heat stress
- wind chill

None of these stresses are inevitable. They depend on the condition of the journey.

Long journeys cause **sleep deprivation** and tiredness in people and no doubt this applies to animals as well. In cattle and sheep, rumination is a prelude to sleep, and rumination will not occur during a journey if the animal is stressed or does not have a full rumen. The prevalence of sleep deprivation from the effects of fasting and long-distance transport in cattle has not been examined, but it could be important because rumination promotes light sleep and because they cannot adopt the sleeping position shown in Fig. 1.2 that is required for deep sleep. Metal lorries often produce noise levels greater than 96 dBA, and these levels can also inhibit rumination and sleep in sheep (Hall *et al.*, 1998).

During long-distance journeys, sheep spend about one-quarter of the time with their heads below the level of the shoulder, and this is thought to be a sign of discomfort and submission to the situation. Cattle assume a similar posture to sheep. They hold their heads either with the poll at the same height as the back, or lowered with the muzzle about 15 cm from the floor (Sutton *et al.*, 1967). This is disrupted when horned animals are close to each other. Instead, the animals at risk raise and lower their heads in an attempt to avoid injury. Cattle normally stand at right angles to the direction of travel. This counteracts the sideways sway of the truck and helps them maintain balance.

The optimum **stocking density** depends on the length of the journey. Animals can be stocked more closely during short journeys as there is less need for them to sit down to take a rest. Normally, pigs do not lie down and rest until 2 to 4 h after the start of a journey. On longer journeys they may need to sit down, and lighter stocking reduces the risk of recumbent animals being trodden on. Stocking density may not affect meat quality, but high densities can cause more physical exertion, as seen in their serum creatine phosphokinase (CPK) activities (Warriss et al., 1998b).

Overstocking cattle trucks can make it difficult to close the tailboard or tailgate, and this can lead to frustration and bullying by the driver during loading. Unloading overstocked vehicles can also be hazardous, especially if the first animal coming off is facing forwards and has to reverse down the ramp. At one abattoir, the prevalence of backward unloading down the entire length of the unloading ramp was 4%. This is not a major hazard, but it can lead to injuries if the animal slips during a turn or the other animals follow on in a rush.

Enforcing standards on livestock transport is difficult and is open to evasion. For example, in one study, 41% of cattle trucks arriving at a beef abattoir exceeded the recommended loading density, which at the time was 379 kg/m^2. Four per cent of the trucks held between 475 and 575 kg/m^2. In addition, 70% of the trucks arriving at the abattoir had horned and unhorned cattle penned together, whilst only 23% of all the cattle were horned. Mixing horned with unhorned cattle in trucks is an inevitable feature of buying cattle through livestock markets.

Shearing sheep before transport enables higher stocking densities but if performed immediately before the journey it adds to the overall stress, and under cold conditions it can increase the risk of exposure, especially where top decks are not fitted with a roof.

In parts of North Africa cattle are sometimes transported to abattoirs as pairs on the back of small open-top trucks. They are tied to the headboard with a horn, neck or head halter. Getting the animal on and off the deck of the truck is hazardous if there is no adjustable ramp. Without the ramp, the animals back off the deck when the head is untied. The risk is that they land on their rumps as they come off backwards. An alternative approach is to lower the tied animal off the vehicle (Fig. 10.3). In South America, adjustable loading and unloading ramps are common. They can be raised or lowered to meet the height of most trucks, and the risk of injuries is lower.

Legislation in Europe now requires **rest periods** during long journeys. Stock are unloaded and watered and they may be fed at staging points such as livestock markets. In other countries it is debated as to whether it is better to complete long journeys quickly, without any interruption, or provide breaks in the journeys. The answer probably depends on the conditions during the journey. For example, if the animals are:

- unable to rest by sitting down, without being trodden on
- unduly cold or hot inside the vehicle
- becoming dehydrated

then it would be reasonable to provide breaks that allow recovery. When lambs were transported for 29.5 h under European conditions with a 1-h rest period in the middle of the journey, the stress response determined from plasma cortisol was lower during the second loading and transport episode (Parrott et al., 1998). However, this does not prove the advantage in breaking up long journeys.

European transport regulations also require that pigs should be able to stand or lie down in their natural positions during transport. This would be achieved with a stocking density of about 235 kg/m^2. However, some authorities consider that there are advantages in stocking more heavily (286 kg/m^2) during short journeys (Barton Gade and Christiansen, 1998). When tightly packed, the pigs give each other mutual support, preventing each other from being thrown around and being struck or bruised during sudden movement of the vehicle. They are also less prone to skin blemishes and producing DFD meat (Barton Gade and Christiansen, 1998; Guàrdia et al., 2005). The difference between these two stocking density standards is shown in Fig. 10.4.

The relationship between journey duration and the prevalence of PSE meat depends on stocking density. At low stocking densities, the prevalence of PSE meat falls as the journey gets longer, but at high stocking densities there is no effect from journey duration (Guàrdia et al., 2004). Reducing stocking density for the sake of avoiding PSE meat would only be advisable for long journeys.

The risk of PSE is lower in trucks equipped with hydraulic lifts instead of static ramps, and can be lower in vehicles fitted with a polyester floor surface. The polyester provides a more comfortable thermally insulated surface and is less slippery and noisy. Iron floors in the vehicle can lead to DFD meat (Guàrdia et al., 2005). High temperatures during transport increase the risk of PSE.

Fig. 10.3.   Manual unloading where there is no suitable unloading ramp.

Rough driving conditions on twisting roads are inevitably stressful, and this has been reflected in plasma cortisol concentrations for pigs (Bradshaw *et al.*, 1996). Calves find 2 Hz vertical oscillation stressful and 8 Hz can also be unpleasant (van de Water *et al.*, 2003). An oscillation of 2 Hz is particularly common on ferry crossings and in trucks passing over rough gravel surfaces (Wikner *et al.*, 2003). Poultry also find low-frequency movement aversive.

In people, oscillations between 0.05 and 0.5 Hz are unpleasant and can lead to motion sickness. In pig trucks, vertical resonance is often between 1 and 5 Hz, and large horizontal resonances are mainly in the 0.01 to 0.06 Hz range. These lower frequencies were due to bends in the road and deceleration and acceleration effects (Randall and Bradshaw, 1998). Travel sickness in pigs can be recognized from retching, vomiting, chewing, foaming at the mouth and sniffing the air while standing (Bradshaw *et al.*, 1996). These signs coincide with a rise in plasma lysine vasopressin

concentration, and can develop during the second hour of a journey.

Transport stress can lead to dark-cutting beef (Gregory, 1998). However, there is one important exception, which occurs in very short journeys. Bulls can produce pale meat with a high drip loss when transported short distances under hot, humid conditions and then slaughtered as soon as they arrive at the abattoir (Nanni Costa *et al.*, 2003). Evidently they are acidotic and heat-stressed at the time they are killed. Pre-slaughter heat stress at high humidity had no effect on meat quality in lambs or broiler chickens (Nagle *et al.*, 2000; Lowe *et al.*, 2002). If cattle become dehydrated during longer journeys, they tend to produce firmer meat (Honkavaara *et al.*, 2003).

Poultry start to lose carcass weight earlier during transport than other farm animals. After 4 hours' feed withdrawal, carcass tissue loss in broilers occurs at 0.24% per hour (Veerkamp, 1978). This is an incentive to minimize journey and holding durations and organize precise scheduling of

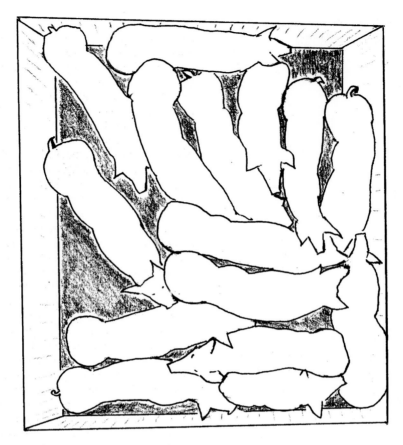

Fig. 10.4. A stocking density of 286 kg per m². There would be 12 instead of 15 pigs if the density were less than 235 kg per m².

the trucks. In some countries, **poultry transport crates** have perforated or mesh floors, instead of solid floors. These assist heat loss during hot conditions, but the birds get dirtier and have higher coliform counts in their plumage by the end of the journey (Buhr *et al.*, 2000). Perforated floors make unloading from dump modules more difficult. During unloading, dump modules tilt and the birds slide out of the transport crates on to a conveyor. With perforated floors, the birds grip the floor with their claws and flap their wings whilst trying to maintain control of the fall. Dump modules with solid floors tip faeces and urine on to the birds during tilting, and so solid flooring is not a perfect solution.

In poorer urban and suburban communities, goats and sheep are often brought to markets and taken away in the boot of a car. This poses risks of heat stress and inadequate ventilation as well as stress from social isolation. In other parts, stock are taken to abattoirs or butcher's premises on foot. This may be no more than a few kilometres to a village slaughter slab, or it may be several hundred kilometres, made over a period of weeks, to a city abattoir. Weight loss is inevitable during the longer treks.

If a vehicle is used for transporting more than one type of animal, the equivalent carrying capacities can be estimated from Table 10.8. However, modern livestock vehicles are becoming more specialized and are now being purpose-built for particular species.

DOAs

The dead-on-arrival (DOA) rates for the main classes of stock in the USA are shown in Table 10.9.

Table 10.8. Different animal-carrying capacities in road vehicles.

Animal type	Carrying capacity units
Cattle	
< 350 kg	4.5
> 350 kg	5.0
Calf	
store	2.5
suckler	1.0
Sheep	
adult	0.8
lamb	0.8
Pig	
slaughter	1.0
sow	2.5
Horse	6.0

Table 10.9. DOA rates in the USA during 2002.

Class of stock	DOA rate (%)
Cattle	0.06
Calves (veal + bobby)	0.86
Sheep	0.03
Pigs	0.25
Goats	0.13
Equines	0.09

In cull sows it is 0.4%, and in cull cows it is 0.3%. The DOA rate plus condemnations for emaciation and injuries in cull cows is 0.6% (USDA, 2006). In the case of spent hens, the DOA rate is just less than 2.2%. Some of the highest DOA rates are in bobby calves. In the USA it is 2.3% and in Australia 0.6%, rising to 1.3% in the worst months. It increases exponentially with journey distance (Cave *et al.*, 2005).

Under hot conditions in Australia, DOA rates for cattle have been about 1%, followed by a further 1% mortality in the holding yards at the abattoir, but these levels would now be exceptional for cattle that are used to being handled (McQueen, 1972). Mortality rates in rangeland cattle mustered by helicopter and then trucked to an abattoir have been in excess of 50%, but, here again, losses are now lower than this, because there is more experience in handling this type of animal. High losses have occurred when cattle have been taken from a drought-affected region.

The DOA rate for pigs depends on temperature conditions and the prevalence of the halothane gene. The *nn* genotype is more prone to dying during transport and holding at the abattoir (Table 10.10). In Canada, the overall DOA rate is low (0.1%) because the *nn* gene is not common (Murray and Johnson, 1998), whereas in EU countries it varies between 0.03 and 0.5%. The *nn* pig is prone to developing a metabolic acidosis when stressed, along with a lethal hyperkalaemia. This type of pig has a higher heart rate during handling (Villé *et al.*, 1993). This is partly due to the greater responsiveness of its sympathoadrenomedullary nervous system, and to a higher intrinsic heart rate.

In broilers the DOA rate usually varies between 0.05 and 0.6%. The target DOA rate should be less than 0.2%, but some allowance may be needed for flocks with ascites. Hot weather also poses a greater risk of deaths, especially in congested traffic (Warriss *et al.*, 2005). Longer journey times, overcrowding in the transport crates and long holding times at the plant before slaughter are also risk factors (Nijdam *et al.*, 2005).

Mechanical harvesters are used by some broiler companies in North America, and there is pressure in some European countries to use them because there are fewer health hazards for catching staff. Most harvesters use rotating rubber fingers that sweep the birds on to a conveyor that takes them to a module loading position. In a recent study with a Danish harvester, the prevalence of DOAs was slightly higher compared with hand-catching (0.35 vs. 0.15%) and plasma lactate in the birds was also higher (Nijdam *et al.*, 2005). There were no differences in subsequent meat quality in the birds that survived, but harvesters can inflict less breast and back bruising (Knierim and Gocke, 2003).

The causes of DOAs in broilers were evaluated recently in the USA (Ritz *et al.*, 2005). About 60% of the deaths were due to handling-related problems and 40% to farm-related problems. Air-sacculitis, septicaemia, ascites and ruptured livers were the largest contributors, and ruptured lungs, head trauma, asphyxia and leg trauma accounted for over half the handling-related deaths (Table 10.11).

The DOA rate in fat cattle and sheep is usually less than 0.02%. However, deaths have occurred when cattle were loaded too loosely or too densely, and in stock that were weak or in poor condition

Table 10.10. DOAs in pigs according to halothane genotype in Canada and Spain.

	Canada		Spain	
Genotype	% of pre-slaughter deaths	% mortality	% of pre-slaughter deaths	% mortality
nn	28	9.2	71	2.3
Nn	25	0.3	24	0.1
NN	47	0.1	5	< 0.1

Table 10.11. Causes of broiler DOAs in the USA.

Cause of death	%
Handling-related problems	
Ruptured liver	25.0
Ruptured lung	15.0
Head trauma	13.5
Asphyxia	13.5
Leg trauma	12.0
Multiple trauma	5.5
Broken bones	3.0
Non-specific/other	12.5
Farm-related problems	
Air-sacculitis and septicaemia	58.0
Ascites	31.0
Runts	7.0
Non-specific/other	4.0

(as in a drought) (Tarrant, 1990). In Australia, high downer plus DOA rates for sheep have been linked to hypocalcaemia and hypoglycaemia (Shorthose and Shaw, 1977).

Holding Animals Before Slaughter

It is usually recommended that pigs should be rested in lairage for at least 2 h before they are put up for slaughter. This gives them an opportunity to settle down, and if they are relaxed they are easier to herd up to the stunning area. It also provides an opportunity for rehydration, and it reduces the risk of PSE meat. In the case of cattle, when a delivery was slaughtered on arrival at an abattoir in New Zealand, subsequent pH_{ult} was 5.81 (± 0.06 SE), whereas animals from the same herd that were rested overnight at the lairage had a lower pH_{ult} (5.47 ± 0.04 SE). Overnight recovery depends on whether the animals settle. If they are active, riding each other, there is a greater risk of high pH_{ult} meat.

Holding stock before slaughter gives them a chance to rehydrate. Dehydration in animals that have just arrived at an abattoir is obvious from their behaviour once they recognize the water trough. In the carcass, it can be recognized from stickiness of the cut surface of the meat and from its dark colour.

Paddocks near an abattoir are sometimes used for holding sick or diseased animals that are not fit for immediate slaughter. They are kept there until it is decided to have a clear-out, or they may be put up for slaughter at the end of each day's kill. Ideally they should be processed every day if they are not receiving treatment or are not likely to be fit for slaughter in the near future. Otherwise, the suffering from the diseased state is prolonged unnecessarily.

It can be difficult to control temperature in poultry processing plant holding areas (Quinn *et al.*, 1998). In a study at two plants in the UK, air temperature in the broiler modules varied between 16 and 31°C, depending on ambient conditions. Protecting birds from heat stress during the holding period can help reduce weight loss (Benibo and Farr, 1985). It may also protect them from blood spot formation in breast meat (Kranen *et al.*, 2000a,b).

Casualty Animals

In most countries the method used for dispatching an injured or diseased sheep or goat at a farm is to cut its neck with a knife. An emerging view is that farmers should be equipped with a captive bolt gun or a firearm for these cases. However, in some countries firearms are disallowed for national security reasons, and neck cutting is the only accepted method.

Pigs that are unable to walk are sometimes left in a pen in the lairage until it is convenient to handle them separately. This may be at the end of the working day. Instead, it is often recommended that casualty animals should be dispatched as soon as possible and without unnecessary handling.

Downer cows at dairy farms can have salvage value for the pet food trade. 'Knacker-men' or 'dog-men' usually insist on slaughtering these animals themselves. This gives them a chance to evaluate whether the animal is likely to be fit for pet food, and it ensures that the carcass is fresh and properly bled. Their preference is for animals that are not diseased and are in good body condition, and injury cases usually fit these requirements best. The main welfare concern with these animals is with delays whilst waiting for the knacker-man to reach the farm.

Ease of Handling

Common stock-handling problems at abattoirs include:

- baulking at the step between the truck and the unloading ramp, where the truck does not have a tailboard
- baulking at steps from the unloading ramp on to the floor of the unloading bay
- reluctance to move down the slope of the unloading ramp
- jamming of animals at the exit of the truck, where half-width doors are used
- mishandling the first sheep to get the others to follow on
- gap between truck and unloading ramp creating a visual cliff effect, where the truck does not have a tailboard
- baulking when stock are required to go into a dark area
- baulking due to a change in floor type or floor surface pattern
- baulking from glare or reflected light ahead of the animal
- arrested flow because of 90° turns
- arrested flow when approaching the forcing pen because of panelled gates
- distractions created by animals or other features in pens alongside a corridor
- unstable floors or platforms
- arrested flow because of narrowing at gateways or entrances to races

Loading and unloading are often the most stressful periods during livestock transport, and stock often have to be threatened or bullied to get them to move in the required manner. The **layout of the loading bay** at the farm, market or abattoir affects the ease of unloading and loading. For example, a sloping vehicle is easy to unload when it gives a clear view for the animals on the vehicle of other animals in the yards below. This encourages them to come off the truck. On the other hand, it is more difficult to achieve a full load when loading a vehicle that is parked on a steep slope running down to the docking bay. Sheep fail to move all the way up the sloping vehicle, and sheep loaded at the lower end are squashed together and may be prone to going down during the journey. Some drivers recognize this problem and redistribute the sheep within the vehicle by braking suddenly just after they have set off.

Three systems are used for loading and unloading stock:

- the tailboard of the truck is used as the loading and unloading ramp
- the truck has a sliding door. There are dedicated unloading and loading ramps at the farms, markets and abattoirs
- a hydraulic vertical lift fitted to the back of the truck

Trucks with sliding doors are common in the southern hemisphere. They have a number of weaknesses compared with trucks that use the tailboard as the ramp:

- narrow entrance – usually half the width of the truck, creating greater opportunity for bruising
- vertical gap between truck and ramp causes baulking
- less versatile – requires dedicated unloading or loading of ramps at the destination
- multi-deck trucks require adjustable ramps at the destination
- faeces accumulate on the unloading ramps at abattoirs during the day, making them slippery
- same ramp may be used for several species – this can lead to arrested flow (e.g. when pigs stop to investigate sheep faeces)

- likely to be fewer docking bays at markets because they are more elaborate and expensive – longer waiting times for trucks

Where a sliding half-door at the rear of the vehicle is used, it is best if the position of the docking ramp is adjustable. If docking ramps are not adjustable, the floor of the trucks may not meet the docking bay ramp. Adjustable docking ramps need to be sufficiently sturdy and stable to prevent wobbling when used by stock. They must join with the edge of the truck and not leave a gap for stock to jump over or slip into. This is usually achieved with unfolding plates or flaps that act as a bridge between the docking ramp and vehicle (Fig. 10.5). Occasionally the flaps are dislodged as stock pass, and this allows a subsequent animal to get a leg stuck in the gap.

Tailboard ramps are usually made of alloy. Their light weight helps keep transport costs down and allows the operator to raise and lower them easily, but they provide a poor grip for livestock. Stock are prone to going down on their knees when moving up steep internal ramps, and sliding when passing down the ramps. These are not usually serious problems, but they can become hazards if there is pressure on the animals from behind.

Most modern livestock vehicles have internal partitions, which allow penning into smaller groups. This can make loading more difficult, because, as soon as the panel gate is closed on a full pen, there are no animals to act as a visual draw for the next batch entering the vehicle. A gate arrangement at the loading bays is critical. There needs to be a set of gates near the ramp, which can create a forcing pen for a small group of sheep. Without this, the person doing the loading has greater difficulty getting them started up the ramp.

The greatest contributors to welfare problems during loading are stock slowness and refusal to move. These behaviours lead to frustration amongst the handlers, especially when there is pressure on them to complete their loading and move the vehicles out. Frustration leads to rough handling. Frustration and bullying during unloading are less common.

A common design problem in markets that hinders unloading is the cross-corridor at the end of the unloading bays. Staff moving up and down the corridor distract animals as they are coming off the vehicles, causing them to double back. Twin cross-corridors with a fully panelled middle partition help avoid this problem.

Some breeds have a reputation for being flighty, and others have a reputation for being stubborn. Both extremes create difficulties, especially when they are unfamiliar with being handled. The handling procedure that causes most difficulty will depend on circumstances. In a study on 18- to 30-month-old Limousin cattle at Lyon abattoir, putting them up for slaughter was the most stressful procedure. They developed more pronounced acidosis during this final stage than at any other point following mustering at the farm (Mouthon *et al.*, 1976). Needless to say, when cattle have been ill-treated they are more shy and their flight zone is larger (Breuer *et al.*, 2003).

Transferring pigs from the holding pens to the stunning point is also a stressful procedure. Pigs that are electrically stunned in **restraining conveyors** have to be reduced to a single file to get them into the conveyor. Common problems in the single-file race are intermittent stopping and starting of pig movement according to the rate they are being removed from the conveyor, excessively long or steep races, and electric goading used for regulating movement along the race. Flow can be made easier by using:

- a double race. Two races side by side increase the chance of having a pig delivered at the restraining conveyor when it is waiting to be filled
- fully panelled walls in the race
- a crowding pen at the entrance of the race

Some handling systems are rough on the animal but are used when there is no alternative.

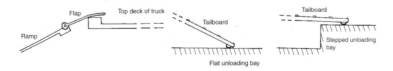

Fig. 10.5.   Three ways of joining the truck to an unloading bay at an abattoir.

In Sweden, reindeer are sometimes restrained by lasso during the harvesting period. The struggle can deplete muscle glycogen, resulting in high pH$_{ult}$ meat and haemorrhagic lesions in the mucosa of the abomasum (Wiklund *et al.*, 1997). Reports from India describe cull buffalo being pulled on to vehicles using a rope around the neck. About one-third of the animals fall or lie down on the ramp during loading, and about 10% go down during unloading. They are then partially lifted and dragged on to the truck whilst applying encouragement from a stick. If they do not go down, loading is quicker if they are hit with a stick compared with pulling them up the ramp with a rope around the neck (Chandra and Das, 2001). Weak animals put up least resistance and are easy to load, whilst male buffalo put up the most resistance. Nervous animals are sometimes blindfolded to make handling easier. If an animal refuses to move, the handlers become frustrated and sometimes resort to thrusting a stick in the rectum or vagina, or poking a stick against the testes or udder. Unloading is difficult if there is no unloading ramp at the abattoir.

Wing flapping during hanging on at poultry processing plants is a source of irritation for handlers, and it can make the work strenuous at turkey plants that process large toms. The flightiness of the birds, lighting conditions in the hanging-on area, and the way the birds are inserted into the shackle influence the duration of wing flapping. Prolonged wing flapping can lower the breast muscle pH$_{15min}$ and this increases the risk of paler breast meat with greater redness (Berri *et al.*, 2005). It can be challenging to hang turkey toms on the line because of their heavy weight. They need to be handled carefully when picked up to avoid limb joint dislocations.

Noise

Abattoirs are very noisy places. This can disturb stock, making it difficult to get them to move in the required direction. Ways that can be used to control noise include:

- avoid metal-to-metal contact on gates and other moving parts
- use welds instead or rivets for metal panels
- use screw compressors in place of piston compressors

- install mufflers or silencers on noisy exhaust systems
- reduce air velocity of exhaust jets
- reduce vibration noise with isolated mountings
- install noise-damping materials where containers are routinely dropped or struck
- use conveyors in place of rollers
- reduce stop–start impact noise with more precise control of line speed
- isolate noise-emitting equipment
- use laminated ('noiseless') steel
- minimize fall height of objects on to hard surfaces
- maintain equipment, e.g. tighten loose parts, balance rotating parts, correct air or steam leaks, replace worn bearings
- maintain better control of fan speed in chillers, or use multi-blade fans

In situations where noise is not a problem, sound can be used as a stock movement aid. One system used for moving sheep is the 'hell's bells', which is a small hand-held set of cymbals that clips on the operator's belt when not in use (Fig. 10.6).

Excessive noise from animals can be a sign of poor handling methods. For example, abattoirs that rely on electric goads when moving cattle to

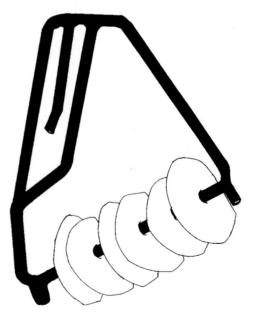

Fig. 10.6. Sheep rattle used in New Zealand (three pairs of cymbals on a hand-held frame).

the stunning pen have higher levels of vocalization in the cattle (Grandin, 2001b).

Mobile Slaughter Plants

In the past, itinerant butchers used to visit farms and smallholdings to slaughter and dress an animal for consumption by the owner's family. This is still practised in some parts and the travelling butchers often bring with them a frame that can be erected for suspending the carcass, or they may rely on whatever equipment is available at the farm. Traditionally, the animal would be bled without stunning, and this method is still preferred in some countries for sheep as the quickest, safest and simplest method. A firearm is often preferred for cattle and pigs.

Mobile slaughter units have been considered as a substitute and as a way of complying with hygienic dressing standards. They are not widely used, but have raised interest amongst reindeer and deer farmers. In Sweden, they have been used for reindeer since 1993. According to EU regulations, the pre-slaughter holding area must be dedicated to the slaughter unit, and this has meant that it cannot be used for routine health and handling procedures such as drafting, drenching and tuberculosis testing. This has been a deterrent for some farmers wishing to use mobile slaughter facilities.

11

Stunning and Slaughter

Pre-slaughter Restraint

When animals are put up for slaughter at an abattoir they are usually confined, separated and restrained. **Separation** is achieved with the animal:

- either free-standing with other animals whilst confined in a small pen
- or free-standing on its own in a stunning pen
- or restrained on its own in a crush or stunning pen
- or restrained in a conveyor

Separation and restraint allow controlled application of the stunning equipment. The animal may try to avoid being isolated from its pen-mates, and it needs to be stunned quickly once it has been separated. Otherwise it may panic and behaviour becomes unpredictable. Stunning an animal which is free-standing amongst a group of animals should involve less stress than complete isolation. However, if the whole group becomes disturbed, this system can be chaotic and applying the stunning equipment is less controlled.

The rate at which animals are stunned can fall behind the speed of the slaughter line if animals refuse to enter the stunning pen or fail to position themselves appropriately for stunning. Refusal to enter is usually due to a front wall inside the pen, giving the impression of a dead end. One solution at a pig abattoir was to have vertical bars at the front of the pen, which were set about 1 m from the wall. This provided sufficient view ahead to stop the pigs baulking. At a beef abattoir, a rotating hazard light set above the pen attracted the animals' attention and there were fewer delays from poor head presentation for captive bolt stunning.

Individual animal **restrainers** are used for small stock and deer. Some have pneumatically operated sidewalls that serve as a V-restrainer. Designs used for pigs include:

- V-restrainer with tilting floor and one lifting side
- V-restrainer with one lifting side and the opposite side pivoted to eject the pig
- lifting V-restrainer with a pivoted side allowing the stunned pig to be dropped on to a horizontal bleeding table

When installing restrainers in pig abattoirs it helps if there is no change in floor material at the entrance; otherwise loading the animal can be difficult.

Supplying animals at an even, unhurried pace to a restraining conveyor and loading them into the conveyor can be difficult. If the rate at which they are being stunned at the other end of the conveyor is slower than the speed of the conveyor, the conveyor has to be started and stopped for each animal. This stop–go situation is highly disruptive for animals in the raceway leading to the conveyor, and in the case of pigs they have to be bullied, often with a goad, to get them to move at the right time. Heel bars are often used to stop pigs backing up raceways. These can be effective, but, if they are set too high above the floor, it is difficult for some pigs to walk over them when moving forwards. One-way gates and backstops are sometimes used in cattle and sheep races.

Head restrainers and head raisers are used for holding cattle during electrical stunning, and they are sometimes used during captive bolt stunning. There are four systems of head capture

©N. Gregory 2007. *Animal Welfare and Meat Production*
(N. Gregory)

and restraint: V-yokes, upright bar yokes, chin lifters and head raisers. V-yokes can be hand-operated with a single moving bar brought towards a static bar, or they can be two hydraulically operated moving bars. In one version the apex of the V is lifted forwards and upwards to raise the chin of the head-locked animal. This allows sticking as well as stunning whilst the head is restrained. Electrical stunning can be through the arms of the V-yoke. Chin lifters are used after the head has been locked in a yoke either to prevent movement during slaughter without stunning, or to allow application of a nose electrode with nose-to-neck-yoke electrical stunning systems. Head raisers are static blocks or buttresses projecting from the front wall that force the animal to raise its head when it is forced towards the front of the pen, for example with a rump pusher (Fig. 11.1). This makes the animal present its head more quickly for a captive bolt stun. Cattle can react badly to being held by the neck, but simultaneous body restraint reduces struggling.

The stunning pen floor needs to provide good foot grip; otherwise there is a risk of untimely slips and falls. This seems obvious but it is not always followed. For example, at one abattoir in

Europe, the skid resistance value for the floor in a calf stunning pen was found to be 22, which is equivalent to the slipperiness of steel plate. About 40% of beef stunning pens in the UK have steel checker plate floors. When wet they are very slippery (skid resistance values between 25 and 30), and there is a risk the animal will slip or slide if it makes sudden movements.

Sheep and pig abattoirs that have high throughputs usually use restraining conveyors (Fig. 11.2). The pressure points from the sides of the restraining conveyor are determined partly by the angle of the two sides. Inappropriate angles have sometimes been blamed for high levels of shoulder blood splash and bruising. This may be linked to struggling whilst restrained. Lambs tend to struggle when they can see an open space at the end of the conveyor. This can be controlled by blocking the view with vertical strips of plastic curtain. Straddle conveyors have been used for pigs but are more common in beef plants.

In poorer countries, sheep and pigs are sometimes restrained before slaughter by tying their feet together and laying them on their side. This is used in slaughterhouses and butchers' premises that

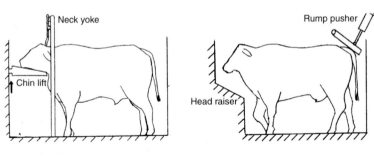

Fig. 11.1. Beef stunning pens.

Fig. 11.2. Sheep conveyors.

are poorly equipped with pens. When restrained in this way, the animal soon gives up and stops struggling provided it is left alone. Where corridors and races are not available, sheep are often herded into a slaughter room by catching one animal and pulling it by one leg into the room whilst someone drives the rest of the group from behind. If the lead sheep goes down, it is sometimes dragged into the slaughter room.

Shackling birds before stunning can be painful if there is excessive pressure on the periosteum of the shank. This is more likely in thick-legged males than in females (Satterlee *et al.*, 2000). End-of-lay hens often have slender legs and the shackle is a loose fit. This increases the impedance between the bird and the shackle and raises the risk of poor current flow through the birds.

Stunning and Slaughter

Unconsciousness means different things to different people. Anaesthetists think of it in terms of reflex behaviour, psychologists in terms of emotions and intellect, neurologists as impaired neural pathways and slaughtermen focus mainly on spontaneous behaviour. The science behind stunning and slaughter blends these approaches in producing practical guidelines. The criteria that have been used most in producing recommendations on how to stun and slaughter are:

- presence of an electroplectic fit, indicating unconsciousness in the context of electrical stunning. This has been very useful in establishing minimum currents that produce an effective stun
- presence of EEG amplitudes and frequencies that are symptomatic of unconsciousness. This has been useful in determining the duration of consciousness during CO_2 stunning and the onset of loss of consciousness following sticking
- absence of primary evoked cortical responses in the brain, indicating interference of specific pathways at a level that is a prerequisite for perception of the stimulus. Evoked responses have been used for comparing the relative effectiveness of different stunning and slaughtering procedures, and for estimating minimum stunning currents for poultry
- behavioural signs such as collapse, seizure activity and absence of brainstem reflexes

Most of the findings given in the rest of the chapter are based on these features.

Concussion stunning

Captive bolt guns are either trigger-fired or contact-fired, and their bolts either are recessed or protrude from the muzzle (Gregory, 1998). The aim is to concuss the animal through transfer of energy to its head, and a successful stun depends on using a gun that fires a bolt at the correct velocity for the size of animal.

Concussion with a captive bolt gun is used mainly in cattle and horses. When positioned correctly and when the right cartridges are used for the class of animal, it produces an immediate and irreversible stun. In cattle, a deep form of concussion is present when the animal shows the following signs immediately after the stun:

- collapse
- breathing is absent
- the muscles in the back and legs are in spasm. The hindlegs should be flexed. If the muscles are flaccid immediately after stunning, this is a sign that the stun is not deep and there is a risk that the animal might regain consciousness
- the eye is not rotated in the eye socket. Instead, the eyes should be directed forwards
- corneal reflex is absent

These signs are associated with immediate absence of evoked cortical responses in the brain. Absence of primary cortical evoked responses indicates failure in neurotransmission at a level that occurs before conscious perception of the stimulus. In other words, it represents a deep stun, as it indicates failure of signals to reach the association cortex where neural activity associated with consciousness is processed. Unlike evoked responses, the spontaneous EEG is not as reliable as an indicator of brain disturbance following captive bolt stunning and evoked responses are preferred (Daly, 1987).

Slow and shallow rhythmic breathing is sometimes seen in cattle after captive bolt and cardiac arrest electrical stunning and whilst they are bleeding out (Daly, 1987). Provided the corneal reflex is absent at this time, the animal is not conscious. Jaw relaxation is another useful sign. This can be tested by prising the jaws apart manually, or it may be seen if the tongue hangs out of its own accord in the suspended carcass.

Guns with low bolt velocities (up to 47 m/s for cattle) are less likely to produce an effective stun, as they impart insufficient energy to the cranium (Table 11.1). The transfer of energy to the head and the depth of stun are improved when bolt diameter is 16 mm or more (Table 11.2). In cattle, shooting in the poll position (caudal to the nuchal crest) is less effective than a frontal shot. In sheep, both frontal and poll shooting are effective initially, but there is a greater risk of recovery of consciousness following a poll shot (Daly and Whittington, 1986). Percussion bolt stunning can be as effective as captive bolt stunning, but because the bolt does not invade the brain, there is less likelihood of intracerebral haemorrhage and an irreversible stun. For this reason, the percussion bolt is accepted by some Islamic authorities as an acceptable pre-slaughter stunning method for producing halal meat. In halal slaughter, the slaughterman must kill the animal through the cut delivered to the neck.

Young bulls are difficult to stun by captive bolt. In a study conducted in the UK during the late 1980s, over 50% of bulls did not have a deep form of concussion (Table 11.3). These animals

Table 11.1. Effect of bolt velocity, shooting position and gun type on the elimination of visual evoked responses (VERs) in cattle.

	Captive bolt				Percussion bolt
Gun position:		Frontal		Poll	Frontal
Bolt velocity (m/s)	47	55	72	55	n/a
Number of animals	7	11	9	8	8
Prevalence of VERs (%)[a]	57	9	11	75	13

[a]During the first 4 s following the shot.

Table 11.2. Effect of bolt diameter and speed on energy imparted during stunning (joules ± SE) and the elimination of visual evoked responses (VERs) in cattle.

	Bolt speed (m/s)			
	47		55	
Bolt diameter (mm)	Energy imparted	VER prevalence (%)[a]	Energy imparted	VER prevalence (%)[a]
12	97 ± 17	50	124 ± 25	13
14	125 ± 18	25	139 ± 25	13
16	158 ± 20	14	186 ± 30	0

[a]During the first 4 s following the shot.

Table 11.3. Prevalence of poor stunning in cattle in the UK.

	Late 1980s		2006	
	Number of cattle examined	% poor stunning[a]	Number of cattle examined	% poor stunning[a]
Steers + heifers	1284	6.6	1290	6.8
Cows	628	1.7	–	–
Bulls	32	53.1	306	16.3

[a]Outright failure to stun, or slaughterman decided to re-stun the animal, or rhythmic breathing shortly after the shot, or positive corneal reflex or rolled eyes.

either received more than one shot, or were breathing rhythmically shortly after the shot, or had a rolled eye.

Following this study it was recommended that the minimum bolt velocity for guns used on young bulls should be 70 m/s. The prevalence of poor stunning in young bulls is now about 16%. The usual recommendation for steers, heifers and cull cows is not less than 55 m/s (Daly and Whittington, 1989). The guns need to be cleaned regularly to maintain these recommended bolt velocities. Otherwise the carbon and silica that accumulate in the breech increase the size of the expansion chamber by limiting the return of the bolt to its correct position, and this reduces the power of subsequent shots. In addition the rubber rings (or 'sleeves') need to be replaced if they show signs of splitting, heat damage or wear. These rings fit around the bolt and provide automatic retraction of the bolt from the head of the animal as well as protection of the bolt from direct impact against the muzzle of the gun (Fig. 11.3).

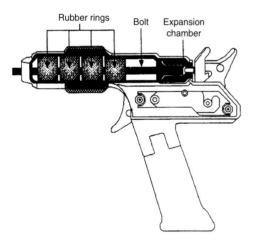

Fig. 11.3. Schematic diagram of a captive bolt gun showing the expansion chamber, bolt and sleeves.

Head restraint can improve shooting accuracy. In an abattoir in New Zealand where an underpowered gun was being used on bulls, the prevalence of repeat stunning was 19% when no head restraint was used, but it was about 2% when the animals were held in a head-bail. However, use of the head-bail should be left to the discretion of the slaughterman because in some cases it is stressful for the animal. If the animal presents its head as soon as it enters the stunning pen, it is better to stun it straight away, rather than attempting head restraint.

Animals that move about within the pen, and even attempt to climb out of the pen, are difficult to shoot. At one abattoir the prevalence of repeat stunning in the highly active animals was 53%, whereas in quieter types it was 19%.

The ideal shooting position in the head is the crossover point between two imaginary lines drawn between the base of each horn and the opposite eye. When shooting accuracy deviated beyond 2 cm from the ideal frontal position, there was a significant increase in the incidence of poor stunning (Table 11.4). Shooting accuracy becomes more critical when using low-powered cartridges. Shooting accuracy is sometimes poorer with contact-firing guns (35% of young bulls) compared with pistol-grip guns (23%).

Sheep are usually electrically stunned. However, the captive bolt may be preferred when there is a risk of blood splash, even though there might be a slightly higher risk of intermediate pH_{ult} meat and cold-shortening (Petersen and Blackmore, 1982; Paulick et al., 1989).

Veal calves are stunned either with a captive bolt or by head-only electrical stunning. When high voltages have been used, the rate of pH fall in the semimembranosus was faster and the meat was paler than for captive-bolt-shot calves (Lambooy, 1985). There was also a higher prevalence of blood splash in the electrically stunned calves.

Table 11.4.   Effect of shot accuracy on the prevalence of poor stunning when using 3-grain cartridges.

	Shooting accuracy – distance from the ideal position	
	< 2 cm	> 2 cm
Number of young bulls	242	90
% poor stunning[a]	11.5	35.5

[a]Outright failure to stun, or breathing shortly after the stun, or rolled eyes.

Spring-powered captive bolt guns are available for rabbits. Skin slip when positioning or firing the gun can lead to inaccurate shooting. Some rabbits vocalize during the shot, even though they are immediately unconscious. It is linked to sudden thoracic compression at the onset of the tonic spasm. It has been suggested that, if the bolt strikes a bone suture in young rabbits, there can be a higher risk of poor stunning.

Pigs and poultry kick violently following concussion, and this can increase the rate of postmortem muscle glycolysis. In veal calves, the kicking following captive bolt shooting is not sufficient to affect the pH, temperature or colour of the meat, but animals that require a repeat shot to produce a reliable stun can have lower pH_{3h} and higher pH_{ult} (Klont et al., 1999).

Electrical stunning

The recommended minimum currents for stunning with a 50 or 60 Hz current are listed in Table 11.5. The standards for sheep, cattle and broiler chickens are sometimes given as 1.0, 1.5 and 0.120 A, respectively. Those currents induce a cardiac arrest when current simultaneously flows through the body, and are higher than the minimum currents producing unconsciousness (Lambooy, 1982; Gregory and Wotton, 1990a; Wotton et al., 2000). The currents given for ducks and geese will induce a cardiac arrest when using a 50 Hz water-bath stunner. The minimum stunning currents for those species are not known. Electrical stunning is not commonly used in deer, but currents of 1.3

Table 11.5. Minimum recommended stunning currents.

	Minimum stunning current (A)
Cattle	1.15
Calves	1.00
Pigs	1.25
Sheep	0.50
Chickens	0.105
Turkeys	0.15
Ducks	0.13
Geese	0.13
Rabbits	0.14
Ostriches	0.40

and 1.0 A are recommended for red and fallow deer, respectively, with the head-only system.

It is usually recommended that the current should be applied for at least 3 s, but this is not a validated standard. The advantage with long application times is that the current rises as impedance is broken down during current flow. For example, during head-only stunning in pigs, electrical impedance starts at about 380 Ω and falls to about 160 Ω within 10 s (Troeger and Woltersdorf, 1988).

Head-only electrical stunning is reversible. In other words, the animal can regain consciousness. This has two consequences. First, the animal has to be killed soon after head-only stunning to ensure that it does not recover. Secondly, head-only electrical stunning can be acceptable to some halal slaughter authorities as it does not kill the animal.

The duration of unconsciousness provided by head-only electrical stunning depends on the current that is used and the length of time it is applied (Table 11.6). If 0.5 A is delivered for 3 s to a sheep, which is the recommended minimum standard (Lambooy, 1982), the duration of insensibility would on average be 62 s. If the time between stunning and sticking at an abattoir is, say, 23 s, and the time to brain failure following sticking is 14 s (Gregory and Wotton, 1984), the required duration of unconsciousness is on average 37 s. So, at that abattoir, the 'average' animal would not regain consciousness and there would be 25 s (62 minus 37 s) margin for error and unusual events. If a lower current is applied briefly (e.g. 0.3 A for < 1 s), there would be a greater risk of recovery of consciousness. This is because the duration of unconsciousness provided by the current (33 s) is shorter than the required duration (23 + 14 s). The corresponding durations of apparent unconsciousness for pigs and calves when using their recommended minimum stunning currents are 58 and 67 s (Anil, 1991; Gregory et al., 1996).

These estimates are based on averages. In sheep stunned with 0.3 to 0.66 A for 3 s, the quickest time to recovery in one of the animals was 41 s. The required duration of unconsciousness (37 s) is still shorter than this period. On this basis, it can be recommended that the stun-to-stick interval should be 23 s or less. It is possible that the required duration of unconsciousness given above is an overestimate, because electrical stunning itself could hasten the time to brain failure following sticking through its exhausting effect on brain metabolism (Bager *et al.*, 1992).

Prompt sticking is no longer important if a cardiac arrest is induced at or shortly after electrical stunning. Blood flow to the brain is brought to a standstill by the cardiac arrest, and so sticking only serves the purpose of releasing blood from the carcass. There is also a practical advantage to so-called cardiac arrest stunning: it produces a still carcass that is easy to handle (Fehrenberg *et al.*, 1991).

There are two risks when using low stunning currents. First, a proportion of animals may not be effectively stunned. Secondly, and more commonly, the duration of insensibility is shorter in some animals. For example, in calves receiving 100 volts head-only stunning, the calf with the quickest time to recovery was substantially faster than for calves receiving higher voltages, even though the average durations of insensibility for the different voltages were about the same (Table 11.7). Similar effects have been observed for lambs, pigs and poultry. Increasing the duration of application of a low current can extend the duration of insensibility (see Table 11.6; Wotton and Wilkins, 1999).

Some electrical stunning equipment used for cattle, sheep and pigs has a low-voltage circuit that senses the impedance between the electrodes before switching to the stunning voltage. The intention is to ensure that the electrodes are in adequate contact with the animal. If the impedance is too high, because of poor contact, the stunning current will not be engaged by the control box. In principle, this is a good system, but if slow switch gear is used the animal may object to the pressure applied by the electrodes and start struggling before it is stunned.

It sometimes happens that an animal has to be re-stunned. For example, if an electrically stunned pig drops from its shackle before it is stuck, it is either shot and then bled or it may be electrically stunned again and then stuck. It has been thought that a second epileptiform episode might be shorter-lasting than the first, but recent research has shown that this is not the case (McKinstry and Anil, 2004). In other words, the duration of insensibility following repeat electrical stunning should be similar to that experienced following a single application of current. Repeat stunning will increase the risk of blood splash (Kirton and Frazerhurst, 1983).

Table 11.6.   Duration of insensibility[a] in seconds (SD) according to current level and duration following electrical stunning in sheep.

Current level (A)	Current duration (s)		
	< 1	3	6
< 0.3	33 (9)	52 (13)	49 (20)
0.3 to 0.66	34 (12)	62 (14)	71 (18)
> 0.66	39 (6)	63 (16)	87 (25)

[a]Based on the time to the onset of head righting in the animal lying on its side following delivery of the current.

Table 11.7.   Effect of stunning voltage on time to recovery in calves.

Stunning voltage (V)	Time to onset of head righting (seconds)	
	Average	Range
100	78	24 to 114
150	63	44 to 83
200	67	44 to 115
250	71	59 to 84

The minimum stunning current for broilers is 105 mA when using either a water-bath or a head-only stunner (Gregory and Wotton, 1990b; Raj and O'Callaghan, 2004b). Layer hens require higher stunning currents than broilers, and even with a successful stun they regain consciousness considerably more quickly (Gregory and Wotton, 1994; Raj and O'Callaghan, 2004b).

When monitoring electrical stunning in a plant, examining the current profile can be useful in recognizing interruption in current flow when the electrodes are applied (Troeger and Woltersdorf, 1988). Examples of current profiles and the problems they depict are shown in Fig. 11.4.

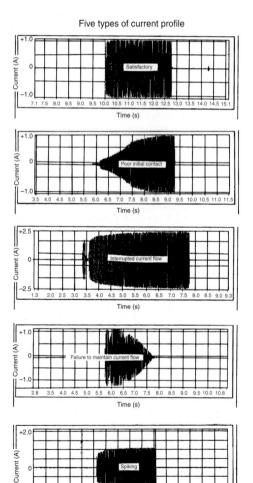

Fig. 11.4.   Current profiles during electrical stunning.

Electrical stunning and meat quality

If a pig is electrically stunned across the head ('head-only') with a high current applied for a relatively long period, the stun can be expected to activate muscle metabolism and lower the pH of muscle at 45 min post-mortem ($pH_{45 min}$). In practice, head-only currents have to be applied for exceptionally long periods to accelerate the early pH fall (Rothfuß et al., 1984; Troeger and Woltersdorf, 1988). Current duration can be 7 to 10 s when using 250 volts and 10 to 15 s when using 180 volts without harming meat quality. However, if the electrodes are applied across the neck, the pH fall in the shoulder muscles might be affected. Since head-only electrical stunning electrodes are quite frequently applied across the neck in pigs, instead of the head, it is probable that electrical stunning has unwanted effects on meat quality (Anil and McKinstry, 1998). Head-to-back stunning would be expected to directly stimulate muscles in the neck and upper latissimus dorsi because of the close proximity of the rear electrode to those muscles. It can result in a lower $pH_{40 min}$ than head-only stunning (Channon et al., 1997b), but this is not invariably the case (Aalhus et al., 1991). Head-to-brisket applications can accelerate the rate of pH fall in the longissimus muscles in comparison with head-only stunning, which confirms that other factors besides direct stimulation of muscle at the site of the rear electrode are important (Channon et al., 1997a). In contrast to head-only stunning, delivering a head-to-brisket current for 10 instead of 4 s resulted in a high $pH_{40 min}$ (Channon et al., 1997b). High-frequency stunning currents (400 and 1000 Hz) are becoming more common in red meat abattoirs, but there is insufficient information on their effect on meat pH.

Electrical stunning has a reputation for producing blood speckle and blood splash in the carcass. Blood speckle is small pinprick haemorrhages in the fascia and connective tissue, often near the site of an electrode. Blood splash (petechial haemorrhages) is bleps (purpuras) of blood either in the fascia or within the muscle. Both are thought to arise from capillary rupture during current flow. There are four theories that could explain how this occurs, but none have been proved. First, it could be due to counteracting muscle contractions during stunning causing localized tearing of the capillary bed, especially at sites where the muscle inserts on to the skeleton (Kranen et al., 2000a). It can be associated with super-contraction of muscle fibres (Leet et al., 1977). Petechial haemorrhages can be induced electrically in tissues that do not possess skeletal muscle, and so this mechanism may not apply in all cases of the condition. Secondly, arterial or venous dilatation at the time of stunning might increase the likelihood of blood splash. This was suggested when propanolol pretreatment reduced and phentolamine increased the expression of blood splash (Shaw et al., 1971). Presumably engorgement of the capillary bed encourages rupture of the vessels when they are placed under pressure. Thirdly, the blood vessels may be unduly fragile in some animals. Blood splash is common in animals that have died from anticoagulant poisoning, and these animals have raised capillary fragility as well as impaired blood coagulation. Raised capillary fragility might explain the susceptibility of unweaned lambs to blood splash (Gregory, 2005a). Finally, during intense generalized contractions, there is probably a rise in venous pressure. If sufficiently severe, this could rupture the capillary bed at sites some distance from those that generate the convulsive activity. In effect, there is referral of pressure to a site where the capillary bed is weak and this is where the bursts occur. Engorgement of the venous circulation at the time of stunning would presumably exaggerate this effect. This explanation would not apply in the case of head-to-back stunning, as this method produces elevated venous pressure and a low prevalence of blood splash (Gilbert and Devine, 1982).

Several approaches can be tested when trying to reduce the prevalence of blood splash. Avoiding rough handling before stunning may help. In veal calves the prevalence of severe blood splash can be greater following electrical stunning in a restraining conveyor when the calves are handled roughly, compared with considerate handling (Lambooy, 1985). Head-to-chest stunning is used in Germany as a way of avoiding blood splash in the loin, as well as broken backs that occur when pigs are stunned with the head-to-back system. The chest electrode is applied after the pig has been stunned across the head. Prompt sticking following stunning can also help reduce the expression of blood splash.

In broilers, if the breast meat is deboned early after water-bath stunning, there is a risk that the meat will go into rigor shortening and be tough, whereas this is a lesser risk with head-only stunning because of the extra muscle activity during stunning.

Head-only electrical stunning in poultry produces immediate wing flapping, which usually subsides to a tonic contraction in 6 s. Although these contractions are short-lasting they are sufficient to lower $pH_{5\,min}$ in the muscle (Savenije et al., 2002). Electrical stimulation during bleeding has a similar effect in water-bath-stunned birds, and it can be used to accelerate glycolysis and allow early portioning, provided steps are taken to avoid heat shortening. It is important to note that the convulsions during head-only stunning are unsightly and most processors would not tolerate them.

Inducing a cardiac arrest at stunning with the head-to-back method can reduce blood splash in lambs as well as pigs, but it does not necessarily reduce **blood speckle** (Gilbert and Devine, 1982). Blood speckle often extends to the hindleg where it is worse when high currents are used (Devine et al., 1983). It is linked to muscle activity at the time of current flow, and has been eliminated by preventing the contractions with curare (Gilbert and Devine, 1982). There may be a genetic component to blood speckle. In one study, blood speckle in the fascia of the loin and blood splash in the ham were more common in Nn than NN pigs (Velarde et al., 2001).

Carcass kicking and convulsions are common with high-voltage stunning in red-meat species, in stunning with anoxia and when pigs and poultry are concussed. In cattle and small stock, it is more difficult and dangerous to shackle a kicking carcass. In poultry, convulsions make the control of blood collection more difficult, and there can be a risk of dislocated wings. There are eight approaches to controlling convulsions (Table 11.8). Normally, they are controlled by regulating the duration of the stunning current or the length of the interval between stunning and sticking. Severing the spinal cord has been used to stop convulsions in cattle and sheep. Severing the spinal cord within 3 min of

Table 11.8. Ways of controlling carcass kicking.

- Pithing with a cane or rod
- Pneumatic pithing
- Severing the spinal cord
- Inducing a cardiac arrest at stunning
- Spinal discharge
- Electroimmobilization
- Prompt sticking
- Sticking before shackling

stunning in pigs had no effect on PSE characteristics, presumably because it was applied too late (Westervelt and Stouffer, 1978).

Pithing with a rod causes a brief rise in blood pressure (of about 20 mmHg), followed by a precipitous fall (Dock, 1940). In pneumatic pithing, a pulse of compressed air is injected into the brain when the animal has been shot by a captive bolt. It is effective at reducing both spontaneous activity and physical responses during sticking. As with mechanical pithing, it raises concerns about distributing bacteria, which are introduced into the brain with the pithing equipment, and distributing macerated brain to the lungs and perhaps the edible carcass via the venous drainage from the brain. For this reason, both methods are disallowed in some countries where there are concerns about BSE. Pneumatic pithing could be acceptable for ostriches, and could find favour if it eliminates skin damage from convulsions. In some countries, severing the spinal cord after stunning is called 'pithing' whilst in other countries it would be called neck stabbing.

Spinal discharge is used mainly in lambs, and a system has been devised for turkeys but is not widely used. In lambs it reduces activity, so the carcass can be worked on immediately after sticking. It reduces oscillation of the carcass whilst suspended by three legs, and this reduces regurgitation of reticulo-rumen contents into the oesophagus before weasand clipping or ringing. Spinal discharge also causes spasm of the oesophageal sphincter, which lasts for at least 30 s following the current, and this controls reflux of digesta on to the neck wound (Gilbert et al., 1984).

High-frequency electrical stunning

The outstanding advantage of high-frequency electrical stunning is that it can reduce the prevalence of blood splash. In the case of pigs, between 1000 and 2000 Hz is usually effective in eliminating shoulder blood splash. The disadvantage is that it can result in a more active carcass.

Early research showed that high-frequency currents produced less initial muscle spiking and a more even contraction during current flow. Muscle tension was lower and more prone to subsiding (Gregory, 2005a). The less forceful muscle contractions probably explain the lower prevalence of haemorrhages associated with high frequencies

in pigs. Some electrical stunning systems now being used in sheep and pigs apply an initial high-frequency current across the head (800 Hz for 2.5 s), followed immediately by a 50 Hz head-to-back current for 2 s to stop the heart. This system can lead to a high prevalence of blood speckle, but this downgrading feature is not as important as blood splash (Velarde et al., 2000).

Blood spots in poultry breast meat have been a common and serious cause of downgrading in the poultry processing industry, but high-frequency electrical stunning (and gas stunning) has helped reduce their importance in recent years. In many countries, high-frequency electrical stunning has replaced 50 or 60 Hz water-bath stunners, and this has produced a great saving in blood-blemished breast meat. These and the other effects of high-frequency currents are listed in Table 11.9.

High-frequency water-bath stunning produces a more active stunned bird. It also results in a shorter-lasting stun (Hillebrand et al., 1996; Mouchonière et al., 1999). In particular, the duration of unconsciousness in the quickest-recovering birds is shorter following higher frequencies (Wilkins et al., 1998; Wotton and Wilkins, 1999). Extra care is needed in checking that the birds remain insensible throughout the bleeding period, and a large sample of birds needs to be examined when making this assessment.

The effect of current frequency on the prevalence of cardiac arrest is shown in Table 11.10 (Gregory et al., 1991). At frequencies higher than 125 Hz there will be fewer birds with a cardiac arrest at stunning. It is only at very high frequencies (500 to 1500 Hz) that there is no cardiac

Table 11.9. Advantages and disadvantages of high-frequency electrical stunning in poultry.

Advantages
- Does not induce a cardiac arrest – can meet halal requirements
- Fewer blood spots in breast meat
- Fewer broken bones
- Fewer red wing tips and red pygostyles

Disadvantages
- More active stunned bird
- More rapid breast muscle pH fall
  - greater cooking loss
  - can be tougher
- Greater risk of recovery of consciousness

arrest, and these are the frequencies that are suitable for halal. Turkey hens are more prone to developing a cardiac arrest at stunning than toms, and so in some cases the frequency setting may need adjusting for the different genders (Mouchonière et al., 1999).

Blood spots in breast meat take many forms, but one of the most common conditions is blood in the tenderloin (pectoralis minor) where the coracoid bone joins the keel. This has become an important defect as more breasts are being sold as fillets. High-frequency stunning does not eliminate this problem but it allows a substantial saving (Table 11.11; Wilkins and Wotton, 2002). High frequencies (350+ Hz) also help reduce the prevalence of red wing tips (Wilkins et al., 1998; Contreras and Beraquet, 2001).

When the frequency of the current is 200 Hz or more, the direct effects of the current on muscle metabolism are usually negligible, provided the current is not applied for unnecessarily long periods. However, the greater physical activity following the current can be associated with more rapid breast muscle pH decline, and this can be linked to greater cooking loss (Santé et al., 2000). When using a pulsed direct current (DC), reducing the pulse width will reduce stimulation of the muscle, but the quality of the stun will also be lower.

The exact frequency–current combinations that provide a satisfactory stun when using high-frequency currents in water-bath stunners have not been specified (Beyssen et al., 2004; Raj et al., 2006a, b). The minimum current required to stun broilers increases with stunning frequency (Contreras and Beraquet, 2001; Raj and O'Callaghan, 2004b). The recommended currents for the head-only system are 100, 150 and 200 mA for 50, 400 and 1500 Hz sinusoidal alternating currents (ACs), respectively. Applying 150 mA for 4 s with a 300 Hz AC produces an acceptable stun in turkeys. Pulse widths less than 0.1 ms need higher peak currents to ensure an adequate stun (Raj et al., 2006c).

Gas stunning

Carbon dioxide, argon and nitrogen are used for stunning poultry and carbon dioxide is used for pigs. The main concern with CO_2 stunning is that it induces breathlessness before the animal is stunned. Set against this, the pig does not need to be restrained in a conveyor, nor do birds need to be shackled before they are stunned. This eliminates some of the pre-slaughter stress.

CO_2 is a potent activator of breathlessness and it is aversive (Raj and Gregory, 1995). However, when used at high concentrations, it produces a quick induction. This was shown in studies on the pigs' EEG. The time at which high-amplitude, low-frequency activity in the EEG increases was used as an indicator of the onset of unconsciousness. During 90% CO_2 stunning, delta (2–4 Hz) and theta (4–8 Hz) activity begin to set in at 6 s following immersion in the gas, and by 30 s they are well established (Martoft et al., 2002). Although the precise onset of unconsciousness is not known, it is thought to begin in 10 to 30 s. Many authorities consider that, even though breathlessness will occur during this period, CO_2 stunning is acceptable because of its other welfare benefits. In particular, when pigs are lowered into a pre-filled well of CO_2 as a group, there is less stress in comparison with forcing pigs individually into a restraining conveyor. If less stressful restraining methods are

Table 11.10. Effect of stunning frequency on the prevalence of cardiac arrest during electrical stunning.

Frequency (Hz)	Current (mA/bird)	% fibrillated
50	112	71
125	114	64
200	111	41
275	114	5
350	109	9
1500	–	0

Table 11.11. Effect of stunning frequency on the prevalence of blood spots and broken bones in turkey breast meat.

	50 Hz	1400 Hz
Blood spots in		
pectoralis minor (%)	38	4
pectoralis major (%)	19	5
Broken coracoids (%)	16	4
Haemorrhages associated with broken bones (%)	21	5

developed for electrical stunning, this outlook might change and CO_2 could become less acceptable.

Meat plants that use electrical stunning often have a higher prevalence of PSE meat than plants using CO_2 (von Zweigbergk et al., 1989). Controlled studies have confirmed that CO_2 stunning results in a slower rate of post-mortem muscle glycolysis. There is also substantially less blood splash in the meat compared with electrical stunning (Table 11.12; Channon et al., 2002). The overall outcome in terms of meat quality will depend on the prevalence of the halothane gene for the pigs entering the abattoir as well as the way the gas is applied. For example, the concentration of CO_2 can influence the severity of PSE in nn pigs (Troeger and Woltersdorf, 1991; Table 11.13), and nn pigs show more physical reaction to the gas (Troeger, 1990). CO_2 stunning will not prevent PSE meat, but it can help maintain a normal $pH_{40\ min}$ in Nn genotypes (Channon et al., 2000).

The handling system used in getting the pigs to the CO_2 stunner is critical. For example, at one plant where electric goads were used and the pigs were moved along a single-file race to a Compact stunner, the prevalence of PSE was 50%, whereas at another plant where the pigs were herded as groups using a semi-automated moving gate and then loaded in batches of four to six pigs into a lift, which was lowered into the gas, the prevalence was only 13% (Franck et al., 2003).

Gas stunning has been trialled with mixed success in sheep. In one study on suckling lambs, it resulted in more tender meat than electrical stunning or no stunning (Vergara et al., 2005). It is used for rabbits in a plant in Spain.

Six gas stunning systems are being used commercially for poultry. They differ in the gas composition they use. Two methods in the UK are: 60% Ar + 30% CO_2, and less than 2% O_2 in air using Ar. The anoxic method is falling from favour because of the severity of the carcass convulsions once the bird is stunned, and because of the cost of Ar. A system developed in the Netherlands uses a mixture of 40% CO_2 + 30% O_2 + 30% N_2 for inducing unconsciousness, followed by 80% CO_2 + 20% N_2 for killing. A system developed in Denmark also uses a two-phase approach with CO_2 in air starting at 5% and rising to 50% to kill the birds. In Japan and Italy, 40% or more CO_2 is applied for over a minute. In the Italian system the birds are lowered into a well containing a gradually increasing CO_2 concentration, with a final concentration of 60% of more. In the Japanese system the birds are suspended on shackles that are conveyed through the CO_2 unit.

Table 11.12. Comparison of CO_2 stunning with electrical stunning on muscle pH and blood splash.

		Electrical stunning	
	CO_2	Head-only	Head-to-back
$pH_{40\ min}$	6.63[a]	6.36[b]	6.20[c]
g blood splash trim	8[a]	101[b]	109[b]

Table 11.13. Influence of halothane genotype and CO_2 concentration during stunning on $pH_{45\ min}$ in the latissimus dorsi and semimembranosus muscles (after Troeger and Woltersdorf, 1991).

	Halothane-positive pigs			Halothane-negative pigs		
$pH_{45\ min}$	60% CO_2	80% CO_2	85–90% CO_2	60% CO_2	80% CO_2	85–90% CO_2
Longissimus dorsi	5.49[a]	5.68[ab]	5.67[b]	6.06[a]	6.09[a]	5.99[a]
Semimembranosus	5.60[a]	5.63[a]	6.05[b]	6.23[a]	6.37[a]	6.32[a]

Means for a halothane type in a row not sharing a common superscript letter were significantly different at $P < 0.05$.

The ways in which the birds are loaded into the gas stunner differ. In the Dutch system the birds are tipped from the transport module on to a belt conveyor, and DOAs are removed at an ante-mortem inspection point. In the UK and Danish systems, drawers are automatically removed from the transport modules and are passed through the gas unit, and the stunned birds are removed from the drawers. The UK and Danish systems involve less handling before stunning and so are less stressful for the birds.

Time to loss of posture is a useful behavioural indicator of failing consciousness during gas stunning (Mohan Raj and Gregory, 1990). It coincides with the onset of a high-amplitude, low-frequency waveform in the EEG during anoxic stunning, and with the onset of suppression of the EEG during $31\% CO_2 + 2\% O_2$ stunning (Mohan Raj et al., 1991, 1992). Time to loss of posture is fastest when CO_2 is combined with anoxia (Webster and Fletcher, 2001). CO_2 in air is not as quick, but the rate of induction is faster at higher concentrations (Table 11.14). In addition, $40\% CO_2 + 30\% O_2$ takes over twice as long to result in collapse in comparison with $70\% Ar + 30\% CO_2$ (Gerritzen et al., 2000). The longer the induction, when using CO_2, the greater the likelihood of breathlessness. These findings, along with other studies, indicate that the most humane kill can be achieved with anoxia (argon alone). However, this method produces violent anoxic convulsions in the unconscious bird, resulting in wing damage. This is a loss for the processor and so the Ar/CO_2 combination is preferred.

The quality advantages of gas stunning in poultry are reduced prevalence of breast muscle and wing haemorrhages, and fewer red wing tips in comparison with electrical stunning (Raj, 1994).

Electrical stunning (200 mA, 60 Hz for 5 s) in turkeys has been associated with higher m-calpain activity in the breast meat compared with CO_2s tunning (40 to $60\% CO_2$ for 60 s), but the effect was not linked to a difference in shear force once the meat had been aged (Northcutt et al., 1998).

Slaughter

The **main slaughter methods** are neck sticking and chest sticking. Chest sticking is done mainly in cattle and pigs, and it is used for sheep in some North American plants. Neck sticking is the usual method for sheep, goats, calves and poultry. With neck sticking the aim is to sever both carotid arteries. The position of the carotid arteries in the neck is shown for sheep in Fig. 11.5 and for chickens in Fig. 11.6.

When animals are given a reversible stun, they should be bled immediately. The aim is to render them insensible by stunning and then deflect blood away from the brain to ensure prompt loss of brain function without recovery of consciousness. The recommended maximum stun-to-stick intervals following low-frequency electrical stunning in pigs, sheep and calves are 23, 27 and 27 s, respectively. These values are based on the reasoning given in the section on electrical stunning in this chapter, and using data on the time to return of neck tension or head righting as the onset of return of consciousness. Working within these values will ensure that the animals do not regain consciousness. They apply to chest sticking for pigs and calves and neck sticking for sheep.

If neck cutting is used as the slaughter method in calves and cattle, the situation can be complex.

Table 11.14. Time to loss of posture and signs of breathlessness in broilers during gas stunning.

Treatment	Time to loss of posture (s)	Number of gasps before loss of posture	Prevalence of birds with gasping
Air	–	–	0
30% CO_2	57.0[a]	16.1[a]	100
45% CO_2	37.9[b]	8.9[b]	100
60% CO_2	35.4[bc]	8.3[b]	100
Ar/CO_2	28.5[c]	8.3[b]	100
Ar	30.7[bc]	0	0

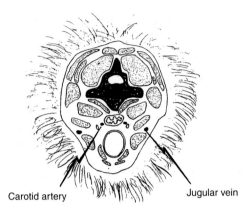

Carotid artery   Jugular vein

Fig. 11.5. Location of the carotid arteries in the sheep's neck.

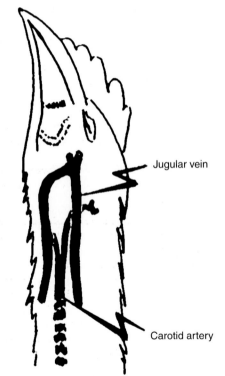

Jugular vein

Carotid artery

In the mid-neck region, the carotids
are enveloped by neck muscle

Fig. 11.6. Location of the carotid arteries in a chicken's neck.

There is a greater risk of recovery of consciousness because cattle and calves have a collateral blood supply to the brain through an occipito-vertebral anastomosis, which can allow extended perfusion of the brain if the carotid arteries do not bleed freely (Bager *et al.*, 1988). The severed ends of the carotid arteries in calves and cattle are prone to developing ballooning. The ballooned ends of arteries contain trapped blood, which disrupts bleeding out (Anil *et al.*, 1995). The ballooning is in fact a false aneurysm and it occurs in 16% of cattle and 25% of calves following neck cutting. The aneurysms increase the risk of resumption of consciousness following reversible electrical stunning, and they extend the period of consciousness following sticking when there has been no stunning (Blackmore, 1984a).

When head-only electrical stunning is used in large cattle for the halal trade, it is sometimes recommended that the neck is cut within 10 s of stunning. The animal is rolled out of the stunning pen on to a bleeding cradle so that it can be promptly bled, instead of being hoisted before bleeding. Even so, it can be difficult to achieve a rapid cut, and in practice in New Zealand it is usually done between 9 and 24 s.

When non-reversible stunning methods are used, the stun-to-stick time is not critical, and it does not need to be specified. It is advisable to stick the animal within 3 min of heart failure; otherwise there is a risk of a poorer bleed-out.

The size of the sticking wound can influence bleeding rate in pigs (Anil *et al.*, 2000). When a long sticking wound (11 cm) was made in the skin, the pigs took 72 s to bleed out, whereas for a short sticking wound (5 cm) it was about 84 s. The longer sticking wound could require more trimming but the unrestricted blood flow with this method should improve confidence in achieving prompt death.

Some poultry processing plants sever the spinal cord during neck cutting. This reduces problems with physical activity following high-frequency electrical stunning. **Spinal cord severance** can be recognized when a bird flips its wings upwards as the blade cuts and stimulates the cord. It is easily achieved by increasing the penetration depth of the rotating blade that cuts the back of the neck. Besides reducing physical activity, it prevents the expression and recognition of conscious activity, and this is a concern from a welfare perspective. It means that it is no longer feasible to identify whether the birds remain unconscious throughout the bleeding period. Using a neck cutter that severs the carotid arteries at the front of the neck would be a preferred option as it would interrupt the flow of blood to the brain. However, this has raised the following commercial concerns. If the front of the neck is cut, it is said that:

- there will be pieces of trachea and oesophagus left in the neck flap after head removal by a head puller. These contaminants will have to be removed by some other means, but this should be feasible with the croppers, final inspection machines and neck flap trimmers used in modern evisceration lines
- more heads are likely to come off during plucking, and, where feathers are a valued co-product, the heads will cause off odours if they are not removed within a day or two
- in some evisceration lines the head is used as an anchor by some machines to maintain tension in the carcass. Machine damage could be higher in headless carcasses

On account of the welfare concerns, one poultry meat buyer in Europe is asking its suppliers to decapitate the birds after high-frequency electrical stunning instead of using conventional neck cutters. This removes any concern about the resumption of consciousness when using high frequencies and a back-of-the-neck cut. Decapitation following electrical stunning does not increase the risk of contamination of the respiratory tract with scald water, and so this is not a justified reason for not decapitating (Buhr et al., 2005). Decapitation following electrical stunning can help arrest physical activity in the carcass, increase breast meat pH$_{ult}$ and reduce the prevalence of red pygostyles, without affecting cooking loss, toughness or meat colour (McNeal and Fletcher, 2003; McNeal et al., 2003).

A little over 40% of total blood volume is usually released at sticking, and a high proportion of the remainder is removed with the viscera. Small amounts of **residual blood in meat** can lead to blood staining of the drip that accumulates in retail packs, and less commonly it leads to blood smears on the surface of the meat when it is cut. Some people find bloody meat objectionable, and so meat processors are looking for ways that keep residual blood content at tolerable levels.

The amount of blood lost from the sticking wound is used as an indicator of **bleeding efficiency**, but it is not closely related to the amount of residual blood in the meat (Warriss and Leach, 1978). The residual blood content in beef is higher when the shot animal is bled horizontally instead of suspended by a hindleg (Hess and Klinger, 1967). Delaying neck cutting following 50 Hz 120 mA electrical stunning in broilers can lead to discoloration of the breast muscle from lividity (Raj et al., 2001).

The main meat safety hazards associated with slaughter are:

- regurgitation of rumen contents on to the neck wound and uptake into the lungs in captive-bolt-shot and stuck cattle
- contamination of the neck with microorganisms introduced with the sticking knife or neck-cutting machine
- contamination of the lungs and possibly the meat with brain emboli in captive-bolt-shot cattle
- contamination of equipment with brain matter following captive bolt shooting in cattle
- contamination of meat by microorganisms present on the bolt of captive bolt guns in cattle
- water aspiration into the respiratory tract from the water-bath stunner in poultry

Less common problems include:

- blood aspiration into the lungs following sticking in water buffalo (Kumar and Parihar, 1998)
- contamination of the edible carcass with microorganisms introduced with a pithing cane or rod

In beef, **electro-immobilization** is used in some countries between captive bolt stunning and sticking to reduce the risk of injury to staff. It makes it easier and safer to shackle the stationary leg before hoisting the carcass. It uses a high-frequency current (2000 Hz) with narrow pulse width (0.1 ms) for about 10 s. This high frequency is safe and has little effect on post-mortem glycolysis. Separate low-frequency electrical stimulation currents can be applied later in the line to hasten pH fall if that is required.

Electrical stimulation of the carcass can improve bleeding efficiency as well as controlling the risk of cold shortening. In some countries it is used as an aid to bleeding, and is called **electronic bleeding**. An extra 1 kg or more of blood is released when using a 5 or 14 Hz 0.5 A current for 35 s. In heavy carcasses, a 5 Hz current is preferred, whereas 14 Hz is used when processors also wish to control the risk of cold shortening.

An electric current can also be used to reduce carcass kicking at the legging stand in the dressing line. This current is applied for 20 to 30 s between sticking and legging, and, where acceleration of pH fall is not wanted, a high-frequency (1000 to 2000 Hz) current is used.

Slaughter without stunning

Some municipal abattoirs in poorer countries operate on the traditional shambles system. A shambles consists of bays either side of a central hall. Each bay is rented by a butcher, who uses it for slaughter as well as dressing the carcasses. The animals are usually led or driven individually to a slaughter bay, and this can be hazardous if they are not easy to handle. A method used at one abattoir in North Africa for controlling unruly cattle is to reduce their mobility by cutting the metatarsal tendons as they are driven to a bay. Eye stabbing is sometimes done to disorient and slow down unpredictable cattle. Cattle with severed tendons usually go down on their own accord within about a minute, but, if they do not, they have to be cast. This is sometimes done by one person holding on to its tail whilst another twists the head. Once the animal is down, the neck is restrained to tension the skin and soft tissues, and the cut is made.

Large numbers of animals are slaughtered for consumption at religious festivals and celebrations. For example, goats are slaughtered at yam festivals, rams at Sallah celebrations, pigeons have a ritual role amongst Yoruba people and cockerels are used in Igbo traditions. There are many other examples. The main religious slaughter methods in industrialized countries that do not usually involve stunning are shechita, halal and jhatka. These methods are used for routine slaughter as well as producing meat for consumption during festivals and periods of religious observance.

The methods used in religious slaughter are not described in explicit detail in religious scriptures. Instead, the Koran and Talmud give guiding principles. Like many other scriptures, these texts are not prescriptive. They give historical accounts, and they portray values and a way of life. They are used by the reader to develop his or her faith by blending the written messages with personal outlook and religious teaching.

Present-day methods used in slaughter according to the Islamic faith have been developed from interpretation of the scriptures, whilst in the Jewish faith the methods are based on repeating the traditional method dating back to times before the scriptures were committed to paper. In both cases, adherence to the slaughter methods and consumption of meat from animals killed by those methods is an integral part of being good Muslims or good Jews. Certain features of the methods have, however, been adapted to commercial constraints. Nevertheless, pressure from people outside those faiths to adopt practices that turn them into 'bad' Muslims or Jews is resented. The situation in Sikh communities is different. According to ancient Aryan Hindu tradition, only meat from an animal killed with one stroke of a weapon causing instantaneous death is fit for consumption. Decapitation (jhatka) is the preferred method. Reciting expiatory verses during the slaughter process is unacceptable. The Sikh community is, however, more flexible in its approach, because the important tenet in deciding whether meat should or should not be eaten is whether it creates trouble in the meat consumer's body and fills the mind with evil. If no ill effects are anticipated, then there is no harm in eating meat prepared by other methods.

It is recognized that during shechita the neck should be cut as a single uninterrupted reciprocating action. In other words, there should be one sweep of the blade in one direction followed by one sweep of the blade in the returning direction, and the travel of the blade throughout this process should be continuous or uninterrupted. A reason for this requirement is that repeated sawing of the neck wound is likely to stimulate free nerve endings in the tissues of the neck and give rise to pain. It may also exacerbate any distress associated with the traumatic experience. A sawing action with the knife is sometimes used by people killing animals for their own use. In order to make a single reciprocating cut and sever both carotid arteries, the blade of the knife should be relatively long. In shechita the length of the blade is twice the width of the neck. The cut is made swiftly with a very sharp shechita knife, and the animal should not show physical reactions, such as leg kicking or tail flicking whilst the cut is being made. In halal slaughter it is common for there to be one non-reciprocating cut. In other words, the knife is inserted into the side of the neck and the soft tissues in the neck including the carotid arteries are severed as the knife is drawn across to the other side of the neck, with much of the skin being cut from the underside outwards. Unlike the method used in shechita, the halal method is subject to variation according to custom or experience of the slaughterman. For example, a minority of halal slaughtermen sever the spinal cord, a minority decapitate the animal, whereas others do not cut the spinal cord and claim that cutting the spinal cord is unacceptable because it reduces physical activity and a good bleed out.

Some people question whether the religious slaughter methods are essential to being good practitioners of the respective faiths. The questions raised are framed in the following ways:

- Do the religious slaughter methods conflict with other religious or moral requirements on humane animal treatment?
- Are the prescribed methods an integral part of personal faith adopted freely by the individual or are they imposed by doctrine?
- Are some features of the prescribed methods cultural rather than religious?

The main focus in debates on religious slaughter has been on the first of these questions. The other questions have not been discussed so openly for two reasons. They are considered offensive, and they are sometimes interpreted as being part of racist and anti-religious attacks. It has been difficult to disentangle motives behind the discussion and the subject matter itself (Klug, 1989).

The legal situation can be summarized as follows. Many countries have a Bill of Rights that allows religious and cultural expression of minority groups provided their activities comply with other laws. Exemptions are usually made for religious slaughter, to allow freedom of expression of religious beliefs. In this situation, the law in many countries allows human rights to take precedence over concerns for the animals.

There are some exceptions to this last generalization. In Australia and New Zealand all halal slaughter is preceded by stunning, and this is done with the agreement of the Islamic authorities, who insist that stunning must not kill the animal. This has to be inspected periodically, by testing whether a stunned animal regains consciousness if it is not bled. In other countries, shechita in cattle has to be followed within a specified time period (usually 10 s) by captive bolt or percussion bolt stunning. The intention is to minimize any risk of distress during bleeding out. When this has been tested, stunning stopped the subsequent rise in plasma noradrenaline concentration (Petty et al., 1994). Whether this was due to a reduction in distress or inhibition of subconscious reflex responses to hypotension is not clear.

The main concern is that the risks of suffering during slaughter without stunning are greater than for slaughter methods that use stunning. There are three hazards. First, cutting the neck of a conscious animal requires a robust restraining method, which in some situations is stressful for the animal. Secondly, the injury discharge when nerves in the neck are severed and subsequent activity in nerves adjacent to the wound would cause a sense of trauma and in some instances there could be pain. Thirdly, the animal may experience distress subsequent to the cut and before it loses consciousness.

Some authorities claim that the animal does not appear to feel the throat cut, and they do not appear to be aware of what has happened (Grandin, 1992), whereas others find this difficult to believe. A more honest appraisal would be that we do not know what proportion of animals experience pain and what they are thinking about in this situation.

Over the years, seven methods have been used for restraining cattle during shechita. They are casting by twisting the head back, casting with a rope, hoisting by a hindleg, restraint in a straddle conveyor or restraining (V-shaped) conveyor, half-inversion in a rotary casting pen, full inversion in a rotary pen and restraint whilst standing upright. Casting by hand can cause the animal to land heavily on a hard floor.

Koorts (1991) compared the prevalence and severity of struggling in over 1500 cattle that were either inverted in a casting pen and subjected to shechita, or were held in the same pen and stunned whilst in an upright position. Animals restrained for shechita were more stressed than for secular stunning and slaughter (Table 11.15). In 15% of the cattle about to be killed by shechita, more than one attempt was made at inverting them in the rotary pen and 4% managed to escape from the pen. If head restraint is applied after an animal has been inverted, there is more vigorous struggling than with head restraint before inversion.

The duration of struggling in an upright pen can be shorter-lasting compared with a rotary pen (1 vs. 11 s), and the number of vocalizations is correspondingly lower (Dunn, 1990). However, it is more difficult to cut the neck when the animal is held in an upright pen, and it is more difficult for the shochet to avoid being covered with the blood.

Table 11.15. Average time spent in the casting pen and prevalence of stressed behaviour before stunning or shechita.

Category	Average time spent in the casting pen before stunning or shechita (s)				Prevalence of stressed behaviour (%)	
	Secular	No.	Shechita	No.	Secular	Shechita
Calm	7	1085	33	511	69.4	31.4
Nervous	14	418	62	868	26.7	53.3
Wild	22	58	81	208	3.7	12.8
Frantic	35	2	99	41	0.1	2.5

Hoisting the conscious animal by a hindleg is used in some beef abattoirs in the Middle East. It is sometimes done with the animal held in an upright pen. When the suspended animal is quiet, it is lowered to rest its head and neck on the ground or a platform and the neck is tensioned and then cut. This method was used in the USA, where hygiene regulations require that animals must not be on the ground when slaughtered. However, it was associated with struggling and vocalization, and has now been replaced by upright pens and slaughter within straddle conveyors (Regenstein and Grandin, 1992).

When the neck is cut, there will be injury discharges in the nerves that are severed. These injury discharges will be brief, and it is likely that they will quickly subside (Gregory, 2005a). Once the severed ends of nerves become depolarized, they are unresponsive. Undamaged free nerve endings in the wound may, however, be responsive, and so the way the wound is managed before the animal loses consciousness will determine whether pain occurs.

The time to loss of consciousness depends on blood being directed away from the brain. In broiler chickens, the time to loss of consciousness following shechita can be assessed from the time to loss of posture after the birds have been cut and immediately placed on their feet. On average this was 14 s (range 8 to 26 s). The implication is that birds could be conscious for up to 26 s following the cut.

In many forms of religious slaughter there must be no residual blood in the carcass. This depends in part on the efficiency of bleeding out following the cut. In sheep, there is no difference in the rate of blood loss between halal slaughter without stunning, halal slaughter following electrical stunning, and sticking following captive bolt

stunning (Anil *et al.*, 2004). There is, however, a greater risk of blood splash following head-only electrical stunning compared with slaughter without stunning (Kirton *et al.*, 1980).

In the absence of stunning there are no tonic spasms in the carcass after the cut has been made. However, the clonic phase can be very vigorous (Petersen and Blackmore, 1982). If this kicking occurs after the animal has bled out, it is not a concern from the welfare perspective, and it can indicate loss of cortical function in the brain. The kicking may contribute to early acceleration of post-mortem muscle glycolysis. For example, in bulls, $pH_{15 min}$ and longissimus dorsi glycogen can be lower in the non-stunned animals compared with bulls stunned by percussion bolt (Önenç and Kaya, 2004).

Meat from bulls slaughtered by the halal method without stunning may have less cooking loss compared with bulls that are either electrically stunned (400 V, 1.5 A, 50 Hz, 10 s) or shot with a percussion bolt before the halal cut (Önenç and Kaya, 2004). This difference can persist even when the meat is aged. In comparison, the greater initial hardness of meat following electrical stunning resolved within 96 h.

When meat quality has been compared in electrically stunned lambs and lambs that were bled without stunning, there were few or no differences (Vergara and Gallego, 2000). However, meat from electrically stunned lambs may age a little faster. In contrast, in unstunned turkeys and chickens, the carcasses undergo vigorous convulsions, which are sufficient to accelerate post-mortem glycolysis in breast muscle (Northcutt *et al.*, 1998; Ali *et al.*, 2005). In turkeys, the wing flapping results in lower breast muscle $pH_{25 min}$ and tougher meat compared with birds bled after water-bath stunning (Alvarado and Sams, 2000). Neck cutting without electrical stunning also results in a lower prevalence of red pygostyles (McNeal *et al.*, 2003). Electrical stimulation of the carcass after halal slaughter can improve meat tenderness in hens (Suguna *et al.*, 1997).

In the case of shechita, broiler breast meat is more tender following shechita compared with electrically stunned non-kosher neck cutting (Mast and MacNeil, 1983). This is because kosher carcasses are chilled more rapidly and they receive less physical stimulation during plucking, and the improved tenderness is incidental to the killing method itself. Evidently, the post-slaughter processing methods help to reduce the toughening effect arising from the convulsions.

The amount of blood lost from chickens slaughtered by shechita is comparable to that in electrical stunning with neck cutting (Kotula and Helbacka, 1966). Decapitating unstunned birds is associated with poor bleeding efficiency.

Fish, crustaceans and frogs

The traditional method used for killing farmed fish is to allow them to die in air (death by **emersion**) or whilst held in crushed ice. When a fish is held in air, the gill lamellae collapse and adjacent filaments adhere to one another. This reduces the surface area for gas exchange and the fish dies from hypoxia. If they are killed in air whilst held in crushed ice, the time to brain failure is protracted (Kestin *et al.*, 1991). In comparison, at normal summer temperatures in Scotland, insensibility occurs within 4 min when rainbow trout are removed from water and allowed to die in air without ice chilling (Wall, 2001). Gilthead sea bream (*Sparus aurata*) take longer to die when held in air. If fish are held together in bins to die in air, they can damage each other during physical activity. This is worst in species that have sharp dorsal spines, which flare as they die.

Death by emersion is not recommended as a humane method because it is slow, and it is considered stressful. The recommended killing methods for farmed fish are concussion stunning followed by bleeding, brain spiking and electrical stunning (Robb and Kestin, 2002).

Electrical stunning is used in some salmonids, and for carp in France and the Czech Republic. Mediterranean farmed tuna that are destined for the fresh market are sometimes stunned electrically to control their movement. A semi-automated stunner has been developed in Norway, which stuns salmon in batches every 70 s. There have been problems with electric shocks when fish have been presented tail-first to the stunner, instead of head-first. A semi-automated electrical stunner has also been developed for trout, but it is prone to producing blood spots in the meat during specific seasons of the year.

The minimum current necessary to stun eels is 600 mA (Lambooij *et al.*, 2002a). This value is based on the induction of an epileptiform EEG in the brain, and it applies to head-only stunning, which is not really feasible where there is a need for high throughputs. A more practical alternative is to pass current through a tank of eels in water, and a satisfactory A/dm^2 is 0.64 (Lambooij *et al.*, 2002b). However, eels have a remarkable ability to recover from electrical stunning, and the stunning current needs to be followed by a second longer-lasting killing current plus partial deoxygenation of the water using nitrogen. This additional procedure can ensure there is no recovery. When eels were killed in this way, their meat was less prone to oxidative rancidity, and was firmer and redder in comparison with the traditional eel-killing method of placing live eels in a bin with crushed salt (Morzel and van de Vis, 2003). The system recommended for salmon, which will stun all the fish whilst minimizing carcass damage, is a 500 Hz current applied for 10 s at a field strength of 50 volts/m (Roth *et al.*, 2004).

Electrical stunning systems have been developed for crabs in restaurants and processing plants. When the electrodes make contact with the carapace of *Carcinus maenas*, a current of 1.5 A for 6 s is delivered and the crab is then placed immediately in boiling water.

In **CO_2 stunning** systems used for salmon, the water in the stunning tank is saturated with CO_2 gas to a pH of about 5.0. The salmon are transferred from a holding tank to the stunning tank in bulk using a pivoted riddle. For the first minute in the CO_2-impregnated water, the fish are hyperactive and show escape attempts. The impression is that it is highly stressful for the fish (Robb and Kestin, 2002). By 2 min they are subdued and by 3.2 min they are unconscious. Although this is criticized because of the initial stress, others support its use because of its speed, efficiency and effectiveness (Wall, 2001).

Salmon stunned with CO_2 usually go into rigor considerably quicker than those stunned by concussion or electricity (Roth *et al.*, 2002). This is because of their greater physical activity. However, electrical stunning can lead to early rigor if the current is applied for too long. Early rigor makes filleting more difficult and meat yield is likely to be lower. **Brain spiking** can delay rigor development by reducing subsequent convulsions, and this is sometimes used in tuna, salmon and a range of species caught by longline. There is resistance against brain spiking in some markets, because it disfigures the head. It can cause an immediate loss of evoked responses in the brain, but is prone to errors in the way it is applied (Robb *et al.*, 2000).

A semi-automatic gill cutter has been developed for salmon that have been immobilized by chilling or partial stunning with CO_2. The number of gill arches that are severed varies between fish according to their size and orientation in the cutter. When unstunned salmon are bled manually, they show head shaking and exaggerated opercular movements. The acceptability of bleeding fish out without prior stunning depends in part on whether they can feel pain. This has been a controversial topic in recent years. Some take the view that fish probably cannot feel pain (Rose, 2002), whereas others consider that elasmobranchs have diminished ability to feel pain but teleost fish probably can (Sneddon *et al.*, 2002).

Less common catching and killing methods used for fish in natural reserves are looping, spear fishing, bamboo traps, explosives and poisoning. Looping is a form of snaring. The fish is caught in a baited nylon loop by its dorsal and pectoral fins. In some regions, spear fishing is with a kukuri knife and torch at night-time.

Blood spots are an important cause of downgrading in salmon. They occur in the blood vessels that branch from the aorta and supply the muscle, and in the worst cases the blood runs along the length of the fillet at the midline. The vessels are thought to rupture during killing, evisceration or filleting, and they become more noticeable after the meat is smoked as this makes the blood darker. High-frequency electrical stunning results in fewer blood spots and broken vertebrae than 50 Hz ACs, but in trout there is a greater risk of recovery of sensibility following the stun (Robb *et al.*, 2002b; Roth *et al.*, 2002, 2004). Blood spots can be seen in salmon that have been stunned by

concussion if they are not bled within 4 min (Robb *et al.*, 2003).

Isoeugenol (AQUIS™) is used as a pre-slaughter sedative for salmon in some countries, and it has also been trialled successfully for eels and some crustacea. Flavour residues in the meat prevent its use in fish destined for the high-value Asian markets.

When **concussion** is applied correctly, it causes immediate loss of evoked responses in the brain (Robb *et al.*, 2000). It also seems to be the stunning method that has the least effect on post-mortem muscle metabolism (Ruff *et al.*, 2002). However, it can disfigure the head, and in particular the eyes. Manual concussion is too time-consuming for small fish, but there are automatic machines that can stun about 5000 fish per day.

Explosives are still used for stunning fish. For example, Nepalese fishermen throw rice or oil cake into a pool to concentrate katle (*Neolissocheilus hexagonolepis*) and tor fish (*Tor* spp.), and then throw in an explosive wrapped in a thick cloth. The stunned and dead fish are collected with a scoop net. The use of explosives has been blamed for unselective destocking in some natural inland water reserves in Africa.

Poisoning is occasionally used as a catching method and it also has the weakness of being unselective. Crushed leaves of *Sapium insigne* and *Agave americana* or bark and roots of *Dalbergia stipulacea* are amongst the preferred natural poisons, and aldrin, γ-benzene hexachloride (BHC), DDT, malathion and thiodine have also been used in the past.

Gill and trammel nets are used more in the fisheries industry than in aquaculture. They are, however, used for catching crustacea for stocking grow-out farms. A gill net consists of netting secured to a buoyed head-rope and a weighted foot-rope, which keep it open in the vertical position. The lobsters or other crustacea become entangled as they try to swim through. Trammel nets have three layers, with a fine-mesh net sandwiched between larger mesh.

The methods used for killing rock lobsters in restaurants and retail shops include 'drowning' in fresh water, boiling, head spiking, chest spiking, splitting and tailing, and electric currents. They may or may not be preceded by chilling. Drowning marine lobsters in fresh water causes an osmotic stress and vigorous struggling and is regarded by many as an inhumane killing method. Marine lobsters are not euryhaline and cannot reduce the loss of Na^+ from the gills sufficiently when placed in fresh water. Instead, it has been recommended that they should be chilled to less than 4°C in order to ensure insensibility before they are killed (Gregory, 2005a). Concerns have been raised that chilling in ice may not provide satisfactory insensibility in cold-adapted yabbies.

Over 5000 t of frogs' legs are produced every year. The most common killing method is to fast them for up to 24 h, then chill them in cold water for at least 15 min, followed by bleeding out when the legs are cut from the pelvis. Other methods are immersion in a concentrated saline solution and electrical stunning.

On-farm killing methods

Animals are killed on farms either because they are injured, diseased, unwanted or needed for home consumption. The main methods are:

- shotgun, rifle or captive bolt
- concussion with a hammer or other blunt instrument
- sticking without stunning
- neck dislocation or decapitation
- gassing

Downer and casualty animals are usually dispatched either by a veterinarian with a captive bolt gun or by a farmer with a rifle or shotgun. This is usually done satisfactorily when in the hands of an experienced operator. Whole herd destruction for health control purposes can be more problematic. In some countries, animals are driven into pits where they are shot from above and then buried. This can conjure images of panicking animals, but that is not necessarily what happens.

One method that worked well during the 2001 foot-and-mouth disease (FMD) outbreak in the UK was to put the cattle through a crush where they received an injection of a sedative (xylazine). The animals were then released into a yard, and as the sedative took effect they sat down. They were then shot with a captive bolt and pithed.

Captive bolt shooting is often preferred for sheep. More than one gun is needed when large numbers have to be dispatched. One gun needs to be rested to allow it to cool whilst another is in use. One of the best bloodless methods for pigs is a pair of electrical stunning tongs, which deliver a 50 Hz current through the head. Current is then passed through the chest to fibrillate the heart. This has been less stressful for the pigs than shooting with a rifle.

Some of the major dairy exporting countries aim for a seasonal calving pattern to maximize milk production from grass. Cows that develop a long calving interval are either culled or induced to calve early, using long-acting corticosteroids or synthetic prostaglandin. The calves either fail to breathe after delivery or they are killed at birth because they are difficult to rear. Clubbing a new-born calf's head with a hand-held hammer can be very effective, but it is not recommended in calves that are more than 2 days old. Older calves should be shot with a rifle or with a captive bolt or shotgun loaded with heavy shot.

Farmed deer in antler are sometimes put up for trophy shooting. A head shot is not accepted if the head is to keep its trophy value, and a chest or neck shot is used instead. Chest shooting is often preferred amongst deerstalkers hunting wild deer because they are more confident that this will kill the animal even though unconsciousness and death may not be immediate.

In poultry farm clearances, birds used to be killed in their sheds with Zyclon B, which is a canister that releases cyanide gas when thrown into the shed. In battery cage units the convulsions in the dying birds caused some cage damage. This product is no longer available, and it has been replaced by other gaseous methods. CO_2-filled skips and trucks were used during the *Salmonella* control programme during 1989 in the UK. The birds were caught by hand and transferred to the CO_2 unit. Some difficulties were experienced in delivering the gas. With vapour-feed systems the manifold was prone to blocking as it froze, and with liquid-feed systems the birds experienced a sudden drop in temperature before they were unconscious. These systems have now been refined, and an in-shed liquid-feed system is considered the most practical option. There is gasping before loss of posture, indicating that there is a period of breathlessness (Gerritzen *et al.*, 2004). An argon–CO_2 system has been developed for single modules filled with birds.

Neck dislocation is a relatively easy method for destroying small birds on farms. It is, however, tedious if large numbers have to be dispatched. In large birds, such as adult turkeys and geese, it is physically challenging when done by hand, and either purpose-built equipment is used, or the neck is held on the floor under a broom handle. Whilst standing on the handle the legs are pulled upwards to break the neck. As with all neck dislocation methods without stunning, there are strong convulsions. Some authorities oppose the routine use of neck dislocation because it does not have a concussive effect and may not produce immediate unconsciousness (Gregory and Wotton, 1990b).

One of the methods used for thinning breeder turkey flocks has caused concern. A video recording made by People for the Ethical Treatment of Animals (PETA) showed an operator walking through a shed of breeder birds. Selected birds were struck on the head or neck with a hockey stick. This method was good in that it did not disturb the whole flock, but some stunned birds that were transferred to a disposal skip recovered and stood up. A cartridge-powered percussion bolt gun is available for geese, and should be suitable for breeder turkeys as well.

12

Meat Quality

The principles that link meat quality with animal welfare have been described already (Gregory, 1998). This chapter gives examples that support those principles and brings together new information on genetic and farm management factors that influence meat quality.

Stress and Meat Quality

Stressing an animal before it is killed can make its meat more tender. Several historical accounts describe ante-mortem treatments used in the past to improve tenderness (e.g. Anon., 1910). Bulls were baited by dogs, pigs were whipped in a similar way to that still used on dogs in parts of Asia, and cattle were chased through the streets by mounted stockmen equipped with goads. Similarly, meat from wild animals that had been severely stressed, such as hares caught in chases, was renowned for its tenderness.

The duration of the stress determines whether the meat will in fact be more tender. Physical activity that lasts for hours before an animal is slaughtered will deplete muscle glycogen. If muscle glycogen is reduced below a critical level, the pH_{ult} of the meat will be raised. In the case of pigs, pH_{ult} is only affected if muscle glycogen concentration at the time the pig is stunned is less than 53 mmol/kg (Henckel et al., 2002). Meat with a high pH_{ult} has higher calpain activity, which imparts tenderness as the meat ages, and it is prone to being dark in colour. It also retains moisture more effectively when stored and cooked. In pigs this type of meat is known as dark, firm, dry (DFD) pork, and in cattle it is called dark-cutting

beef (DCB). High pH_{ult} meat from stressed animals does not keep as well as meat with a normal pH, and it has a poor shelf life when sold fresh. In these ways, stress benefits some meat quality features and has a detrimental effect on others.

Sometimes, relatively mild stresses can upset the suitability of meat for a particular market. At an export abattoir that specialized in supplying kobi beef to Japan, feedlot cattle produced meat that lacked sufficient brightness in colour because of a high prevalence of intermediate pH_{ult} cuts. This developed because the animals did not sit down on the hard concrete floor in the holding yards on the night before slaughter, and so their postural muscles were not sufficiently rested when the animals were slaughtered.

Stress before slaughter can affect meat quality in other ways. If an animal is stressed briefly just before it is slaughtered, it may die whilst its muscle is acidotic. In this situation, there is a risk that the meat will be pale and have poor water-holding capacity (WHC). Pig breeds and strains that are double recessive for the halothane gene (nn) are prone to accelerated post-mortem muscle acidification and producing pale pork which has a poor WHC (PSE meat).

PSE and DFD are now recognized features in poultry meat but they are not thought to be causing consumer dissatisfaction. They can be recognized from the criteria shown in Table 12.1, and, in the case of PSE, the prevalence is claimed to be high. For example, in turkey meat in the USA, the prevalence of $L \geq 53$ is 40% (Owens et al., 2000a). Poultry breast meat is naturally pale in colour, and so high L values are not thought to be abnormal. If complaints do arise in the future about PSE

©N. Gregory 2007. *Animal Welfare and Meat Production*
(N. Gregory)

poultry meat, they are more likely to be connected with its texture. PSE chicken meat can be abnormally soft and almost mushy in texture (Zhang and Barbut, 2005b). This situation is different from PSE pork, which is often chewier in texture. DFD broiler meat can be more prone to developing thiobarbituric acid reactive substances (TBARS) when stored, and this could give rise to complaints about off flavours (Young *et al.*, 2003). However, DFD broiler meat can be ideal for producing processed products because of its ability to retain water during cooking (Zhang and Barbut, 2005a; Table 12.2).

The DFD condition has been responsible for carcass condemnations by meat hygiene services. In 2004, about 9% of turkey carcasses in Canada were condemned for 'cyanosis'. This abnormality was misdiagnosed, and it was due to an excessive dark colour caused by the DFD condition. Cold stress during transport was probably a contributory factor (Mallia *et al.*, 2000a, b). At the other extreme, high temperatures can produce PSE in turkeys but this would not normally be recognized during routine meat inspection (Babji *et al.*, 1982).

Modern Problems with Meat Quality

Meat consumers are becoming more discerning about the appearance of fresh meat quality and their buying patterns are affecting demand further up the supply chain. For example, in European markets there is good demand presently for double

Table 12.1. Classifying criteria for PSE and DFD broiler meat.

PSE	L > 53	pH < 5.7
Normal	L > 46 and < 53	pH > 5.9 and < 6.1
DFD	L < 46	pH > 6.1

L, measure of surface reflectance and paleness in meat.

Table 12.2. Cooking loss and softness in chicken meat.

	PSE	Normal	DFD
Cooking loss (%)	27.96[a]	25.77[a]	21.32[b]
Fracture force (N)	2.50[a]	4.16[b]	4.95[b]

muscling in cattle because it provides large cuts of lean meat. Similarly in pigs, leanness fetches a premium. In general, leanness is associated with lower levels of marbling fat within the muscle (intramuscular fat) as well as smaller amounts of subcutaneous and intermuscular fat. Marbling fat can, however, enhance the flavour and juiciness of the meat. Meat with limited marbling is likely to have a dry texture, and this has been one of the primary eating quality faults with pork in recent years. It has arisen from selection against fat deposition during growth, resulting in excessively lean meat (Fernandez *et al.*, 1999a). This has been managed by incorporating the Duroc breed in sire lines, which imparts superior marbling and succulence whilst producing acceptably lean carcasses. Meat consumers are reluctant to buy chops with a high level of marbling based on their appearance in a retail display cabinet, but this behaviour is susceptible to change once a consumer realizes the benefits that marbling has for eating quality (Fernandez *et al.*, 1999b).

Retail displays aim at minimizing differences between cuts and portions of meat. Otherwise, the purchaser may think that something is abnormal or wrong with a particular cut. Two-toning is becoming a more common problem in this respect. It can be due to differences in muscle fibre type within the same cut and to fat accumulation in a muscle masking the normal muscle pigment. These are natural effects, which cannot be avoided, but in pigs the colour contrasts are sometimes very obvious. In fish, two-toning occurs in the form of stripes where fat accumulates when the fish reach sexual maturity. Patches of pale muscle also occur in breeder fish where pigments have been mobilized for deposition in the skin and eggs. Both conditions lead to downgrading of breeder fish from the aquaculture industry.

Young fish do not have these problems but they can produce fillets that have poor cohesion. This becomes a problem when the meat is sliced thinly. The fillets show gaping between muscle blocks, and the fillet may break up and fall apart during cooking. This condition is exaggerated if there has been excessive contraction during rigor, causing the myotomes to separate from each other. Stress during catching, inadequate temperature control during processing and autolysis also contribute to this problem.

Blood splash, blood speckle, blood spots and ecchymoses are problems when the residual blood is visible at the surface of a meat cut. They are connected with the way the animal is stunned and bled, and are common causes of downgrading in both red and white meat species. In salmon, blood spots can occur where blood vessels branch from the aorta. The problem arises from engorged blood vessels, which rupture during evisceration and filleting. This is often worst during the summer when water temperatures are high.

Rearing Conditions

There are conflicting views about the effect growth rate has on subsequent meat tenderness. Theoretically, situations that enhance muscle catabolism through elevated calpain or reduced calpastatin activities could improve meat tenderization post mortem. Alternatively, if the animals are very slow-growing, or even losing weight, background toughness may be raised because connective tissue and collagen will be a larger component in the muscle. Overall, the majority of the evidence suggests that there is little or no effect of pre-slaughter growth rate on meat tenderness if the meat is aged (Perry and Thompson, 2005). The relationship may, however, depend on the cooking method that is used (Listrat et al., 1999). When beef was cooked very lightly (to 55–60°C internal temperature), meat from fast-growing cattle was tougher.

Theoretically, if an animal is slaughtered during a period of compensatory growth, its meat could be more tender. If it is fed generously after a period of weight loss and growth rapidly compensates, the meat could have superior background tenderness to meat from animals grown throughout the rearing period at a more constant rate, provided calpain activity is still elevated. However, there is limited evidence that has tested whether this is a real effect.

In general, feedlot beef can have a less 'grassy' flavour than pasture-fed beef. This distinction is sometimes quite subtle and may not always be detected. Feedlot beef can also have a slightly paler red colour than grass-fed beef, but there is no difference in the tendency to develop surface discoloration when left in retail display cabinets. Meat from spayed feedlot heifers has the same tenderness as beef from maiden heifers, but meat from once-bred heifers can have more variable tenderness without, on average, being more or less tender (Field et al., 1996).

In lambs, regular exercise during the rearing period is conducive to a leaner carcass with more tender meat in the hindleg (Garrett et al., 1999). In steers, regular exercise during the finishing period can enhance redness in the meat (Dunne et al., 2005b), but it may also increase two-toning (Dunne et al., 2005a).

It has been suggested that it should be feasible to produce pigs outdoors in colder climates using a once-bred gilt system. The gilts would produce a litter in the spring with the piglets reared to slaughter weight outdoors. Replacement gilts would be retained over winter for farrowing the following spring. When this system was tested in Sweden, it was found that overall growth performance and instrumental meat quality were similar in comparison with conventional indoor finishing systems (Stern et al., 2003).

Meat from red deer has poor colour stability compared with beef. It is prone to metmyoglobin formation during retail display, and this limits its retail life. This has been particularly noticeable in deer raised indoors and fed a pelleted ration. Pasture-raised venison has better colour stability, presumably because of the potentially higher concentrations of vitamin E in the meat (Wiklund et al., 2006).

Turkeys reared in hot conditions might be prone to producing breast meat with a higher than normal pH_{ult}, but it is not usually serious enough to affect overall quality (Owens et al., 2000a).

Castration

Castrating male animals is considered less necessary than it used to be, and the growth advantage of the entire male outweighs its other disadvantages. Nevertheless, many countries still castrate pigs, lambs and cattle. In Australia, Eire, New Zealand, Spain and the UK most pigs are left entire, and in Denmark castration is no longer used on every farm. Castration is mainly done where there are concerns about boar taint, and this is associated with slaughtering at heavy weights.

The off odours and flavours associated with boar taint are linked to androstenone and skatole in the fat. The risk of skatole odours is increased if the boars are kept at high stocking densities and they lie in their own manure (Hansen et al., 1997). Androstenone contributes to both boar odour and flavour, whilst skatole and its derivatives are quite volatile and so they are more likely to create odour problems during cooking.

The way the pig meat is presented determines whether boar taint is noticeable (Bañon *et al.*, 2003a). This is shown in the detection thresholds listed in Table 12.3. Dry cured meats contain oxidized fats, which help mask boar taint. Cooked ham is often eaten cold and so volatile off odours are also less obvious in this type of product. However, there are ethnic and gender differences in sensitivity to boar taint, and awareness of the odour is not always predictable (de Kock *et al.*, 2001).

When castration increases the fat content of pig meat, salt uptake during dry-curing is reduced and saltiness can be less pronounced (Bañon *et al.*, 2003b). In other words, boar meat can carry a saltier flavour. In Canada, boar meat has been shown to be marginally tougher and paler, but this is not a widely recognized feature (Sather *et al.*, 1999). When boar carcasses are leaner, there can be less marbling. Including the Duroc in the sire line helps to reduce the adverse effect this has on eating quality (Channon *et al.*, 2004).

The Meishan breed reaches sexual maturity earlier than other breeds, and it is favoured by some breeding companies for female lines because of its prolificacy. The cross-bred progeny also express precocious puberty and the males are prone to developing boar taint by the time they reach market weight. In this situation, including sugarbeet pulp in the grower ration can help to correct the taint (Whittington *et al.*, 2004).

In cattle, castration helps to increase fat content of the meat. This is not always sufficient to improve juiciness, but there can be less cooking loss in meat from steers compared with bulls (Destefanis *et al.*, 2003).

Table 12.3. Threshold values for the organoleptic detection of androstenone and skatole in pig meats.

Pig meat product	Threshold concentration ($\mu g/g$)	
	Androstenone	Skatole
Cooked meat	0.5	0.10
Cooked ham	1.5	0.75
Dry-cured ham	2.0	0.12

The flavour difference between ram and castrate lambs is subtle and not easily distinguished. Ram lamb meat can have a sweaty odour, which is due to 4-methyloctanoic and 4-methylnonanoic acids in the fat (Sutherland and Ames, 1995). In Hispanic countries a substantial number of ram lambs are slaughtered at light weights (e.g. 10 kg carcass weight). It would be possible to slaughter ram lambs larger than this without incurring meat quality defects for this particular market (Sañudo *et al.*, 1996). There would, however, need to be changes in the way the carcasses are cooked and presented if these lambs are to be slaughtered at heavier weights.

Goat meat from intact males has a reputation amongst some consumers for its rank odour. However, objective trials have shown no difference in flavour from castrates up to the age of 310 days (Madruga *et al.*, 2000).

Cattle

The beef quality defects that are linked to the way the animal is grown and managed before slaughter include:

- dark colour (e.g. DCB)
- poor keeping quality
- strong flavour, for some markets

DCB is particularly common in bulls, but bulls do not invariably produce dark-cutting meat. Considerate handling and individual penning help minimize the exercise that leads to glycogen depletion in muscle and high pH_{ult} meat. Many cattle markets that specialize in selling bulls for slaughter keep the bulls in stalls during the holding period. This helps limit exercise and DCB. In addition, some abattoirs that process bulls pen the animals individually in races with sliding gates between consecutive animals. This also helps limit mounting behaviour and muscle glycogen depletion. However, when the processing line throughput is high, there is a greater risk of high pH_{ult} bull beef because the handling is likely to be rushed (Young *et al.*, 2004). This effect has been less obvious on rainy days, presumably because bulls are less active in wet conditions. Hot weather, in contrast, increases the risk of high pH_{ult} meat.

Long transport journeys before slaughter can increase the risk of DCB in bulls, but muscle glycogen depletion during these journeys can be controlled in other ways (María et al., 2003). Resting Simmental bulls for 36 h in the lairage, without feed and water, helps reduce DCB (Kuzmanvić and Elabjer, 2000).

The objectionable dark colour in uncooked high pH$_{ult}$ meat is lost when it is cooked, even when it is fried. It can have a redder colour because the myoglobin is less prone to heat denaturation and this gives the impression that the cooked beef is more rare. Beef that has a low residual glycogen content and normal pH$_{ult}$ has better juiciness but the residual glycogen content does not influence flavour (Immonen et al., 2000b).

High pH$_{ult}$ meat is often used in manufactured products such as ground-meat patties. Large quantities are shipped around the world to meet the demand for the burger trade. For example, in New Zealand, bull calves are not usually castrated, and about 60% of the beef produced in that country has a high pH$_{ult}$. Much of it is exported frozen to the USA for burgers. High pH$_{ult}$ beef loses less weight as purge (drip) on thawing (0.1%) compared with beef with a normal pH$_{ult}$ (2.0%), and this is a significant financial saving for meat importers (Swan and Boles, 2002).

There is a perception that raising cattle on pasture is more welfare-friendly than raising them in either a feedlot, yard or barn. There is probably an overlap in people's minds between naturalness and welfare-improved status. Bulls reared intensively in tie-stall housing have produced more tender beef with superior flavour for the north-west European palate compared with pasture-fed bulls (Vestergaard et al., 2000). Supplementing the feed with concentrates for 17 days before slaughter helped to increase muscle glycogen and reduce the effect of stress in raising meat pH$_{ult}$ (Immonen et al., 2000a). The size of the protective effect from the high-energy feed was commercially significant (0.65 pH units), but it would no doubt be less than this if the bulls were penned together in the lairage.

The flavours that can cause criticisms amongst beef consumers are 'grassy', 'milky–oily', 'soured dairy' and 'fishy' flavours in pasture-raised beef, and, less commonly, oxidized flavours in feedlot beef that has a relatively high concentration of polyunsaturated fatty acids. The oxidized flavours can develop when beef is stored for long periods, but their formation can be inhibited by increasing the concentration of antioxidants in the meat (Reverte et al., 2003; Gatellier et al., 2005). It is not possible to give precise recommendations on how

to avoid oxidized flavours without considering the fatty acid composition of the feed, the level of antioxidants in the feed and the storage period used for the meat. However, both the grassy and the oxidized flavours can be masked quite easily with the commercially available beef flavouring agents that are used in restructured steaks.

Belgian Blue bulls are now used widely in Europe for crossing with dairy Holstein cows to improve carcass value in the slaughter progeny. The cross-breds have higher dressing percentage, with a high yield of lean meat in the double-muscled hindquarters. However, the meat has a lower intramuscular fat content and water-holding capacity, and it is slightly paler in colour (Mojto et al., 1998). There are conflicting reports about the tenderness of beef from double-muscled cattle. Some studies suggest that it is tougher because of the high proportion of ligament proteins, whereas others indicate that it is softer and has a lower background toughness because of its lower connective tissue content (Oliván et al., 2004).

Some breeds, such as the Limousin, have a reputation for erratic behaviour in markets and abattoirs and being difficult to handle. The advantage with this breed is its tender meat in comparison with Holsteins, partly because of its lower collagen content in muscle, but the difference in tenderness is less obvious if the meat has been aged (Monsón et al., 2004). Cull cow beef has a reputation for being very dark in colour. This applies to beef cows as well as dairy cows, and it is independent of any effect from pH$_{ult}$ (Fiems et al., 2003).

Veal

The veal market has been in decline over the past 10 years, partly because of the reputation gained by veal patties in passing on E. coli O157, and also because of its declining animal welfare image. Nevertheless, white veal still commands a premium price in select markets. Paleness is the main factor determining its price in Dutch and French markets, and pale veal is associated with low iron status and anaemia in the live animal, and not withholding feed on the day of slaughter.

Some veal calf units are managed intensively using labour-saving methods, whilst in others the farmers spend time making contact with the animals. Contact with the animals has been shown to have some small benefits for the quality of the meat, but growth rate is not affected. Calves reared by considerate farmers were less reactive during handling in the lairage before slaughter. They were less prone to slips and falls and to striking gates when being moved. The prevalence of bruising was no better, but pH_{ult} in the meat was lower. The cumulative stressors before slaughter evidently had a less demanding effect on muscle glycogen stores, and as a result the meat was slightly less red (Lensink *et al.*, 2001a).

Sheep

Meat quality defects are less common in lamb than in beef and pork. However, problems with carcass grading are important. Excessive fatness and small eye muscle area are common complaints, and where lambs are sold early to avoid over-fatness penalties there is often insufficient marbling. The Texel, Suffolk and Poll Dorset breeds are used as terminal sires to correct deficiencies in muscle development.

There has been interest in using callipyge sire lines for the same purpose, but the meat in some cuts from cross-bred callipyge lambs is tough because of higher calpastatin activity (Goodson *et al.*, 2001). The carcasses have to be aged for over 24 days to achieve the same level of meat tenderness as non-double-muscled lambs (Kuber *et al.*, 2003). However, the ageing period can be less than this if the carcass is hung by the hip instead of the leg tendon.

Lambs that are sold for slaughter directly from the ewes do not experience the growth check associated with weaning and so they have a commercial advantage. They tend to have paler meat than lambs that are weaned and then finished on concentrates and straw before slaughter, but this is not a serious weakness (Vergara and Gallego, 1999). A more important concern is their susceptibility to blood splash.

Meat from feedlot lambs fed concentrates and dry roughage (oat hay) is more prone to lipid oxidation when ground to make patties, compared with meat from pasture-raised lambs (Rhee *et al.*, 2003). Grass-fed lambs can have a more pronounced liver flavour, and feedlot lamb is paler and it can be juicier and more tender if finished at a higher level of fatness (Priolo *et al.*, 2002).

In parts of the Middle East and Australia, salt-tolerant plants such as saltbush (*Atriplex nummalaria*) are being encouraged to help regenerate saline land.

When lambs graze on *Atriplex*-dominant land they can produce meat with a strong cooked aroma, but it is not an unpleasant smell (Hopkins and Nicholson, 1999).

High pH_{ult} lamb is prone to producing more purge in retail packs and to greater weight loss during cooking (Bond *et al.*, 2004). The greater fluid loss is assumed to be due to leakiness of the myofibrillar proteins arising from their early breakdown by proteolysis.

Pigs

The main pig meat quality problems are PSE, DFD, boar taint, bruising, meat dryness, two-toning and rancidity in stored products. Both PSE and DFD can be linked to pre-slaughter stress. In a survey of five EU countries during the 1990s, the prevalence of PSE meat was 24%, whereas in Canada it was somewhat lower at 13% (Murray and Johnson, 1998; Warriss *et al.*, 1998c). Differences between nations are largely due to different pig genotypes, but differences in climate and pre-slaughter handling methods may also contribute.

Pigs are sent for slaughter when they fall within a target weight range. The pigs in a rearing pen are not all sent at the same time, but instead they are drafted on a weekly basis. Pigs that grow most slowly may be mixed with other pigs, to allow a pen to be restocked with a new batch. When mixing does not occur, the last pigs to be sent for slaughter produce meat that has more drip and cooking loss (Beattie *et al.*, 2002). The reason for this is not clear, but it could be linked to their slower growth.

High stocking densities during rearing have been linked to a reduced risk of DFD and paler meat (Hamilton *et al.*, 2003; Guàrdia *et al.*, 2005). In PSE-prone pigs, high stocking densities have been associated with less drip loss from their meat.

Feeding a ration containing carbohydrate that has a low digestibility in the foregut, but is fermented in the hind-gut, has been used to lower muscle glycogen pre-slaughter and control PSE meat production (Rosenvold *et al.*, 2001). The effect this has in satisfying hunger and reducing physical activity during the pre-slaughter period has not been examined.

Feeding pigs a sugar solution whilst they are in the lairage has been used in the past to improve liver yield and quality. It lowers the pH_{ult} in most muscles and protects against stress-induced muscle glycogen depletion (Fernandes *et al.*, 1979). Whether it will reduce two-toning has not been tested, but this is worth exploring as two-toning is a common complaint amongst pig meat processors.

Although acute stress under hot conditions can lead to PSE meat in *Nn* and *nn* genotypes, exposure to high temperatures throughout the rearing period in *NN* genotypes can lead to higher pH_{ult} meat (Rinaldo and Mourot, 2001). This was found to be the case in Large White pigs kept in the French West Indies (Table 12.4).

In Spain, the prevalence of PSE meat is higher during hot weather in the summer. The overall prevalence is 6.5%, but it also tends to be higher following short journey times to the abattoirs, especially when stocking density is low (> 0.4 m^2/100 kg) (Gispert *et al.*, 2000; Guàrdia *et al.*, 2004). Reducing the stocking density for the sake of reducing PSE meat would only be advisable for long journeys.

The risk of PSE was lower when vehicles were equipped with hydraulic lifts in place of loading and unloading ramps. Vehicles with non-slip comfortable floor surfaces resulted in less PSE than aluminium and iron floors. Resting the pigs for 1 h or more in the lairage also helped reduce the development of PSE meat (Owen *et al.*, 2000).

In temperate regions, PSE meat can be worst in November and December, when abattoirs have high kill rates in preparation for the Christmas trade (O'Neill *et al.*, 2003a). Outside this period, DFD meat is more common during winter. It is also more common when pigs are transported on long journeys at high stocking densities (< 0.4 m^2/100 kg) or are held for long periods in the lairage (9 h) (Gispert *et al.*, 2000). Much of the stress associated with short journeys occurs during loading and the start of the journey. Extending the journey for up to 3 h helps the animals recover from that acute stress, but it increases the risk of DFD meat through physical exhaustion (Pérez *et al.*, 2002).

Pre-slaughter mixing reduces the prevalence of PSE and increases the risk of DFD, provided the pigs are rested and slaughtered after they have stopped fighting (Sather *et al.*, 1999). When exercise occurs as pigs are transferred from their holding pens to the point where they are slaughtered, muscle creatine phosphate (CP) and ATP may be depleted but muscle glycogen stores may be unaffected (Henckel *et al.*, 2002). This has been sufficient to precipitate PSE meat, even when the pigs are stunned with CO_2, which is normally associated with a low prevalence of PSE meat (Støier *et al.*, 2001). Using an electric goad when moving the pigs to the slaughter point can be sufficient stress to increase the expression of PSE meat (D'Souza *et al.*, 1998; van der Wal *et al.*, 1999). The stress need only last for 1 min to produce this effect (Table 12.5).

There are some subtle effects from pre-slaughter handling that do not seem to be mediated by pH decline in muscle. In pigs that are not prone to developing PSE meat (*NN*), pre-slaughter exercise can increase meat redness in the absence of any effect on pH_{ult} (Rosenvold and Andersen, 2003). Walking pigs 125 m to an abattoir, instead of transporting them in a vehicle, resulted in a substantial increase in drip loss from the loin after slaughter, without causing any marked effect on meat colour or the rate of muscle pH decline post mortem (de Smet *et al.*, 1998). The mechanisms explaining these effects are not obvious.

Table 12.4. Effect of rearing conditions on meat quality in Large White pigs.

		Tropical climate	
	Control 20°C 75% RH	Cool season 25 ± 4°C	Warm season 28 ± 3°C
Longissimus dorsi $pH_{45\,min}$	6.21	6.21	6.37
pH_{ult}	5.52[a]	5.60[ab]	5.71[b]
Moisture loss (%)	21.4	21.4	20.7
Biceps femoris $pH_{45\,min}$	6.05	6.06	6.17
pH_{ult}	5.58[a]	5.63[ab]	5.88[b]
Moisture loss (%)	24.8[a]	22.7[ab]	21.4[b]

Means in a row without a common superscript letter were significantly different at $P < 0.05$.

Table 12.5. Effect of 1 min of handling stress immediately before electrical stunning on subsequent meat quality.

Longissimus dorsi	Mean ± SE	
	Not stressed	Stressed
$pH_{45\,min}$	6.66 ± 0.04^a	6.39 ± 0.04^b
Temperature$_{45\,min}$ (°C)	39.4 ± 0.2^a	39.9 ± 0.2^b
Exudate$_{24\,h}$ (mg)	30.6 ± 3.8^a	51.2 ± 3.9^b

Means in a row with a different superscript letter were significantly different at $P < 0.05$.

With the possible exception of pigs in Canada, the halothane genotype of the pig usually has a greater influence on meat quality than pre-slaughter management (Costa *et al.*, 1999). Meat from *Nn* genotypes has the same problems as *nn* genotypes (low $pH_{45\,min}$, more drip, paler colour and less tender), but to a lesser degree and so it is more easily tolerated (Fernandez *et al.*, 2002a).

The rapid acidification of meat due to become PSE will favour a more rapid metmyoglobin formation. This could contribute to the brown-grey colour of fresh PSE meat. The greater paleness of meat from *nn* pigs does not persist with cooking (roasting), but it is less juicy (Moelich *et al.*, 2003). The greater toughness of PSE meat might be partly due to its lower μ- and m-calpain activities (Claeys *et al.*, 2001).

Cooked ham is now sold mainly as pre-packaged slices, and this trade is handicapped by PSE meat. During slicing at the processing plant, PSE zones in the ham crumble easily, making holes and splits in the slices, which affect the appearance of the product. In France, it is said that as much as 15 to 20% of hams experience this defect, and this is adding pressure on breeders to reduce the prevalence of PSE meat.

Processed hams are inferior in other ways when made from PSE meat. Cooked hams tend to be drier, tougher and stringier, and dry-cured hams have poorer cohesion between muscles and two-toning is more obvious between the muscle groups. Dry-cured PSE hams are also less salty in flavour than non-PSE hams because they take up less salt during curing, but they can be softer and more tender (Fernandez *et al.*, 2002a).

It has been suggested that severe pre-slaughter stress could enhance warmed-over flavours (WOFs) in meat, but this may not be as important as previously thought (Juncher *et al.*, 2003). WOFs have not so far been shown to occur under experimental stress conditions even when the stress resulted in PSE meat (Byrne *et al.*, 2001). Part of the reason may be that mild acidic flavours associated with PSE meat could mask the WOFs. In spite of this, when cooked hams were prepared from PSE meat, they were more prone to developing WOFs, as well as producing more drip and cooking loss (O'Neill *et al.*, 2003b). When the stress has been more protracted, resulting in DFD meat, WOFs have not been a problem, and this could be because high pHs inhibit lipid oxidation (Juncher *et al.*, 2001).

Fast chilling can help retard the rate of pH fall post mortem, but relatively aggressive chilling regimes are required to produce this effect (e.g. holding pork sides at −20°C for 2 h prior to boning at 3 h post mortem). In *nn* genotypes, chilling is usually applied too late to prevent the critical combination of high temperature and low pH that precipitates the PSE condition.

When the PSE condition in a batch of meat is bad enough to downgrade the carcasses to the sausage and minced meat trade, yield is not inevitably worse, but financial return will suffer because of the lower value of those manufactured meats. When PSE meat was used in making Chinese pork sausages, it introduced an unwanted acidic flavour and formed a less appealing texture (Kuo and Chu, 2003).

Rendement Napole pigs

The Hampshire breed is favoured by some pig breeders as a terminal sire because of its low prevalence of PSE and high value in producing tender and juicy meat. There is a subpopulation within the breed derived from the so-called Rendement Napole (*RN⁻*) mutation, which has about 70% more muscle glycogen. It is present as macroglycogen and these pigs have faster rates of glycogen re-synthesis after it has been depleted by exercise (Essén-Gustavsson *et al.*, 2005). The dominant *RN⁻* allele is associated with low pH_{ult} values and a low prevalence of DFD pork. This meat has poor WHC when fresh, and reduced processing yield when cured. Thawing losses are also higher but both fresh and processed products are more tender (Byrne *et al.*, 2003). These effects can also be seen in the heterozygotes (Andersson *et al.*, 2003).

The high-drip–high-tenderness combination of the *RN⁻* pig is unusual. The greater drip in the PSE condition is usually associated with tougher fresh meat (Stalder *et al.*, 1998). The greater tenderness of *RN⁻/–* is largely due to a faster rate of tenderization post mortem, and a lower tension during rigor adds a smaller benefit (Josell *et al.*, 2003). *RN⁻* carriers are also slightly more prone to developing WOFs. Like PSE-prone genotypes, the *RN⁻/m⁺* heterozygote is liable to produce dark meat when mixed before transport and then allowed to rest at the abattoir before slaughter.

Organic and free-range pig meat

There are differences in eating quality between conventional and organic or free-range meat but the differences do not seem to be consistent. This unpredictability could be due to differences in climate, feed or leanness of the carcasses. The following highlights some of the more consistent effects.

In parts of Italy and Spain, some pigs are grown under free-range conditions with access to woodland. Their meat can have a higher lipid content and produce less drip as well as less cooking loss, and it is paler but more red in colour (Pugliese *et al.*, 2005). In the Netherlands, free range pigs produced meat which also had less drip (Lambooij *et al.*, 2004). In general, these are positive attributes.

Rearing pigs outdoors in cold conditions can increase the risk of two-toning through an increase in the proportion of aR fibres in particular muscles giving them a more intense red colour (Lebret *et al.*, 2002). When meat from outdoor sows is cured, the intense redness diminishes, but some two-toning persists and this can be an unwanted feature (Lebret and Guillard, 2005). On the positive

side, pigs reared outdoors under cold conditions are likely to have greater marbling in their meat.

The effect that outdoor rearing has on the oxidative stability of marbled meat depends on the feed the pigs received, and in particular on the balance between unsaturated fats and antioxidants such as vitamin E. Access to pasture promotes the consumption of both unsaturated fatty acids and vitamin E. It is not easy to predict the outcome in terms of susceptibility to WOFs or rancidity, and it is unwise to give generalizations that apply to all outdoor systems. The outcome may also be influenced by the fat content of the meat, as low-fat meats are often more prone to oxidative rancidity. In Spain, when pigs of the traditional Iberian breed have been grown on pasture, their meat has had variable oxidative stability or higher TBARS values when irradiated (Estévez *et al.*, 2003; Cava *et al.*, 2005). It has also had lower water-holding capacity, juiciness and tenderness (Olsson *et al.*, 2003). Liver pâté made from these pigs had lower polyunsaturated fatty acids (PUFA) and non-haem iron, and it should be less prone to oxidative deterioration (Estévez *et al.*, 2004).

Poultry

Rearing broilers under free-range conditions results in leaner birds, and their breast meat can have a lower pH_{ult}, water-holding capacity and tenderness (Castellini *et al.*, 2002). The meat may be more susceptible to lipid oxidation if it is subjected to strong oxidizing conditions, but usually there has been no effect on TBARS values (Jahan *et al.*, 2004).

Withholding feed for 8 h or more before slaughter helps to make broiler breast meat paler and less red (Smith *et al.*, 2002). Heat stress before slaughter causes weight loss, and the breast meat is slightly less red in colour but can have a poorer WHC (Petracci *et al.*, 2001; Sandercock *et al.*, 2001). Pre-slaughter transport for as little as 3 h can be sufficient to cause an increase in DFD features in turkey breast meat (Owens and Sams, 2000).

The way broilers are managed when they are being slaughtered can affect the rate of post-mortem glycolysis in the meat and its subsequent quality. However, there are several processing features that can either exaggerate or mask this outcome. The potentially important processing factors are the interval between neck cutting and scalding, the scald water temperature and duration, and the severity and duration of plucking. A rapid pH fall, short neck cut-to-scalding interval, high scald water temperature, long-lasting scalding period, harsh or prolonged plucking and delayed chilling will all increase the risk of heat shortening and firm or tough meat. The outcome depends on the circumstances at each processing plant.

A slow pH fall in turkey breast muscle can pose a risk for cold shortening where early rapid chilling is used, but, more importantly, it is likely to pose a hazard for rigor shortening in early-deboned chicken and turkey meat. Here again, the overall risk depends on circumstances, and advice has to be tailored to the conditions at each plant.

Deer

One of the most striking features of venison is the consistency of its quality. It is probably one of the more uniform fresh meats on the market. The main quality defect is spoilage, but this is not necessarily linked to high pH_{ult}. The prevalence of high (> 6.2) pH_{ult} longissimus dorsi meat has been put at 6% for reindeer, 1.5% for red deer and 1% for fallow deer. The other main quality defect is its unique flavour, which some people find unpleasant. Reindeer meat is gamy, and it has been likened to the odour of wet dog hair.

Fish and Crustacea

Some of the more important quality features in fish meat that need to be managed before or at the time the fish are slaughtered are:

- taste and off flavours
- texture (gaping, mushiness, chewy)
- freshness
- pigmentation
- fat level in the meat and fat distribution
- drug and chemical residues

Of these, texture and fat level can be influenced by the ways the fish are handled and fasted just before slaughter. Freshness is influenced by the way the fish are chilled as they are being slaughtered.

In trout, **pre-slaughter fasting** helps:

- reduce physical activity, making it easier to handle and transport the fish without causing injury

- empty the gut, and reduce faecal contamination of the surface of the fish
- reduce the fat content of the muscle

It can improve meat firmness, but it cannot be relied on alone to ensure that there will be no fish with soft meat.

Salmonids need to be fasted for up to 3 days to empty the gut. It takes longer to deplete excessive amounts of fat in the muscle that develop when they are reared on high-fat feeds. Normally, salmonids are fed high-fat rations because the high energy content promotes rapid growth. The benefits from reducing the oil content of muscle with pre-slaughter fasting are less greasy mouth feel, and less expressed oil in vacuum-packed smoked fillets. The exact period needed to deplete fat in muscle will also depend on water temperature and the activity of the fish. At some trout farms fasting is combined with exercise by forcing the fish to swim against a fast-flowing current in a narrow raceway. The exercise period needs to be sufficient to shift substrate utilization from carbohydrate towards long-chain fatty acid oxidation, and this takes days rather than hours. An alternative approach is to feed the fish a **low-fat finisher ration** during the weeks before slaughter.

Feeding low-fat rations for long periods can result in a saltier taste to the meat, especially if it is smoked (Regost et al., 2001). It also makes the meat firmer in texture, darker but less red, with less exudate during cooking. An abnormally low fat content in the meat, as occurs in chronically undernourished fish, can result in watery meat, which is less juicy after cooking. The higher fat content of farmed salmon compared with wild-caught salmon makes it more suitable for sashimi, and smoked salmon benefits from a relatively high fat content (Robb et al., 2002a).

In Norway, salmon are usually fasted for 2 weeks before slaughter in winter, and for 1 week in summer. Pre-slaughter fasting has less effect on eating quality or appearance in salmon than in trout (Erikson, 2001). It can increase the intensity of the red colour in salmon, especially when the fat content of the meat is reduced, but this does not always have a consistent effect. Long fasting periods (3 weeks) can lead to a more acid flavour, but this disappears if fasting lasts more than 8 weeks.

Fillet yield and meat flavour should also be considered when deciding on the optimum length of fast. With long fasting periods there is a risk of reduced yield. Fasted trout can develop a muddy flavour, and in some of the worst cases there has been a putrid odour. In addition, fasting can make the skin colour paler, but this is not usually sufficient to cause downgrading (Grigorakis and Alexis, 2005).

Intense **physical activity** before slaughter will deplete white muscle ATP, CP and glycogen. Fish muscle does not have high concentrations of glycogen, and many species can be taken to complete physical exhaustion quite quickly. Associated with this, there is accumulation of lactate and the muscle becomes acidic. In Atlantic salmon, H^+ or lactate may contribute a preferred fresh, acidic flavour to the meat (Einen and Thomassen, 1998). In general, lactate is removed from fish muscle slowly compared with other meat species, and it is particularly slow in fish that are unused to exercise. There are, however, pronounced species differences in muscle lactate accumulation during exercise (Table 12.6), and this may explain differences in meat acidic flavour. The differences in species' tolerance to exercise depends on their ability to buffer lactic acid as well as their glycogen storage capacity.

Any procedure that causes excessive muscle activity shortly before or during slaughter can hasten **rigor** development and pose a risk for meat gaping (Pottinger, 2001). If the fish are physically stressed during catching, rigor is likely to develop quickly and it can be more intense. Excessive tension during rigor introduces a risk of meat gaping, where adjacent blocks of muscle separate from one another. When rigor occurs too quickly, the fish may be filleted whilst they are stiff and when this is done automatically there is often poor yield because of poor presentation to the machine. Alternatively, rigor is allowed to resolve and the fish is filleted once it has gone limp. In salmon it usually takes a day for rigor to resolve and this additional holding period sacrifices some of the shelf life. From a butchery perspective, it is usually best to fillet fish before rigor sets in, even though there will be more drip loss (Thompson *et al.*, 2002). Pre-slaughter stress and excessive muscle tension at rigor are also recognized problems in cod, blue tilapia, sturgeon and carp processing.

In some circumstances pre-slaughter stress can result in a softer meat texture post-rigor. This is partly due to disruption of the structure of the meat by more forceful rigor contractions, and partly to increased post-mortem proteolysis. Stress-induced softness is a problem if it gives the impression that the meat is no longer fresh. In the case of Atlantic salmon, the fish have to maintain a 'fresh' appearance in the market for between 4 and 14 days after slaughter. Minimizing handling stress before slaughter is recognized as one way of maintaining this aspect of shelf life.

At the other extreme, if the fish are rested at the time of slaughter, and then are frozen soon after slaughter, there is a risk of toughness through thaw rigor. Colour intensity can also be influenced by the rate of post-mortem muscle acidification. When pH decline is rapid, the paleness can be increased and colour intensity reduced.

In general, post-mortem tenderization of fish muscle is undesirable because the meat seems to be unnaturally soft and this reduces its perceived freshness.

Impacts of Genetic Improvement on Meat Quality

Meat lines are being genetically selected for rapid growth, high feed conversion efficiency and large muscle mass. The effects this has on meat quality are beginning to emerge, and it is now recognized that they depend partly on the genotype that is chosen at the outset. Each species, and sometimes each breed, needs to be considered separately.

Some of the improvements in growth performance and conformation have had unwanted effects on meat quality. Animals with large muscles often have poor marbling in their meat and this is leading to problems with meat dryness. This has been a feature in beef, pigs and turkeys, but an additional problem in turkeys is the high cooking loss of its breast meat.

Table 12.6. Peak muscle lactate concentrations in a range of farmed fish species during severe exercise (after Kieffer, 2000).

	Typical maximum muscle lactate concentrations (µmol/g) following exhaustive exercise
Atlantic salmon	45
Rainbow trout	41
Yellow perch	24
Largemouth bass	20
Starry flounder	10
Shortnose sturgeon	6

Selection for growth rate in cattle, pigs and broiler chickens is resulting in paler meat. In the case of pigs, this is linked to the PSE condition, where selection is being directed towards stress-susceptible genotypes. Turkey sire lines are being selected inadvertently for stress-induced glycogen depletion in their muscle and DFD meat.

Some of the traditional European beef breed societies claim that meat quality is superior in their genetically improved pedigree stock. However, the differences are not as pronounced as some of them might hope. When two African breeds (Afrikaner and Nguni) were compared with a range of temperate cattle breeds under comparable conditions, their meat had less marbling but when oven-roasted it had similar eating quality (Strydom et al., 2000). At the other extreme, genetic selection for double muscling is leading to less marbling and paler, less red meat and, because of the lower level of marbling, there is a risk that the meat has less succulence and flavour. In temperate cattle breeds not expressing double muscling, selection against fatness could be leading to tougher meat, as in some breeds the genetic correlation between the two traits is strong (Table 12.7). However, simultaneous selection for growth rate should counteract the trend towards toughness. In subtropical beef breeds selection against fatness may be less closely linked to meat toughness, whilst selection for growth rate could be leading to paler meat (Reverter et al., 2003). In addition, the heritability of meat quality traits is slightly higher in subtropical compared with temperate breeds (Johnston et al., 2003).

It should be possible to improve marbling genetically when an appropriate marker is discovered. Marbling score can be highly heritable in beef, but this depends to some extent on breed ($h^2 = 0.71$; Splan et al., 1998). The h^2 for tenderness is not as strong (0.26 to 0.35), and meat flavour and juiciness are not heritable.

In pigs, intramuscular fat content and meat tenderness have a high heritability ($h^2 = 0.4$ to 0.6), as does meat redness ($h^2 = 0.72$; Oksbjerg et al., 2004). It might be hoped that insufficient redness could be corrected genetically by influencing the fibre types in the muscles. However, genetic selection for a high proportion of slow-twitch type I fibres in the longissimus dorsi muscle did not influence the rate of post-mortem muscle glycolysis in that muscle or subsequent meat colour in the biceps femoris (Lefaucheur et al., 2000).

The halothane gene and the RN^- gene are two major genes that have a direct effect on meat quality. Both arose as mutations. In the case of the halothane gene (nn), the effects are mediated through the rate of post-mortem muscle pH decline. The h^2 of $pH_{45 \ min}$ is usually in the range 0.14 to 0.37. It is difficult to predict the likely responses in $pH_{45 \ min}$ to genetic improvement in growth and carcass composition in populations with an unknown prevalence of the n/– condition. In Large White pigs, which normally have a low prevalence of n/–, it is thought that selection for growth rate and fatness is more likely to raise $pH_{45 min}$, whilst selection against fatness will have little effect (Sonesson et al., 1998). The halothane gene has a large effect on WHC, whereas the effect is smaller for the RN^- gene.

The h^2 of most other pig meat quality traits is generally low (Verbeke et al., 1999). The h^2 of pH_{ult} is about 0.21 (0.07 to 0.39), and in NN and Nn pigs it is invariably low. Pig meat quality is best improved by breed substitution, rather than by selection within breed. For example, insufficient marbling in lean pig genotypes has been controlled by using the Duroc breed in developing meat sire lines. This has been highly successful, and dry meat with insufficient succulence is not such a problem as it was in the past. If this were taken too far, such that very high levels of marbling were present in the ham, there could be negative effects on flavour and the odour of cooked ham slices, but these repercussions are not likely at present (Fernandez et al., 2000). Breeds like the Belgian Landrace, which have heavily muscled hams and a high lean meat yield, do not necessarily have superior eating quality. They have poorer meat flavour, tenderness and juiciness in comparison with the Large White (Verbeke et al., 1999).

Breast meat has been getting paler in recent years and the reason may be genetic. The h^2 of meat paleness in broilers is high (0.50). It has not reached the stage where it is a consumer complaint, but it is causing processors inconvenience when trying to match breast meat colour in retail display packs. Pale breast meat can indicate other problems such as excessive amounts of purge in the pack and this adds to the negative image. In broilers, genetic selection for growth rate and size of breast meat can lead to paler, less red breast meat and a large abdominal fat pad

Table 12.7.   Some examples of genetic correlations between production performance and meat quality traits.

Animal type	Parameters	r_g
Cattle – temperate breeds	Growth rate and longissimus dorsi L	0.49
	Growth rate and longissimus dorsi a	−0.68
	Growth rate and pH_{ult}	0.05
	Growth rate and longissimus dorsi tenderness	0.28 to 0.70
	Fat depth (P8) and longissimus dorsi tenderness	0.57
Cattle – subtropical breeds	Growth rate and longissimus dorsi L	0.36
	Flight time and longissimus dorsi tenderness	0.37
Pigs	Growth rate and ham L	0.43
	Growth rate and loin L	0.74
	Growth rate and drip loss %	0.78
	Ham size and ham a	−0.60
	Meat yield and loin L	−0.46
	Meat yield and longissimus dorsi myoglobin	0.28
	Meat yield and longissimus dorsi glycogen	0.35
	Meat yield and WHC	−0.70
	FCR and longissimus dorsi L	−0.59
	FCR and drip loss %	−0.58
	FCR and $pH_{45\ min}$	0.40
	Feed intake and $pH_{45\ min}$	0.66
Broiler chickens	Growth rate and breast meat L	0.51
	Growth rate and breast meat a	−0.25 to −0.30
	Growth rate and fat pad weight	0.46
	Breast meat weight and L	0.37
	Breast meat weight and a	−0.29 to −0.48
	Abdominal fat % and pH_{ult}	−0.54
Turkeys	Growth rate and breast meat L	−0.41
	Growth rate and breast meat $pH_{20\ min}$	0.55
	Breast yield and breast meat L	−0.24
	Breast yield and breast meat $pH_{20\ min}$	0.62

a, measure of surface redness in meat.

(Le Bihan-Duval *et al.*, 1999). Theoretically, this could be from one of three causes. They are:

- an increase in the PSE condition
- lower haem pigment content in the meat, associated with
  - either a lower capillary density and residual haemoglobin content
  - or a lower myoglobin content

The evidence indicating which of these is most important is conflicting. When pale broiler and turkey breast fillets were selected at a processing plant and compared with normal fillets, it was found that the paler meat had a lower pH_{ult} and cooking yield and produced more drip (Owens *et al.*, 2000b). This implies circumstantially that paleness is commonly due to the PSE condition. In contrast, when broilers were genetically selected for growth rate and breast meat yield, their pectoralis major $pH_{15\ min}$ and pH_{ult} were higher than normal whilst the meat was paler, less red and had a lower iron content (Berri *et al.*, 2005). This suggests that genetic selection has increased paleness through some other mechanism, and perhaps it is a lower haem pigmentation in the meat.

The situation in turkeys is different. Selection for high growth rate and breast yield is linked to

DFD characteristics instead of paleness (Le Bihan-Duval *et al.*, 2003). Fast-growing turkey genotypes are prone to producing high $pH_{60 \, min}$ leg meat. This is not due to pre-slaughter glycogen depletion from their greater weight-bearing (Fernandez *et al.*, 2001). In fact, faster-growing strains have greater glycogen storage capacity in their ilio tibialis muscles. Breast meat in these lines can be darker than normal, and it is prone to producing more drip during storage, but the difference in breast meat quality from traditional lines is small. A more striking feature in fast-growing turkey lines is the uniformity of their breast meat compared with traditional strains. For example, there was less variation between birds in the $pH_{20 \, min}$ for the pectoralis major in a fast-growing line. The implication is that there could be greater uniformity in quality in the modern fast-growing type, and this could be an advantage in retail display packs. A low $pH_{20 \, min}$ in turkey breast muscle is associated with a more acceptable paler and slightly redder colour when the meat is cured with nitrite (Fernandez *et al.*, 2002b), but it also produces more drip and has a lower processing yield. It is likely to be less preferred in terms of flavour and texture.

There is limited information on the genetic relationships between stress-related behaviours and meat quality. In broilers, there are strain differences in meat $pH_{15 \, min}$, which are probably due to differences in the severity of wing flapping when they are hung on the shackle line at processing plants (Debut *et al.*, 2003). Birds from the strain with faster growth rate flapped their wings less. A nervous temperament in subtropical cattle breeds (measured from their shorter flight time when released from a cattle crush) has been linked genetically to tougher meat (see Table 12.7). This could be due to a greater prevalence of intermediate pH meat.

Other points worth noting are:

- in cattle, culling for dystocia could lead to indirect benefits for meat tenderness (Splan *et al.*, 1998)
- in Australian Merino sheep, genetic selection for wool fineness is leading to darker meat ($r_g = 0.66$; Fogarty *et al.*, 2003)
- selection for fast growth in rabbits could be leading to harder, more chewy and poorer WHC meat (Ramírez *et al.*, 2004)
- selection for growth rate and leanness in ducks has resulted in slightly darker and redder breast meat, but so far the effect on overall acceptability has been negligible
- triploid Atlantic salmon produce larger fibres in their muscle. This imparts a softer texture to the meat, but it has a higher proportion of soluble collagen and is more prone to gaping (Bjørnevik *et al.*, 2004). In addition, the meat may be redder, with a low pH_{ult}

13

Implementing Effective Animal Welfare Auditing Programmes

Temple Grandin

You manage the things that you measure. Maintaining a high standard of animal welfare during stunning and handling can be achieved by using a simple objective numerical scoring system. This prevents practices from slowly becoming poorer without anybody realizing it. Numerical scoring has been extremely successful in improving animal welfare. It is used by many meat-buying customers to audit animal welfare at slaughter plants. Data on the improvements are in Grandin (2000, 2005a, 2007a). The author has worked on implementing objective scoring in many different countries. The best plants are audited by both their internal quality assurance staff and external auditors from a major customer. A typical effective programme is one or two annual audits from a customer and a minimum of a weekly audit by the plant's own staff.

Effective Audits

To be effective, standards have to be clearly written so that different auditors will give a plant a similar score. Words such as properly, adequate and sufficient must be avoided. These terms are too subjective. What one person may consider proper handling, another person may consider abusive. An example of a clearly worded guideline is: 'All the animals must have enough space so that they can all lie down at the same time without being on top of each other.'

Objective numerical scoring of welfare uses the HACCP (Hazard Analysis Critical Control Point) principles that are used in food safety. The principle is to measure critical control points that measure the outcomes of good practices. A good system uses a small number of well-chosen critical control points that are able to measure a multitude of problems. This is an outcome-based scoring system where things that can be directly observed can be assessed. This is much more effective than audits that are largely based on paperwork. To pass the audit, an acceptable score is required on all of the critical control points. This method produces a numerical score. The use of a numerical score enables plant management to determine if a practice is improving or deteriorating. This is the same principle as tracking bacterial contamination on meat. One hundred animals are scored for each critical control point in large plants or 1 h of production in small plants. The six critical control points are:

1. Percentage of animals rendered insensible with one application of the stunner. All species of animals, including poultry.
2. Percentage of animals rendered insensible before hoisting, skinning, scalding or plugging of the oesophagus. Must be 100% to pass an audit. For pigs and poultry, the audit is failed if a pig or bird enters the scalder showing any sign of sensibility. In poultry plants, the audit is failed if a red, discoloured bird is found that does not have a throat cut. It is likely that the bird was scalded alive.
3. Percentage of cattle or pigs that vocalize (moo, bellow or squeal) during handling and stunning. Score this in the stun box and race leading to the stun box.

©CAB International 2007. *Animal Welfare and Meat Production*
(N. Gregory)

4. Percentage of animals that fall down during handling. Body touches the floor during handling. All mammals. Score in all parts of the facility, including the unloading ramp and stun box.
5. Percentage of animals moved with no electric prods. All mammals.
6. Poultry only – percentage of birds with broken or dislocated wings.

There are also acts of abuse which would be an automatic audit failure.

1. Dragging downer, non-ambulatory animals that cannot walk or throwing animals.
2. Beating, blinding, breaking tails or cutting tendons.
3. Poking animals in sensitive areas such as in the eyes, rectum, nose or ears.
4. Deliberately slamming gates on animals.
5. Deliberately driving an animal over the top of another animal.
6. Hoisting or dragging sensible mammals with chains, shackles or ropes.
7. Puntilla or electrical immobilization that does not induce insensibility. The puntilla is a short knife used to sever cervical vertebrae to immobilize the animal. Electrical immobilization that does not induce insensibility is extremely aversive (Lambooy and Van Voorst, 1985; Grandin *et al.*, 1986; Pascoe, 1986; Rushen, 1986).

Setting Critical Limits

A perfect score on every critical control point is impossible, so limits for a minimum acceptable score must be set. The following critical limits have been used since 1997 for cattle, pigs and sheep (Grandin, 1998a) and since 2001 for poultry. The latest standards are in Grandin (2005b).

Stunning critical limits

Captive bolt stunning

For a minimum acceptable score, 95% of the animals must be rendered insensible with one shot. An excellent score is 99%. All animals requiring a second shot must be rendered insensible prior to hoisting, bleeding or any other invasive procedure.

Electric stunning

1. Correct placement of the stunning tongs (electrode) to ensure that the current passes through the brain of 99% of the animals (Fig. 13.1). For water-bath systems for poultry, 99% of the birds must be positioned in the water bath so that the current will pass through the brain.
2. Hot wanding – 1% or less. Hot wanding occurs when the electrode is energized before the tong or other device is firmly pressed against the animal. Score a hot wand if the animal vocalizes or pulls away in direct response to stunner application. For poultry, score pre-shocks at the water-bath entrance as hot wanding.

CO_2 and other controlled atmosphere stunning methods

Animals must be able to stand or lie in the gondola or container without being on top of each other. Score an overloaded gondola if there is an animal on top of the others with no space for it to stand or lie down: 4% maximum overloaded gondolas or containers. Use this standard for pigs and poultry.

Insensibility critical limits

All animals must show no signs of return to sensibility prior to hoisting. There is a zero tolerance for starting any dressing procedure such as scalding, skinning or leg removal on an animal showing signs of return to sensibility. This standard applies to both standard slaughter with stunning and ritual slaughter (kosher or halal) without pre-slaughter stunning. See Chapter 11 for signs of sensibility following captive bolt and percussion bolt stunning.

For CO_2 and other controlled atmosphere stunning, there is a need to develop a scoring system for monitoring animal welfare during induction. Windows should be provided so that the induction phase up until the animal or bird loses posture and falls over can be observed. The audit would be failed if more than 1% of the birds or animals attempt to escape from the container by jumping up on the walls. Flapping or struggling continually before they fall over and lose posture would also result in audit failure. This would be a minimum standard. Standards need to be developed for the critical limits on gasping, head shaking and other signs of possible aversion to the gas.

Fig. 13.1. When a head-to-back cardiac arrest stunner is used, the head electrode must be placed as shown right behind the ear or on the forehead. The head electrode must never be placed on the neck.

Vocalization critical limits

Cattle

Vocalization (moos or bellows) is scored during handling and in the stunning box or other slaughter point. Vocalization in cattle is an indicator of stress (Dunn, 1990). Grandin (1998b) found that 99% of cattle vocalizations occurred in response to an aversive event such as missed stuns, slipping on the floor, electric prod use or excessive pressure from a restraint device. This standard applies to both conventional slaughter with stunning and ritual slaughter without stunning. In plants with single-file races, vocalization is scored in the race and in the stun box or ritual slaughter restraint box (Fig. 13.2). In plants that do not have races, vocalization is scored any time an animal is being moved by people and at the stun/slaughter point. The critical limit is 3% of the cattle vocalizing. If the plant uses a head-holding device for either conventional or ritual slaughter, the standard is 5%. It is yes/no scoring on a per animal basis. An animal is scored as either a quiet animal or a vocalizing animal. Plants can easily achieve this standard (Grandin, 1998b, 2005a, 2007a). Data collected by the author in ten beef plants that conduct ritual slaughter indicated that this standard can be achieved when the animal is held with a well-designed head-holding device (Fig. 13.3). The audit is failed if any cattle vocalize after hoisting.

Pigs

Squealing in pigs is associated with stressful handling and painful procedures (Warriss et al., 1994; White et al., 1995). Unlike cattle, it is difficult to determine which pig is squealing. A sound level meter could be used by the quality assurance department to monitor sound level. This works well within a plant, but works poorly for comparing one plant with another plant due to differences in plant design and the number of pigs in a room. For comparisons between plants, squealing of individual pigs can be scored in the following areas. In plants with races, squealing in the restrainer is scored. In plants with gas stunning, each individual pig is scored when the gondola is loaded. In plants that stun pigs on the floor, squealing should be scored on each individual pig as it is being handled for stunning. It is yes/no scoring on a per animal basis.

Fig. 13.2.   Well-designed curved race constructed from concrete block that has been plastered with grout to form a hard smooth surface. This is an easy construction method for developing countries. Vocalization scoring is done in the single-file race and in the stun box. Note that the overhead lamp has been moved off the centre line of the race to reduce reflections on the wet floor and improve animal movement.

Fig. 13.3.   Plants that use a head holder must have a vocalization score of 5% or less of the cattle vocalizing to pass the animal welfare audit. Causes of vocalization are often excessive pressure on the neck, slipping on the box floor or holding the animal too long before stunning or ritual slaughter.

A pig is scored as either silent or vocalizing. The limit should be 2%. The audit is failed if pigs vocalize after stunning.

Sheep

Do not use vocalization scoring for sheep. Sheep are an ultimate prey animal and they often do not vocalize even when they are stressed and in pain.

Goats

A vocalization standard needs to be developed.

Poultry

Do not use prior to stunning. The audit is failed if a bird vocalizes (clucks or cackles) after passing the backup bleeder person.

Electric prod (goad) use critical limits

Cattle, sheep, pigs, goats and all other mammals – 75% must be moved with no electric prod. An excellent score is 95% moved with no electric prod. Flags or other non-electrified aids should be the primary driving aids (Fig. 13.4). If the electric prod is replaced with abusive driving methods such as beating, dragging, breaking tails or other abusive practices the audit is failed. A yes/no scoring is used: the animal is either moved with an electric prod or not moved with one. Since it is very difficult to determine if the switch on an electric prod is pushed, electric prod use is scored if the animal is touched with the electrical device. Plants can easily achieve the standard (Grandin, 2005a, 2007a).

Falling critical limits

Cattle, sheep, goats, pigs and all other mammals – the maximum acceptable score is 1% of the animals falling, from unloading trucks until the animal is stunned or ritually slaughtered. A fall is scored if the animal's body touches the ground. Plants can easily achieve this standard (Grandin, 2007a).

Poultry handling

Handling of poultry is assessed by counting the percentage of birds that have broken or dislocated wings. Broken wings should be scored with the feathers on to avoid confusing handling damage, which is a welfare issue, from damage caused by the feather removal equipment. The standard is

Fig. 13.4.   Electric goads have been replaced with flags to move cattle out of the crowd pen and into the single-file race. The handlers are moving the animals quietly with no yelling. Calm animals are easier to handle than excited agitated animals. Cattle and pigs should be moved in small groups and the crowd pen should be filled half full. There is a dark shadow across the race entrance caused by bright sun. This may cause baulking at certain times of day. To reduce contrast the sun may need to be blocked by either a wall or shade cloth. The person standing near the front of the pen should move away so that the cattle do not see him.

1% (per bird basis) in lightweight birds and 3% (per bird basis) in heavy birds over 3 kg. The industry can easily achieve this standard (Grandin, 2007b). Data from 2006 restaurant audits of 22 poultry plants that process heavy chickens weighing 3 kg or more showed that all of the plants had broken wing scores of 3% or less. Six excellent plants processing heavy chickens had 1% or less broken wings. When the audits first started, most plants had 5 to 6% broken wings but their scores became better when they improved their handling practices.

Objective Scoring Improves Welfare

Some animal welfare advocates have complained that allowing 1% of the animals to fall or 5% to vocalize is not very good welfare. In plants that have really good internal and external auditing, the actual numbers are much lower. In 1996 before the restaurant audits started, the worst plant had 32% of the cattle vocalizing and the average was 7.7% (Grandin, 1997, 1998b). Since the audits started, the average vocalization percentage is under 2% and the worst plant is 6% (Grandin, 2005a). Reductions in falling were even more striking. In 2005, data from the most heavily audited plants, which were audited by both McDonalds and Wendys, were examined. Over 3000 cattle and pigs were scored in over 30 plants, and 0% fell down. Before the welfare audits were started, 100% of the cattle and pigs were moved with an electric prod in many plants. Today the average prod use is under 20%. In most plants, the only place the electric prod is used is at the entrance of the stun box or restrainer. During the first 4 years of audits, 17.5% of the cattle were moved with an electric prod (Grandin, 2005a). Young fed cattle can be moved more easily than old cull cows. The average for young fed beef was 15.2% electric-prodded and 29% for old cull Holstein cows (Grandin, 2005a). More recent data collected from the 2005 audits of 72 beef and pork plants indicated that only one plant failed on electric prod use and 23% of the beef plants used electric prods on 5% or less of the cattle. Cattle stunning scores averaged 97.2% on the first shot and 91% (60 out of 66 beef plants) passed the insensibility audit (Grandin, 2005a). In 2005, 100% of the beef cattle were rendered insensible prior to hoisting in 42 plants.

When performance is measured, a plant can determine if they are getting better or getting worse. In setting these standards, the critical limit had to be made high enough to force poor plants to improve but not so high that it would be too difficult to improve. I am often asked by animal activists whether or not the standards should be more strict. I have resisted changing them because there are differences in the ease of handling different types of animals. Difficult-to-move animals or bad weather would cause too many plants to fail. Plants have been audited with these standards for 8 years and many of the plants are continually improving. Data from a major restaurant company showed that, in 2001, 21% of the beef plants and 33% of the pork plants failed on one or more of the five critical control points. In 2005, the failure rate was reduced to 2.5% of the beef plants and 0% of the pork plants.

Cost to Implement Improved Welfare

There are benefits to improved animal handling and stunning. Safety for employees is a sufficient reason for eliminating abusive handling practices of cattle and other large animals. Cruel practices such as shackling and hoisting live cattle prior to ritual slaughter are extremely dangerous. Installation of modern restraint equipment, which holds the animal in a comfortable upright position, greatly reduced accidents and injuries to employees (Grandin, 1988). Other benefits are reduced bruises and better meat quality. Stressful handling and overuse of electric prods shortly before stunning will greatly increase meat quality problems such as PSE (pale, soft, exudative) pork (D'Souza et al., 1999; Hambrecht et al., 2005). Multiple shocks from an electric prod will greatly increase lactate levels in the blood and the percentage of downer pigs (Benjamin et al., 2001). I have worked with plants that were able to export 10% more quality pork to Japan after they improved handling practices.

In the vast majority of plants in the USA, Europe, Australia and South America, major capital costs, such as building totally new handling systems, were not required. In the USA and Canada, only three plants out of over 75 pork and beef plants had to rebuild the entire animal handling facility. The other plants had to make lots of little improvements in the facility and improve employee training and maintenance of equipment. Improvements in equipment usually cost under $2000 and in many plants under $500. The most common simple improvements were installation of non-slip flooring in the stunning box, changes in lighting to improve animal movement and installation of solid panels to prevent approaching animals from seeing people and moving equipment (Grandin, 1982, 1996, 2005a). In poultry, the reduction of broken wings was accomplished with better management of catching employees and improved equipment maintenance.

Management of the plant is a big factor. In three plants, audits were failed until the plant manager was changed. Management is a much more important factor than line speed. Electric prod scores for cattle were very similar at different line speeds ranging from 50 to over 300 cattle per hour (Grandin, 2005a). Good welfare can be maintained at high line speeds if the plant is designed and staffed for the speed. The worst problems occur when equipment is overloaded or the plant is understaffed. The main reason that three plants had to be remodelled was due to either undersized equipment or the wrong type of equipment for their line speed.

Troubleshooting Guide

The first step to improving poor audit scores is determining if the failure is due to an employee problem or an equipment problem. The next step is to improve training and supervision of employees. If the scores do not improve then there is probably an equipment problem. The troubleshooting guide will enable the company to correct problems easily and economically. It will cover both causes and ways to fix problems that cause unacceptable scores. Understaffing is a major cause of problems. For example, it is difficult to do a good job of stunning if the stunner operator has to do too many other jobs. In any job in the plant, increasing either the line speed or the number of jobs a person has to perform can keep being increased until the employee reaches a 'breaking' point. When this point occurs, performance will immediately deteriorate. The principle is similar to overloading a bridge. It can either hold the load or not hold it.

Captive bolt stunning (all species) – problems and corrective actions

1.   Poor maintenance of the gun is a major cause of audit failure. Implement a daily maintenance and cleaning programme (Grandin, 2002b). The use of a test stand to measure bolt velocity is strongly recommended.

2.   Air pressure is too low in a pneumatic stunner for effective stunning. Most stunners will require a dedicated air compressor. Air accumulator tanks will only work in very small plants that process four or five cattle per hour.

3.   Damp cartridges are a major cause of audit failure for cartridge-fired stunners (Grandin, 2002a). Cartridges must be stored in a dry environment such as the office or in a sealed container.

4.   Agitated animals make stunning difficult.

a.   Improve handling practices so animals will enter the stunning area quietly. Reduce electric prod use and stop employees from yelling and whistling.

b.   Stunning is difficult because the animal is slipping on the floor. A non-slip floor grating should be installed. Weld 2 cm diameter steel rods in a 30 cm × 30 cm square pattern to the floor of the stun box. The rods should be welded flat against the floor and not overlapped.

Electric stunning (all species) – problems and corrective actions

1.   Insufficient amperage to induce insensibility. The minimum amperage settings are 1 A for sheep, 1.25 A for pigs and 1.15 A for cattle. Old or large animals may need higher settings. See other chapters for more details.

2.   Electrodes are in the wrong position to pass the current through the brain. To correct this, retrain the employees on correct electrode positioning. The electrodes (tongs) may have to be redesigned or adjusted to facilitate correct positioning in both manually operated and automated systems.

3.   If the animal's body surface is too dry, stunning may not be effective. Wet either the animal or the electrodes to improve conductivity.

4.   Dehydrated animals are harder to stun. This is especially a problem in old animals or animals that have travelled a long distance. Providing water during long trips and in the stockyard (lairage) pens at the plant will help prevent dehydration.

Gas stunning (pigs and poultry) – problems and corrective actions

1.   Gas concentration is either too low or the wrong gas mixture is used. For pigs, 90% CO_2

is recommended. For poultry, see Chapter 11. Another problem that can occur is that the gas is not evenly distributed in the chamber. This may be due to a design fault in the chamber or the ventilation system in the plant may be sucking the gas out of the chamber. Correcting this problem may require the expertise of an engineering professional skilled in the ventilation system.

2. An overloaded undersized machine is one of the worst problems with gas stunning. As a plant increases production, the machine may become overloaded. Plant managers who are purchasing a gas stunning machine should purchase a large enough machine to handle future increases in production. An overloaded machine has to be replaced. Specific signs of an overloaded machine are:

   a. Animals are not rendered insensible because the exposure time has been decreased by speeding up the conveyor.

   b. Gondolas or containers are overloaded and pigs or birds do not have enough room to stand or lie down without being on top of each other.

Insensibility – problems and corrective actions for both conventional and ritual slaughter

1. Stunning not done correctly to induce insensibility. Refer to other parts of this chapter or Chapter 11.

2. Stun-to-bleed interval is too long when head-only reversible electric stunning is used. The stun-to-bleed interval for head-only reversible electric stunning should be 15 s or less (Blackmore, 1984b).

3. Insufficient blood flow after bleeding. This was a major cause of audit failure in many pork plants. The employees should be trained in more effective bleeding methods. In pork, making a larger wound improved blood flow and corrected return to sensibility problems (Grandin, 2001a). In plants doing ritual slaughter, insensibility will occur more rapidly when the cut is performed on a calm animal (Grandin, 1994). A rapid fast cut with a very sharp knife is usually more effective. The knife should be long enough for the end of the knife to remain outside the neck during the cut. The wound should not be allowed to close back over the knife during the cut.

Vocalization – problems and corrective actions for both conventional and ritual slaughter

1. Excessive electric prod use is a major cause of vocalization. Train employees to use a flag or other non-electric aid as their primary method for moving animals (see Fig. 13.4). They should only pick up the electric prod when it is needed to move a stubborn animal and then put it back down.

2. Missed stuns – see stunning recommendations to correct this problem.

3. Falling or constant small slips in the stun box, restrainer or lead-up race may cause vocalization. The remedy is the installation of non-slip flooring.

4. Excessive pressure from a head holder or other restraint device causes pain. Install pressure-limiting devices on hydraulic or pneumatic-powered equipment. A head-holding device will need to have its own separate pressure control. It must be set at a lighter pressure than the pressure control for heavy gates and other parts of the apparatus that would require more pressure. There is a problem with the restraint device if animals vocalize in direct response to application of the device.

5. Sharp edges stick into the animal. Surfaces that contact animals should be smooth. A small sharp edge may stick into the animal. It must be found and removed.

Falling – problems and corrective actions for both conventional and ritual slaughter

1. Slick, slippery floors are the primary cause of animals falling. There are a number of ways to provide non-slip flooring.

   a. Non-slip floor gratings made from 2 cm diameter metal rods in a 30 cm × 30 cm square pattern. Do not overlap the rods. They must be cut and welded to form a flat mesh. Gratings work best in high-traffic areas such as stun boxes, races, crowd pens, scales and unloading areas.

   b. Groove the floor with a concrete grooving machine. This can be done by a concrete contractor or the machine can be rented. This method is recommended for large areas of existing slick floors.

   c. Recommendations for new floor construction for cattle and other large animal – groove the floor in pens and races in a 20 cm × 20 cm square or a diamond pattern. The grooves should be a minimum of 2 cm deep by 2 cm wide (Fig. 13.5). For pigs, sheep and other small animals, smaller grooves can be used that are spaced closer together. Imprinting the pattern of expanded metal mesh into the wet concrete works well. A rough broom finish is not recommended. It wears out too quickly.

d.   Recommendations for unloading ramps. For new construction, a 20° or less slope is recommended. A steeper slope may be acceptable in an existing plant if animals do not fall during unloading. Grooving alone is usually not sufficient for concrete unloading ramps because it wears out too quickly. On concrete ramps, stair steps are recommended. The dimensions of the steps for large animals should be 10 cm high with a 30 cm to 45 cm tread length. For pigs and sheep, a smaller rise of 8 cm is recommended. On metal ramps, the cleats should be spaced with 20 cm of space between them for cattle and 10 to 15 cm apart for sheep and pigs. The animal's feet should fit easily between the cleats. If the cleats are too far apart, the animal will slip.

Electric prod use – problems and corrective actions

1.   Poorly trained employees – employee should be trained in behaviour principles of handling animals (Grandin, 2007b). Employees should not be allowed to yell and whistle at animals.

2.   Overloading of the crowd pen that leads up to the single-file race makes handling more difficult. Cattle and pigs should be moved to the crowd pen in small separate groups and the crowd pen should be filled half full. Sheep can be handled in large groups as one continuous flow.

3.   Animals baulking and refusing to move are a major cause of excessive use of electric prods. This is a facility problem that must be corrected. To improve the flow of animals, distractions that attract their attention must be removed from the facility. Refer to Table 13.1 to troubleshoot distractions that must be removed. To find the cause of baulking, people should get down in the race and see what the animals are seeing. Table 13.2 will help the company solve problems with stun boxes and restrainers. A calm animal will point its eyes and ears towards distractions. A calm animal can be used to help locate distractions.

Fig. 13.5.   Modern stockyard pen with good non-slip flooring. Non-slip flooring is essential for all species. To pass the welfare audit, only 1% of the cattle, pigs or sheep can fall throughout the entire facility, including the unloading ramp and stun box.

Table 13.1. Troubleshooting guide for finding and removing distractions that cause animals to baulk, turn back or refuse to move.

Distractions that cause baulking	How to improve animal movement
Race or stun box entrance too dark	Add indirect lighting that does not shine into the approaching animal's eyes (Fig. 13.6). If sunlight makes the entrance look dark, a shade may need to be installed to block the sunlight
Seeing moving people or equipment up ahead	Install solid sides on races and install shields for people to stand behind. Block the animal's view of moving equipment that causes baulking. Experiment with large pieces of cardboard
Air blowing into the face of an approaching animal	Change ventilation so that air will not be moving at the stun box entrance
Excessive noise	No yelling, silence hissing air exhausts and install rubber pads on clanging equipment
Reflection on metal or wet floors	Adding or moving lamps will usually eliminate reflections. A person must get in the race and view from the animal's eye level to determine if the reflection is gone. Do many experiments with a portable lamp. Existing lamps may need to be moved
Small objects that cause baulking	Remove loose plastic, clothes hung on fences and chains dangling in races
High colour contrast	Paint facilities all the same colour and dress people in clothes that are similar in colour to the walls
Floor drains or changes in flooring colour (Fig. 13.7)	Move the drain from the area where animals walk or make the flooring surface look the same

More information on animal movement can be found in Van Putten and Elshof, 1978; Grandin, 1982, 1996; Tanida *et al.*, 1996.
Information from many specialists is in *Livestock Handling and Transport* edited by T. Grandin (2007b).

Issues in Developing Countries

In most cases, developing countries should be able to fix their facilities and practices so that they can get acceptable scores on an audit. Specific problem areas are covered in this section.

Lack of unloading ramps

In some countries, animals are unloaded at the abattoir by pushing them off the back of the vehicle. Ramps can be easily built in developing countries by local welders and concrete masons. Local people are skilled in welding and concrete work. Figure 13.8 shows an easy-to-construct unloading ramp constructed from concrete blocks. Stationary ramps built at the abattoir are preferable to mechanically complicated devices such as hydraulic

tailgates on trucks or tailgate ramps that fold down. Tailgate ramps require springs or hydraulics that are expensive and difficult to obtain in developing countries. Portable loading ramps on a trailer could also be easily built for use on farms in areas where the animals are loaded on to trucks. An old car axle and part of a car frame could be used for the trailer and the ramp would be built on it. By using part of an old car, the local welder does not have to make complex moving parts. Many local people are very clever at building things.

Tame intensively raised animals versus extensively raised animals

A single-file race or stun box is not required if a few halter-broken animals that have been trained to lead are processed. A tame animal can be led on to the slaughter floor and stunned. Small animals

Fig. 13.6.   A lamp installed on a dark stunning box entrance improved animal movement into the box. It provides indirect lighting and must not shine into the eyes of approaching animals. Experiment with different lamps and different positions. The electric goad in this photo is only picked up when an animal refuses to enter.

Fig. 13.7.   Pigs refusing to move over a white plastic strip. Drains or changes in floor colour or texture may cause animals to stop. A uniform colour and surface work best.

Table 13.2. Solving problems with stun boxes, head holders and conveyor restrainers.

Problem	Possible causes	Remedies
Animal refuses to enter	See Table 13.1	See Table 13.1
	Visual cliff effect in a conveyor restrainer. Animal can see that the conveyor is above the floor	Install false floor to provide the illusion of a floor to walk on (Grandin, 2003)
	Hold-down rack touches animal's back while entering	Raise hold-down rack
Animal becomes agitated or vocalizes	Slipping on the floor	Install non-slip flooring
	Excessive pressure applied by a restraint device	Reduce pressure. There is an optimal pressure for holding an animal. It must not be too loose or too tight
	Held too long in the restrainer	Stun or ritually slaughter within 10 s
	Sudden jerky motion of the apparatus	Install speed-reducing devices so that the device moves with a steady, slow movement. Hydraulic or pneumatic controls should have good throttling control. This enables the operator to move a head holder or other restraint device with smooth movement like a car's accelerator pedal (Grandin, 1992)
	One side of a V-conveyor restrainer runs faster than the other	Both sides should run at the same speed
	Hold-down rack does not fully block the animal's vision on a conveyor restrainer	Experiment with cardboard in different positions to block the animal's vision to lengthen the hold-down
	Animal sees people and other moving things in front of the stun box	Install a solid shield in front of the stun box

Fig. 13.8. Well-designed truck unloading ramp built from masonry. A ramp like this can be easily built in developing countries because many local people are skilled at laying concrete blocks and doing masonry work. This photo shows a ramp for unloading the top deck of a truck. A separate lower ramp would be used for loading the lower deck.

such as sheep or goats can be held by a person and either ritually slaughtered or stunned. Simple systems that may be adequate with tame intensively raised animals that are accustomed to people will be highly detrimental to animal welfare if they are used with extensively raised animals. Systems that may be adequate for tame Middle Eastern sheep are not adequate for imported Australian sheep that are not accustomed to handling. For these sheep, a single-file race is required. For small plants, a simple race made by a local welder could be used. The sheep could be lined up in the race and removed one at a time by a person. For large Middle Eastern plants, a conveyor restrainer would need to be installed at the end of the race.

Effect of blood on animal behaviour

In some systems, especially in developing countries, the animal can see another animal being bled or walk over a floor covered with blood. The author has observed that cattle will voluntarily enter a kosher restraint box that is covered with blood, provided that the previous animals had remained calm. If a previous animal had been jammed in malfunctioning equipment for 10 to 15 min, the other cattle would refuse to enter. Research shows that a stressor applied for 15 min, such as multiple shocks from a prod, will trigger the secretion of stress pheromones (Vieville-Thomas and Signoret, 1992; Boissy et al., 1999). If animals remain behaviourally calm, then having blood on the floor may not affect welfare. If an animal becomes agitated for several minutes, the floor may need to be washed before the next animal. Observations indicate that stunning an animal in the sight of other animals has almost no effect on behaviour, but the sight of a decapitated head may cause the next animal to panic. They do not appear to understand what has happened provided that the fallen animal's body remains intact.

Practices that should be banned

Highly stressful cruel practices such as shackling and hoisting of fully sensible cattle and other large mammals should be banned. Other practices that should be eliminated are cutting tendons, poking out eyes, throwing animals, breaking tails and dragging animals. Pushing animals off a truck that has a platform more than 45 cm high should also be stopped.

People engage in many cruel practices because they do not have an alternative. Eye gouging and tendon cutting are most likely to occur if cattle or some other large mammal which is not accustomed to handling is handled on an open slaughter floor. Large plants can easily build stun boxes and

races from locally available steel or concrete to handle these wild, extensively raised animals. This will also improve safety for employees. In small plants, cattle could be blindfolded with a dark opaque cloth and non-slip flooring is essential. Blindfolding with a completely opaque material will keep cattle calmer (Mitchell *et al.*, 2004). In some countries, cattle are tied up by the legs for transport on trucks that have no sideboards. Sideboards and non-slip flooring can be easily constructed and they would also provide great advantages for hauling other types of freight in the truck (Fig. 13.9). This cruel practice should be stopped and sideboards should be constructed.

Auditing on the Farm

The principle of using the HACCP system can also be applied to on-farm audits of animal welfare. Some existing welfare audits contain too many questions and it may be possible to pass an audit even if a very serious problem is present. A bad farm may be able to pass because they obtain enough points on record keeping, training of employees and other factors. Record keeping and training of employees are important, but an audit should have

Fig. 13.9. An innovative rancher in a developing country installed this excellent non-slip floor grating and side panels on his vehicle. This is a good example of work by a local welder that greatly improved animal welfare.

a point system where enough points are taken so that serious problems will result in an automatic failure. For each species, there are certain directly observable bad conditions such as neglected health problems, beating animals or high ammonia levels in indoor housing which should be an automatic failure. Some of these items, such as handling, lameness and body condition scores, would be audited with numerical scoring and a high standard would be required similar to the slaughter audit. Other items, such as beating animals or other obvious abuses, would be an automatic failure with zero tolerance.

Major Critical Control Points for Animals and Poultry on Farms

For a minimum animal welfare standard, all 12 items must be passed. Some welfare programmes will have additional standards that specify certain types of housing such as no-sow stalls. or straw bedding.

1.   Handling scoring for truck loading and veterinary procedures (Grandin, 1998b; María *et al.*, 2004). Similar to scoring in the slaughter plant. The following critical control points would be audited: percentage of animals falling, percentage moved with no electric prod, percentage of animals moving calmly at a walk or a trot and the percentage hitting fences.
2.   Euthanasia equipment inspection – fail if not available or in poor condition.
3.   Body condition scoring – critical limits will need to be determined on the percentage of skinny animals that will be allowed. Use published body condition scoring methods.
4.   Lameness scoring – critical limits need to be determined for the percentage of lame animals that will be allowed. Animals, including poultry, that have difficulty walking have poor welfare. Use published lameness scoring methods. Scoring of swellings and other lesions on the feet and legs can also be used.
5.   Air quality in mechanically ventilated or enclosed buildings. Recommend 10 p.p.m. ammonia for an excellent rating and 25 p.p.m. maximum.
6.   Life support backup in mechanically ventilated buildings during a power failure. Acceptable methods are generator, automatic curtain openers, automatic telephone dialler or a person living at the site. Not required for naturally ventilated buildings.

7.   Obvious neglected health problems in animals that should have been euthanized. Examples would be advanced cases of cancer eye in cattle, large infected wounds or advanced fly strike in sheep. Should be a very low percentage of the animals.
8.   Stocking density – must be within a specified guideline.
9.   Cleanliness of the animals – use published scoring systems to assess soil on animals from mud, manure or wet bedding.
10.   Abnormal behaviour – standards will need to be developed for each species. May be difficult to measure.
11.   Heat stress and cold stress – standards will need to be determined. Some of the problem areas are shade for feedlot cattle, heat stress on ships and heat or cold stress in unacclimatized animals brought in from another region. This will be difficult to measure on a numerical scale. Standards need to be developed. For example, in areas where summertime temperatures exceed 37°C, cattle in feedlots must have shade. Death losses of a large number of animals or poultry due to heat or cold stress would be grounds for failure.
12.   No acts of abuse – similar to the slaughter audit.

Transport and Farm Measurements that can be Measured at the Slaughter Plant to Assess Handling and Welfare

**1.   **Bruises on carcasses and broken wings in poultry. Figure 13.10 shows a simple gate modification to reduce bruises.
**2.   **Injuries consistent with abuse or neglect.
**3.   **Body condition score.
**4.   **Lameness score.
**5.   **Neglected health problems.
**6.   **Lesions on feet, laminitis and overgrown hooves.
**7.   **Dirty animals and poultry.
**8.   **Producers who have higher than average numbers of non-ambulatory downer animals.
**9.   **Percentage of animals or birds dead on arrival (DOA).
**10.   **Percentage of carcasses with PSE or dark cutters.

Holding transporters and producers financially accountable for bruises and other damage will greatly reduce bruises. Grandin (1981) found that cattle sold on a live-weight basis where the plant pays for bruises will have twice as many bruises compared

Fig. 13.10.   To prevent bruising and injury, the bottom 45 cm to 60 cm of this heavy vertical slide gate has been replaced with a curtain made from flexible conveyor belting. Since the animals cannot see through, they do not attempt to walk through it when the gate is closed.

with a system where producers who sell in the carcass are held financially accountable for bruises. In Brazil, bruises on cattle were reduced from 20% of the cattle to 1.3% after a supermarket started auditing the bruises and held transporters accountable. Bruising moved back to 9% when the frequency of audits was reduced (M.J.R. Paranhos de Costa, personal communication, 2006). Insurance policies on death losses should only cover catastrophic losses. If every dead animal due to carelessness is covered, the producer or trucker has no motivation to change bad practices. Financial incentives to reduce losses and damage are one of the most effective methods to improve both welfare and meat quality.

Conclusions

Implementation of an auditing programme that uses objective numerical scoring will greatly improve animal welfare. Another benefit will be reduced bruises and severe meat quality defects such as PSE and dark cutters. To be effective, the programme should audit critical control points such as the percentage of animals stunned correctly on the first attempt, percentage rendered insensible, percentage that fall, percentage moved with no electric prod and percentage that vocalize. Numerical scores enable a plant to determine if their practices are improving or deteriorating.

14

The Future

This chapter considers some likely trends in meat production and animal welfare over the next 10 to 15 years. Some of the points are speculative and are simply suggestions. Others are logical extensions of existing trends.

Changes in Meat Consumption (2008 to 2020)

In the 1980s, people in developing countries consumed about one-third of the world's meat supply. Now, they are eating almost half of the world's meat production, and by 2020 it is expected to be about two-thirds. Meat consumption is destined to rise mainly in those developing countries that are finding new wealth (Table 14.1), and the increase is occurring mainly in high- and middle-income households. When a low-income family in a developing country improves its income, it typically switches from eating maize, rice and beans towards poultry, red meats, fruit and vegetables.

This additional demand for meat could help set new directions in world meat trading. It will present growth opportunities for meat-exporting countries that do not have a subsidized farming industry, and the aim will be to capture new markets provided there is a satisfactory margin for profit. To begin with, the focus will be on China and South America (see Table 14.1). In China alone, the anticipated increase in demand is equivalent in size to the total demand of the US market.

Per capita meat consumption in existing industrialized countries is not expected to change to the same extent. There may be changes in consumption patterns and these will presumably continue along existing trends. Those trends are towards rapid access to food, recognizing safe foods through branding, perceived healthy eating and value for money. In the processed food sector, value for money is becoming just as important as cheapness, and value is based on consumer appreciation from previous experience.

Much of the increase in meat production will be supplied by intensive livestock farming methods. Over the past 20 years, world meat production has been increasing mainly in the intensive livestock sector (Table 14.2), and this trend is expected to continue. Growth in pig and poultry production will probably outstrip that of beef and lamb. Nevertheless, overall beef and lamb consumption in Asia is predicted to rise, but at a much lower rate than pig and poultry meats.

The issues that intensive livestock farming in emerging industrialized nations such as China will be facing will presumably be similar to those presently confronting livestock farming in the developed world. In the absence of agricultural subsidies, industrial efficiency is an overriding prerequisite for remaining competitive. Presently, industrial efficiency is focusing on:

- value-for-money livestock feeds
- industry concentration
- working within the constraints imposed by waste nutrient management
- livestock genotypes that are appropriate for the feed source, the environmental conditions and the needs of the meat processing sector

Some examples of the pressures that are confronting industrialized livestock farming, and will

©N. Gregory 2007. *Animal Welfare and Meat Production*
(N. Gregory)

Table 14.1.   Actual and projected meat consumption in different parts of the world (million tonnes) (from FAO, 2001).

	1983	1993	2020
China	16	38	85
Other East Asia	1	3	8
South-east Asia	4	7	16
India	3	4	8
Other South Asia	1	2	5
Latin America	15	21	39
West Asia and North Africa	5	6	15
Sub-Saharan Africa	4	5	12
Developing World	50	88	188
Developed World	88	97	115
World	138	185	303

Table 14.2.   Sources of recent increases in world meat supply.

System	% increase
Industrialized livestock production	4.3
Mixed farming	2.2
Extensive grazing	0.7

at some stage need to be considered in countries embarking on intensive farming methods, are as follows:

- water shortages may force water-dependent sectors such as large-scale dairying and aquaculture to concentrate in particular regions
- the infrastructure used in managing nutrient wastes adds to the cost of production, and one way the additional cost is being controlled is through economies of scale. Large units or consortia help justify capital expenditure on waste nutrient management, whereas small independent livestock farms are sometimes forced out of business because of environmental regulations
- large livestock farming enterprises can be vulnerable to financial insecurity created by high levels of borrowing and capital investment. A 'bad' year for production can be catastrophic financially

To summarize, the likely trends in livestock farming over the next 10 to 15 years are as follows. There will be:

- an increase in demand for meat, which will be met largely by intensive farming systems
- greater world trade in meat and livestock feeds in response to the increase in demand for meat
- a competitive advantage for industrialized farming systems that adopt technologies that improve the efficiency of converting livestock feed into meat
- pressure to increase productivity in environmentally sustainable or acceptable ways

At the national level these trends will need to be managed in ways that are consistent with international regulatory standards. The next section considers the potential importance of animal welfare in those regulatory standards.

International Regulatory Standards on Trading in Meat

Meat-exporting nations are nervous about the prospect of international regulatory standards on animal welfare. The standards could harm their export trade. Animal advocacy groups have recognized this situation, and they are using it to promote their aims. Attention was initially directed at the World Trade Organization (WTO). The animal advocacy groups criticized WTO rules for promoting trade liberation at the cost of moral standards in livestock production. The rules prevented one country from banning meat imports on the grounds of unethical production methods in an exporting country. It was argued that animal welfare laws applying to production methods in one country should also apply to imported commodities. Otherwise there is the risk that the standards will be futile, and even counterproductive.

WTO rules are framed in very broad terms, and the way they are interpreted is crucial. They permit non-tariff trade barriers against a commodity if the commodity or its production methods are potentially harmful to people in the importing country. Justified non-tariff trade barriers do not presently extend to alleged unethical production methods. Take, for example, the US ban on importing shrimps caught by trawlers that take sea turtles as a by-catch. The WTO dispute panel ruled that this was an unlawful ban. Based on this experience, it seems that animal welfare cannot be used as a trade barrier in its own right. The position that the WTO is defending is that, if one country imposes ethical standards on another through trade barriers, it is working against the principle of legitimate trade liberalization. The interpretation of legitimacy could be a key issue.

The focus has now shifted away from the WTO to the Office International des Épizooties (OIE). The OIE is seen as an appropriate body for establishing international standards on animal welfare. It is not a regulatory authority, but it helps develop standards for international trade in animals and animal products. If the OIE is successful in producing harmonized standards on animal welfare, contravention of those standards could be legitimate grounds for denying trade access. Production standards could be upheld as trading requirements, provided the WTO recognized that they were accepted standards, and provided it recognized the OIE's position in formulating accepted animal welfare standards. This status, along with that of nations wishing to withdraw from an endorsed OIE standard, may require dispute panel adjudication to establish the rules. The WTO's position will still be influential as it has the power of arbitration.

The Global Market and Antiglobalization

In future, animal advocacy groups may direct their attention at companies that promote industrialized livestock farming. Their aim will be to prick the moral conscience of those companies on animal welfare issues. The potentially sensitive animal welfare issues connected with meat production have been described in this book, and in general terms they can be divided into four categories. They are concerns about:

- large-scale livestock production systems
- intensive farming methods
- livestock management during large company financial crises
- cultural and national differences in consideration given to animal welfare

The animal advocacy groups will no doubt target those issues that are open to change. If the attention they engender is not welcomed, the situation could become polarized, and the antipathy they generate could be channelled into anticorporatism.

Anticorporatism is an important component of the antiglobalization movement. It originated as a concern about the power that conglomerates have over local communities. That power is seen in the following:

- the way some companies monopolize the supply of important ingredients or resources, such as feed or day-old chicks
- the displacement of traditional practices by new technologies imposed by a company. For example, there may be reliance on a particular genotype for cost-effective use of intensive production methods
- the control over the marketing of the livestock farmers' produce, along with price-setting
- the displacement of decision making at the local level. This leads to loss of ownership of responsibility, which in turn results in a sense of domination and loss of self-worth amongst farmers

Two examples show how these concerns develop in practice:

- In 1998, shrimp farmers in Indonesia started demonstrating, and eventually rioting, in protest against unfair pressures imposed by a corporate company that controlled the supply of fry and feed and the marketing of finished shrimp from their farms. The unfairness arose from the 1997 downturn in the Asian economy. The shrimp enterprise in the corporate company was not directly affected by the downturn, and in fact its trading conditions benefited from the lower value of the Indonesian currency. Instead, it was pressure from within the corporate company that precipitated problems in the shrimp sector. The company owned a bank, and the economic crisis meant that the shrimp plus other businesses had to support the banking division in its time of need. This was done through lowering the price given to farmers for the finished shrimps. If a farmer complained, the corporate company could withhold supplies of feed or the next batch of fry for stocking the farmer's ponds. This was an example where a company that dominates an industry exerts unfair leverage through control over both the supply of an essential raw material and the marketing of the product.

- The second example also comes from the 1990s. A vertically integrated broiler chicken company in the USA failed to control the way its staff were unloading chicks and catching finished chickens at the contractors' farms. The farmers felt powerless to improve the situation or complain about the way the birds were handled, because of the domineering position of the corporate company. Here again, the corporate company had control of the supply of raw materials and marketing of the product. The situation came to a head when undercover video footage of the handling methods was shown to senators, and questions had to be answered at a higher level. This was an example of animal cruelty through irresponsibility on the part of a corporate business. No doubt there will be other cases in the future, but whether they are common will depend on whether corporate companies come to recognize their influence in people's and animals' lives, and act responsibly. Companies that focus on stability through motivation, instead of manoeuvring for absolute control and short-term competitive gains, may have longer-term prospects.

Antiglobalization is a broader issue than anticorporatism. There are five key concerns behind the antiglobalization movement. They are:

- Unaccountability. It is felt that trade negotiations are becoming dominated by largely unaccountable government agents, and the negotiations are taking place behind closed doors with insufficient attention paid to the needs of local agricultural economies.
- Undemocratic processes. National democracy is subordinate to WTO-sanctioned controls.
- Dependency on large companies. Farming is becoming dependent on transnational companies for technology and inputs.
- Dependency in food supply. People are becoming too dependent on centralized food manufacture, some of which is subsidized, and food is imported when it could be produced locally.
- Erosion of self-worth. Individual influence is disappearing, and this is leading to less self-esteem through constrained responsibility.

Secondary concerns are that farming is becoming less integrated within the national economy, and farmers are only valued by corporate food companies for their land resource and labour.

It seems unlikely that these concerns will fade away. Indeed, concerns about animal welfare could fuel the debate.

Organic farming

Organic farming is seen by some as a symbolic and even practical way of reversing some of the unwanted trends towards industrialized farming. Organic and free-range products have been a growth sector in the vegetable and egg markets over the past 8 years. Organic meat and animal welfare-friendly meats are not as prominent in the marketplace. The reason is simple. Interpretation of organic meat and animal welfare-friendly meat labels has less meaning compared with 'free-range' eggs. Free-range eggs stands for eggs produced by hens that are not kept in cages. Animal welfare-friendly meat does not have a comparable condensing symbol.

The standards used in organic farming vary between the organic farming associations. Harmonization and regulation of those standards would be a small step towards certification of animal welfare standards. In this way some animal welfare standards could be integrated with organic farming standards and this could accelerate the adoption of animal welfare improvements.

The present lack of harmonized standards for organic production is said to be inhibiting exports from developing countries. Developing countries wishing to access the more lucrative organic standards markets overseas have to negotiate import permits with individual nations or customers, or rely on foreign certification services from importing countries. This is often too difficult for some of the developing nations.

Another obstacle is that organic meat has a relatively low turnover. This means that the packaging methods need to ensure a long shelf life. Long-life packaging is not always consistent with a 'natural' image.

Developing countries

One of the biggest dilemmas in animal welfare for the future is in deciding whether external investment in developing countries should embrace intensive farming systems. A simple way of providing more meat for the growing world population is to promote intensive chicken and pig production, especially in Asia and South America. This would reduce the price and improve the availability of meat for medium- and high-income groups, and in the longer term for low-income groups as well. However, some observers feel that if this is poorly directed it could support and help proliferate poor animal welfare standards. Finding the right balance may be difficult, especially when relying on commercial incentives to sustain unsubsidized systems.

In poor countries that do not have the same short-term economic growth potential as China and Brazil, the focus could be on resource management. Managing underfeeding and overstocking and matching resources to needs during droughts are three of the most serious animal welfare problems these countries encounter. Many strategies have been tried, including:

- educational programmes on seasonal cropping of livestock
- introducing livestock taxes or grazing fees to control stock numbers
- enforced culling of animals judged to be inferior genetic stock
- controlling access to rangeland through regulatory grazing associations
- enclosing rangeland, and replacing it with ranching systems
- providing transport, markets and abattoirs to create outlets for surplus stock

Some of these were successful, albeit temporarily, in improving resource management. For example,

during the 1972/73 Sahelian drought, ranches established by the French protectorate managed to withstand the worst effects of the drought whilst surrounding areas suffered severe losses. However, many programmes have benefited only a minority of people.

A recurring problem is that the stocking rate does not establish its own sustainable level without periodic catastrophic losses. Well-meaning vaccination programmes against livestock diseases promote survival, which in turn leads to higher stocking rates. Financial or infrastructure incentives also encourage pastoralists to keep more stock, and this can add to stocking pressure. The rise in numbers is checked the next time a drought or other natural disaster strikes. Instead, the emphasis needs to be on educational programmes aimed at maximizing profit through timely cropping of stock before droughts or the seasonal downturn in feed availability takes effect. The incentive to destock has to be framed in terms of minimizing losses whilst maintaining milk supply for the household and manure for arable crops. In some communities, encouraging the transfer of decision making to women may help this change, as women have a greater interest in using animals for food production instead of wealth assertion.

Many rural communities in developing countries do not have banking facilities, and livestock are the only means of accumulating wealth. Goats are equivalent to a 'current account', and cattle serve as the 'savings account'. Animals also provide security against crop failure and currency fluctuations. There need to be alternative forms of security, investment and wealth accumulation if overstocking is to be controlled by reducing stock numbers in pastoral regions. Encouraging reliance on a monetary system would only be the answer where the people are confident about the stability of the financial system. Inflation and currency fluctuations can erode and even destroy wealth accumulation. In addition, there needs to be confidence in the stability and fairness of regional and national government, to make alternative investments attractive. In some cases the answers may lie in capturing new opportunities at the local and regional level, but in districts that are under-resourced and impoverished it is difficult to see any better alternative to keeping a herd of breeding livestock.

Providing emergency relief is a necessary but often too-late remedy for natural disasters. The scale of stock losses during these disasters emphasizes the seriousness of the problems and the ineffectiveness of aid programmes. During the 1982 drought in Botswana, cattle numbers in some parts fell by over 50% (White, 1993). Death from starvation was the main cause of losses. During the 1998 flood in Bangladesh, 4000 cattle, 7000 goats and over 313,000 chickens died (Wahra et al., 1999). Assisting temporary translocation or slaughter of stock is often the only feasible course of action. In Australia, sheep farmers recognize that when a drought is coming they have to act promptly to avoid financial ruin. In their case, the infrastructure, in terms of roads, vehicles, abattoirs and refrigeration, is available to allow realization of the value of the animals. This is not necessarily the case in developing countries, and a sustainable urban abattoir industry is an integral part of rescue and salvage operations.

In some countries, ranching could emerge where other pastoral systems have failed. For example, when transhumance becomes more difficult through restrictions on freedom of passage, the pastoral land will either be enclosed or revert to subsistence communal use. Ranching has been linked to feedlots in some poorer countries where there is a market for this type of beef, and with sound management they could be successful.

Keeping livestock in urban areas will probably hold its present importance, especially in poorer communities where there are few employment opportunities. Presently, about 91% of city households in Bangladesh keep either livestock or poultry, and in Kenya 1.5 million livestock are kept in the major towns (Ghirotti, 1999). Relatively high livestock densities also exist in some cities in the Middle East. Containing contagious animal diseases is a major concern.

The animal welfare problems associated with urban livestock farming include:

- Livestock owners taking on more than they can manage. For example, the owner may have a regular job as well as trying to manage a flock of rooftop broilers or sows and their progeny in a backyard. Normally this may be manageable, but, when there is a problem that requires time and regular inspection, the problem may escalate.
- Trying to manage whilst minimizing costs. This can range from inadequate health care to inappropriate feeds.

- Ignorance: not knowing how the animals should be fed or cared for. For example, how to avoid parasites.

Some onlookers have suggested that mobile phones will be a great help to rural poor people. It could give them immediate access to market information including price movements for livestock, and help them manage feed shortages. Whether this comes about depends on access to mobile phones and on making use of the information in decision making. For example, market information will be of little use if infrastructure is inadequate to allow delivery of the animals in good time.

In emerging market economies, small-scale farming is at risk of being relegated to subsistence or lifestyle farming because of controlled access to the retail supply chain. Small-scale farming is being sidelined because wholesalers and supermarkets think that it is 'not worth bothering with'. This applies less to meat than to crops because animals are often collected into larger slaughter groups at livestock markets. These markets will be important in future in serving small livestock and mixed farming systems. Their benefits to society need to be weighed against the risks they pose for distributing animal diseases.

Livestock numbers are increasing, whilst at the same time the amount of grazing land is decreasing. This combination is leading to localized overgrazing and it is probably increasing the risk of seasonal underfeeding. For example, in Rajasthan more than 75% of rural householders reported that the present grazing grounds are inadequate and are progressively deteriorating (Bharara et al., 1999). In northern Nigeria there is less seasonal transhumance because of land awards and closure of migratory routes. Nomadic and semi-nomadic peoples are adopting a more sedentary existence, and the former interdependence of nomads with sedentary agriculturalists is disappearing (Omotayo et al., 1999). Less transhumance means that stock are being kept in tsetse fly zones throughout the year, and there is greater dependence on by-products from arable farming. Arable aftermath will probably play an increasing role in feeding stock based in settled communities, and both agricultural and pastoral communities should benefit from this, provided good relations can be maintained. Overstocking will inevitably change the plant composition and this in turn will impact on which animals can be sustained. For example, when tall species of the

Hyparrhenia genus are replaced by *Pagame*, which is more resistant to trampling, livestock keepers may have to change to small ruminants in place of cattle. Similarly, in Australia barley grass species are taking the place of *Phalaris* cultivars in regions that are overgrazed. New strategies are needed for managing sward composition in heavily stocked semi-arid regions, to avoid underfeeding.

Subsidized Production

In EU countries, up to 40% of farm income comes from subsidies generated by the Common Agricultural Policy. Until recently payments have been coupled to production on a commodity basis. For example, a sheep farmer received a payment according to the number of breeding ewes that were on the farm on a particular day of the year. Now the payment is being decoupled from production, and instead it is based on the amount of land that is managed by the farmer. It is anticipated that, once the new system is established, European farmers will receive €300 per ha per year for simply owning or leasing land. This change is consistent with the need to avoid international trade conflicts arising from alleged unfair subsidization of commodities within a free-trade market. However, it will lead to increasing focus on how the farm is managed. It is envisaged that farmers who fail to take adequate responsibility for conservation, amenity, environmental management and animal welfare will not be given the subsidy. This could have three repercussions for animal welfare. First, there could be more inspection of farms and farming standards, and subsidy payments will be withheld if the standards fall short of expectations. Secondly, farm enterprises that do not attract a worthwhile subsidy will not be inspected under the scheme. This could mean, for example, that egg producers will not receive regular visits from inspectors who determine whether payment should be made or withheld. Thirdly, the subsidy could evolve into a payment for service scheme where farmers are asked to demonstrate how they have improved or maintained standards of conservation, amenity and welfare. In all but the second case there will be more inspection of farming practices, including animal welfare.

The previous system, which was coupled to the number of livestock on the farm, raised some animal welfare problems. For example, some sheep farmers retained ewes that should have been culled, to enhance their subsidy payment. Dead ewes that failed to survive the winter were a common sight on moorland in some situations. Whether making payments conditional on acceptable welfare standards works will depend on how it is implemented and how the standards are evaluated.

Future Issues in Specific Livestock Sectors

Over the past 40 years the livestock industry has tried to economize on feed costs by using cheap feeds. This has taken several forms, including feeding dried poultry manure to ruminants, feeding pig and poultry manure to fish and incorporating surplus offal into a range of animal feeds. This trend has been checked by animal and public health regulations aimed at reducing hazards associated with salmonellosis and TSEs. In future the emphasis could be on making better use of crop residues and co-products from processed plant material. Those sources will be safer, but they will not be as useful in meeting demand for protein or nitrogenous supplements. Legumes could play an increasing role as sources of protein in animal feeds, with the added advantage that they have a low demand for nitrogenous fertilizers.

Breeding and rearing livestock will no doubt continue to be a good way for smallholders to increase their wealth. However, it may become more demanding in terms of labour requirements. As heavily populated regions become even more crowded, livestock farming methods will have to adapt. There will be greater reliance on cut-and-carry and tethering systems for **cattle, sheep and goats**, instead of free grazing. Confinement and tethering can introduce injury hazards if they are not properly managed, and require greater time, care and diligence on the part of the owner. Lack of time or labour will be a constraint in successful family farming activities. As children attend school for longer they are not available as unpaid labour, and so moving stock to water or between grazing patches, or providing fodder will be duties that have to be taken on by adults. In some sectors this change may be welcomed, because there has been growing concern about child exploitation, especially in parts of Asia. For example, in Bangladesh children spend up to 13 h a day at the hatcheries collecting fry to stock shrimp farms.

Future trends in **pig** production will include:

- emergence of industrialized pig farming in Asia
- refinement of food safety measures
- larger litter sizes, leading to stronger justification for teeth-clipping piglets and the use of farrowing crates

As litter size increases there will be a corresponding decrease in average birthweight and increase in the variability of birthweight. The question will be how to manage the increasing number of undersized piglets. Small piglets often consume less colostrum during their first feed, and so their immune defence system can be compromised. In future, a research focus will probably be on developing efficient ways of maximizing growth and survival in the newborn undersized piglet.

No doubt, there will be technical innovations that will make pig farming and management easier. For example, the focus for monitoring temperature in pigs could shift away from measuring air temperature towards recording skin temperature using infrared sensors. This will be simpler and it should relate more closely to the thermal comfort of the pigs.

The phasing out of feed antibiotics is likely to result in later weaning in the pig industry. There will be fewer farms using 21-day weaning, and segregated early weaning will become less attractive because of health-related risks. This will reinforce the need for effective vaccines in controlling respiratory disease.

World **poultry** meat consumption is expected to increase above its present 31% share of the total meat market. Much of the growth will be in Asia, which already consumes over one-third of the world's poultry meat production. About 70% is raised in China where the preference is for chicken wings, leg quarters, feet, gizzards and necks. Breast meat is considered too plain and dry. The fast-food market in China prefers drumsticks and the middle part of the wing. There will be two opportunities in this growing market. First, there could be more emphasis on developing fast-growing lines that suit Asian demand for leg and wing meat. These could be quite different from lines that have high breast meat yield, and the birds should not have the same physical problems that lead to poor gait. Secondly, there could be interest in preparing and presenting breast meat in ways that are more attractive to the Chinese palate.

Some traditional Asian markets will continue to have their unique features. For example, there is little sign that there will be an immediate change from the traditional wet markets for poultry, where birds are slaughtered and sold individually by shopkeepers for home consumption on the same day. No doubt the 'yellow bird', which is preferred in the south of China, will hold on to its present popularity. It is a relatively slow-growing broiler chicken and is raised in three stages. They are often brooded in brick buildings, and then transferred to floor pens before finishing in cages. Moving the birds is inevitably labour-intensive, but this is considered worthwhile as they are renowned for the flavour of their meat and command a good price.

In broiler strains that have a high breast meat yield, there is now considerable research evidence that shows there are benefits for bird welfare from limiting growth during the second week of life. These benefits include fewer leg disorders, improved resistance to heat stress and fewer deaths from ascites and SDS. The commercial benefit is improved feed conversion. This win–win situation for welfare and profits should encourage widespread adoption of early growth restriction.

The range of species that satisfy the demand for meat will change only slightly. Most of the meat we eat comes from animals that are bred and fattened specifically for the meat market. There are animals that are eaten when they reach the end of their working life, but they are a minority, as are harvested or hunted wild animals. Reserves of **wild animals** are declining and there will inevitably be less harvesting and hunting. As they become scarce, offtake may be limited to recreation and people who take wild animals specifically for their own consumption. Before that stage is reached there could be an intermediate period when hunting pressure escalates as the value of game and bushmeat rises because of its scarcity. Some highly valued harvested species could be farmed. We can see this happening in the aquaculture industry. For example, cod fishing is giving way to cod ranching and, if captive breeding can be managed successfully, it will lead to cod farming.

The sector that will see the greatest technological changes is undoubtedly the **aquaculture** industry. Aquaculture has been benefiting from declining stocks in the marine fisheries. Fish produced on farms now accounts for about 30% of overall fish meat consumption, and in the short term this will continue to rise. The aquaculture industry is almost ubiquitous, and is even becoming important in remote regions such as the Pacific Islands, because of declining marine fisheries.

There could be greater emphasis in using entire natural reserves as intensively managed fish farms. This will help to increase the scale of production whilst providing a more 'natural' image for fish farming. The concerns that exist in marine fisheries about by-catch and dwindling stocks of native fish will presumably extend to some of these larger inland natural reserves. These issues will need to be managed to ensure perceived standards of sustainability. The reserves will either have to be restocked regularly, or the catch will need to be controlled, and the natural breeding grounds protected. Lake Malawi is a good example of a fishery reserve that is being overfished and breeding habitat is being lost through a laissez-faire approach to management. Regulations on mesh size and out-of-season fishing have been ignored and fish stocks in the lake are falling.

By-catch in managed natural reserves could be reduced by:

- regulating the type of gear that can be used
- regulating the time an area can be fished
- regulating the area that is fished
- improving the selectivity of the gear
- reducing the amount of discarding by making use of most of the by-catch

In most fisheries there are few economic incentives to adopt more selective gear. It would have to be introduced and enforced through regulations. Even so, there are often ways of circumventing regulations, such as using loose knots in the mesh and allowing the mesh to close under drag.

Restocking and replenishing inland waters with wild-caught or cultured fish present the following welfare concerns for the translocated animals:

- stress of transport
- predation on release
- competition through initial overstocking

There may also be concerns about loss of genetic integrity of indigenous stock, and disruption of some features of the ecosystem.

Continued overfishing of marine fisheries should help sustain a successful and diversifying aquaculture industry. Even though the two sectors specialize in different species, there is sufficient overlap to make them competitors in future markets. The marine fisheries' loss will be the aquaculture industry's gain.

Future Markets and Trade Barriers

Past experience shows that international trade in meat increases as people get wealthier. Between 1961 and 2003, total global meat production increased from 71 million t to 271 million t, and international trading in meat increased from 3.5 million t to 26 million t. In other words, global trade increased from about 5% of overall meat production in 1961 to about 10% in 2002. During this period, there was a strong correlation between the increase in meat trade and GNP per person.

If this trend continues, then in future a growing proportion of meat will be produced for sale overseas. In the past, tariffs and quotas have constrained exports, but there has been regional relaxation of the constraints through trade agreements such as the North American Free Trade Agreement (NAFTA), the Southern Cone Common Market (MERCOSUR), the Caribbean Community (CARICOM), the EU, the European Free Trade Association (EFTA), the ASEAN Free Trade Area (AFTA), the Andean Community, the Common Market for Eastern and Southern Africa (COMESA) and the Southern African Development Community (SADC). We are now reaching a stage where people are asking whether nations should continue with such trade concessions, or whether we should unanimously adopt WTO rules, which regulate both tariffs and quotas and render regional trade agreements superfluous.

Presently, most new international trade agreements focus either on relief from tariffs or setting quotas for trade access. In future, there could be more attention to technical barriers to trade, food safety being a primary concern. Surveillance methods for food-borne pathogens will undoubtedly improve and this will increase the opportunity for imposing trade barriers. The evidence required when introducing a food safety trade barrier will itself become more regulated, for the sake of transparency and fairness.

Trade barriers have already been used against importing farmed fish, and this may become more common in the future. A notable case was the importation of Nile perch from Lake Victoria into EU countries. The ban originated in the 1990s when *Salmonella* were found in some of the products. This was shortly replaced by another ban because of pesticides found in the lake and fish. Other contaminants that are frequently monitored in fish include heavy metals, veterinary drugs, agrochemicals and harmful algal bloom (HAB) toxins. HABs are not a new problem but they are becoming more common. In industrialized countries they are monitored by chemical analyses, but developing countries still use mouse-bioassay methods, which are criticized as an inhumane use of laboratory animals.

Future Marketing Methods

In some countries there has been talk about phasing out **livestock markets**. It is argued that, if there were no markets, contagious disease outbreaks could be contained more easily. In addition, it is said that the additional handling in putting animals through markets incurs stress and injuries. The counter-argument is that livestock markets are not responsible for breakdowns in border control of diseases, and why should domestic auction markets suffer for a political decision on trade liberalization between countries? The advantages and disadvantages of livestock markets are summarized in Table 14.3.

Procuring animals through livestock markets is important for countries that have a large number of small abattoirs. Small abattoirs cannot afford to spend time recruiting and selecting stock directly from farms. Instead they source through livestock auction markets. Markets are particularly helpful in communities that have a large number of dispersed small-scale farmers, as it is easier for livestock buyers or abattoir agents to visit a market instead of a large number of smallholdings. Selecting and drafting stock, when a buyer visits a farm, involve extra handling, and so the apparent disadvantage of

Table 14.3. Advantages and disadvantages of livestock markets.

Advantages
- Encourage competitive buying and fair prices
- Enable small abattoirs to source animals
- Act as a venue for enforcing animal health control measures
- Social benefit for the farming community

Disadvantages
- Additional handling stress and injuries compared with direct sale to an abattoir
- Can encourage spread of contagious diseases

additional handling when selling through a market may not be as significant as some people think. The system that involves the least handling is direct sale, sight-unseen, to an abattoir, but this does not always suit the discriminating abattoir that requires particular types of stock or does not want to be treated as a disposal route for inferior animals. These abattoirs have to employ buyers, who visit the farms to select the animals they want. Specialization in the type of animal needed makes procurement more complicated, and employing a buyer to source the right animals at the right time through centralized markets is an efficient approach from the meat processor's perspective.

Livestock Transport

In future, EU legislation will define a long-distance journey as greater than 8 h. About 10% of livestock transport in Europe is long-distance, and from 2007 drivers will need specific authorization and a Certificate of Competence when providing a long-distance service. The new requirements will enable enforcement of higher standards for vehicles used in long journeys, including a temperature warning system, additional mechanical ventilation when needed and recording of journey information. Vehicles will be inspected for cracks in alloy floors, broken battens on tailboard ramps, gaps between the tailboard and truck floor, step heights on to and off the tailboard and slopes of the ramps. The overall aim is to improve standards for the animals experiencing long-distance journeys.

Vehicles that have hydraulic lifts and moving floors have some advantages in terms of animal handling. Lifts are potentially useful for young calves, which are prone to going down on both knees when failing to step on to a tailboard ramp, and for pigs that are difficult to load. However, they need to be fitted with safety cut-out mechanisms to reduce the present risk of injuries to operators.

Higher fuel prices in future will have the following effects:

- Vehicles carrying large numbers of animals will be more cost-competitive.
- Vehicles not filled to capacity will be less cost-effective.
- Stock owned by different people are more likely to be carried on the same load. This will result in:

- longer waiting times;
- longer journeys;
- greater control needed in counting stock on to and off trucks, and keeping batches separate.

In the marine fish farming industry there could be a move towards slaughtering and processing fish on well-boats, which are presently used for transferring fish from sea pens to the on-shore processing facilities. The commercial benefit will come from avoiding problems associated with early rigor. The welfare benefit will be reduced pre-slaughter stress. Blood collected at slaughter will have to be stowed on the boats, and under hot conditions quality will be ensured if the fish are held in ice before offloading.

Water

The main areas of concern about water quality and water use in meat production are:

- insufficient water in arid and semi-arid regions
- contamination of waterways by livestock and aquaculture
- land and water degradation with salt through irrigation

Water resource agencies in arid and semi-arid regions are concerned about rising salinity levels in soil and the upper aquifers. There are a number of causes of the problem. First, water extraction rates from boreholes is increasing and, as the upper reserves diminish, boreholes have to be sunk deeper and the salinity of the extracted water inevitably rises. Where this water is used for irrigation, the soil surface accumulates salt through evaporation. Secondly, changes in vegetation in the landscape, away from deep-rooting trees towards pasture or cropping, in combination with irrigation, have meant that there is a continuous column of water between the soil surface and deeper salt-enriched zones. This is allowing salt water to rise to the surface from below. The knock-on effects for animal survival and welfare are indirect, but they will become obvious where conditions deteriorate. The stock-carrying capacity on land affected by salt burden inevitably declines, and this is first seen in seasonal problems with underfeeding. Salt loading in drinking water could become worse in regions with high salt

levels in the upper aquifers, and young stock and lactating and pregnant animals are the first affected. Young stock are most likely to refuse salt-contaminated water.

Aquaculture is facing a range of environmental issues connected with water quality, and regulatory constraints are already having an impact on the growth of the industry in particular regions. In Thailand, there was a period when rice fields were converted to more lucrative shrimp farms. However, this trend has now been halted by the government because of problems with waste-water disposal from the shrimp farms. In India, the growth of shrimp farming is being limited by contaminants from its own operations.

Meat Quality

The fresh meat trade is becoming more demanding in the quality and appearance of the meat it sells. Conditions such as two-toning, residual blood, colour brightness and cohesion in thinly sliced products are increasingly important. One way of meeting this trend is to be more rigorous in downgrading. The less than perfect meat will end up in manufactured products, and the present move to producing more manufactured product fits well with that trend.

PSE will have to be managed by genetic selection to hold it at tolerable levels. This is the approach presently used in Denmark, where the aim is to prevent the prevalence exceeding 2% in loins and 4% in hams. If similar objectives are set in other countries, there will need to be a reduction in heterozygotes (Nn) in the slaughter generation. Genetic selection may be extended to other meat quality traits, but companies taking such steps must be aware of possible downstream effects. For example, if marker-assisted genotyping is used to select against beef toughness, it will probably result in reduced flavour intensity, but there could be improved marbling (Wulf *et al.*, 1996). The overall merits and disadvantages need to be assessed.

In future, PSE will assume greater importance in turkey meat and possibly chicken. Turkey meat has moved away from the whole bird market towards portions and further processed products. This has increased the importance of water-holding capacity and myosin gel strength. PSE poultry meat has a poor water-holding capacity and forms soft gels when used in cooked processed meat products (Woelfel *et al.*, 2002). The solution is not immediately obvious, but it may lie in greater use of polyphosphates and tempering before slicing the products, rather than changes in live bird management.

Meat dryness could become a more common complaint as meat becomes leaner. Exploiting breeds and lines with higher levels of marbling could be one solution, or the meat could be injected with an emulsion or marinaded. In hot climates, most Christmas turkeys are injected with an emulsion and sold frozen. This is a very effective solution for the dryness problem, and it gives an opportunity to add flavour enhancers. However, it is less appropriate for the fresh or portion turkey trade and, since more turkey meat is being sold as unfrozen portions, the opportunity for injecting emulsions may be limited.

In general, the eating quality of organic pork has fallen short of the original expectations (Grunert *et al.*, 2004). The intention was to position this meat on its production characteristics, but it turned out that this has few or unclear benefits in terms of eating quality for the consumer. Without those benefits, consumer interest could collapse. However, experience with other meats shows that organic brands benefit in the aftermath of a food scare, and so the future for these products could depend on how food safety issues are managed in the future.

Food Safety

One of the effects of a food scare is that consumers pay less attention to price when they resume consumption of at-risk foods. This occurred following the BSE crisis. Consumers showed a preference for beef of known provenance, and they looked for quality labelling and outdoor or free-range production emblems. Stall-fed beef in particular experienced poorer consumer interest compared with free-range beef in the aftermath of BSE (Mesías *et al.*, 2005).

As explained in Chapter 10, the focus in pigs and poultry needs to be on controlling *Salmonella* and *Campylobacter*. With the uptake of traceability systems, blame for food poisoning outbreaks may be directed at the farming sector. Test cases may establish the growers' potential culpability, and proof of negligence will be a deciding factor.

The outcome from such cases could help determine where future emphasis is directed on how to manage food-borne pathogens. If farmers share in the responsibility for a serious outbreak, reducing the risk of cross-contamination between farms will become a more important goal. There will also need to be a better understanding of the mechanisms that raise shedding and infection rates. The stress mechanisms that increase *Salmonella* excretion during transport and pre-slaughter handling are not understood. However, they are linked to the following:

- Feed withdrawal before transport and slaughter, resulting in a rise in gut pH and proliferation of *Salmonella*.
- Increased shedding from *Salmonella* carriers, partly through increased defecation frequency.
- Increased susceptibility to infection, with subsequent excretion from a larger population of infected animals. This may be linked to reduced activity of macrophages and leucocytes. The hope is that gut flora and the shedding rates of specific bacteria could be managed in ways that benefit rather than compromise the welfare of animals.

There may come a time when simple biosensor tests can be applied to meat which identify whether the animal that produced that meat had an infection at the time it was killed. Acute-phase protein biosensors could be used in this way, and they could form part of a trade barrier imposed by importing countries. Acute-phase protein responses are non-specific responses to infectious disease, and so they would not necessarily be present when there is either a public health risk associated with the meat or when there is an animal health risk associated with importing the product. Nevertheless, their screening could be applied in a disruptive manner, for example as justification for detaining product for further tests.

In future, greater attention will probably be given to residues in fish meat and fish meat products. Presently, there is concern about formaldehyde residues in shrimps, and there is growing awareness of the risk of antibiotic residues in farmed finfish and shrimps. The image of the aquaculture industry is at a formative stage, and it needs to decide whether it should continue to support the use of antibiotic top dressings as growth promoters, as well as the development of genetically modified strains of fish.

Genetics and Bioengineering

The acceptability of transgenesis in livestock production will depend on the recognizable benefits to society and the animals. Society is not likely to favour genetic manipulations that only enhance revenue for a limited sector. There must be wider benefits, and early developments in this area could decide whether transgenesis is accepted or debunked.

Applying molecular biology wisely could hold some far-reaching benefits for society and meat production. It could open up opportunities for evading particular animal diseases, avoiding genetic antagonisms that have hitherto been intractable, and improving traits that have been difficult to select genetically. This includes selection for improved muscle mass and meat quality. For example, lines of transgenic mice have recently been developed that have a dominant transinactivation of myostatin on the Y chromosome (Pirottin *et al.*, 2005). This means that males are produced which exhibit muscular hypertrophy, whilst the females are non-transgenic. If, or when, this is applied to cattle, bull calves could be produced with double muscling for the meat industry from the same parents that produce females with high-yielding dairy features without double muscling. It is understandable that there are some nervousness and feelings of distaste about such prospects and about what they might lead to in the future. This happens whenever there is a potential development that impacts on existing standards or expectations. This was the case when AI was first introduced in the dairy industry, and looking back at some of those concerns is interesting. For example, in 1943 there was a debate in the House of Lords in the UK where the fear was expressed that AI for cattle would lead to the breakdown of family life, as it could lead to some women choosing to have children without marriage.

Since Macer's surveys on attitudes to foods derived from GM organisms (described in Chapter 1), public attitudes have been manipulated by fears expressed in the media. The concerns have broadened into fears about:

- harming existing wild types or crops through interbreeding with introduced GM plants
- whether GM foods are safe to eat
- the reality of the claimed advantages of GM crops and animals
- mistakes during the development of GM foods, leading to pathogens or detrimental genotypes that happen to escape control (Laros and Steenkamp, 2004)

Fear of GM foods is greatest among people who are strongly concerned about nature, and less amongst consumers who have faith in technology. The USA is less opposed to GM foods compared with Europe. People's judgement in the USA is based on a balance of the perceived benefits and risks, and there is greater awareness of potential benefits, such as longer shelf life, greater sensory appeal and reduced allergenicity, through application of GM and biotechnologies. Within the USA, the greatest resistance to GM technologies is amongst people who are more religious (Hossain and Onyango, 2004).

No doubt there will be continued interest in conserving indigenous breeds and making use of their resistance to particular disorders. This could either be by conventional breeding methods, by embryo transfer or possibly by gene insertion. Some developments could have benefits for the welfare of future generations of animals. For example, introducing enhanced foot-rot resistance could be a long-term solution where it has not been possible to eradicate this debilitating condition.

Broiler chickens in the 1950s took about 14 weeks to reach market weight, whereas today they can reach 2.6 kg body weight in 6 weeks. No doubt growth rate will be increased further by genetic selection, as will milk yield in dairy cattle and prolificacy in pigs. There will be a limit to these advances, and they will probably develop through changes in emphasis. For example, new opportunities may be recognized on how to improve meat quality, and this will set genetic selection in a different direction. For example, it should be possible to select for marbling in lean lines of pig. The h^2 of the intramuscular fat content of pig meat is high (0.50), and the genetic correlation with overall fatness is low.

Some farmers are supplementing their income by supporting pharmaceutical, bioengineering and biomedical companies in developing new technologies. Often this is little more than supplying the company with eggs or animals raised under high herd or flock health conditions. In other cases, the farm is being used as a site for multiplying special lines. In future some of these enterprises may attract attention and ethical concern. For example, there is interest in developing transgenic donors for animal-to-human xenotransplantation.

At present, effort is being directed at overcoming xenograft rejection by developing animal donors with transgenic modifications that render their organs compatible with the human immune system. The complications for the donor animals depend on the modification being made, but increased susceptibility to septicaemia, impaired vision and lower fertility have been identified as potential risks in some instances for xeno-donor pigs (Dahl *et al.*, 2003).

Some Questions for the Future

One way of managing future problems in animal welfare is to legislate. Animal welfare legislation and codes of practice can either focus on outcomes or they can be prescriptive. Prescriptive requirements define, quite closely, how something should be done or how something should be provided. For example, 'the stocking density for broiler chickens must not exceed $34 \, kg/m^2$'. An outcome-based requirement sets a more general goal and it is up to the individual to find a way of meeting that goal. For example, 'care must be taken to avoid damage to layer hens during removal from cages'. The trend in code-writing has been towards prescriptive standards, and the risk is that future codes of practice will come to resemble standard operating procedures rather than guidelines. Some see this as a necessary precaution. Others view it as unfortunate, as the purpose should be to direct rather than manage stock people and farmers.

Recurring questions about the standards are the following:

- Who should write them? In other words: (i) Should they be written by industry and modified or endorsed by an independent body that is sanctioned by government? (ii) Should they be treated as technical standards that are written by technical experts? (iii) Should they be formulated by consensus? The answers to these questions depend on how the democratic principle is applied. Clearly there has to be a point where the view of society is taken into account, especially on moral issues, such as whether it is right to keep dry sows in stalls. In such cases, society's expectations may be a deciding influence, but it can be difficult to gauge society's view. All too often, society's view is assessed from media comment, and there is growing awareness that this reflects a biased and distinctly undemocratic standpoint.

- Should infringement offences be introduced for animal welfare misdemeanours? In other words, should a person be issued with an immediate fine for specific acts that are considered callous or harmful? On-the-spot fines are issued in some cities for allowing a dog to foul the pavement. It would be a small step to extend this enforcement method to excessive use of the whip during horse racing or excessive use of a stick at a market.

- Should veterinarians be warranted? In some countries a veterinarian can be warranted to raise a charge for an alleged animal welfare offence. Those vets usually volunteer to take this responsibility on, or it may be part of their duty if they work for the government. Should all vets have this authority? If so, who compensates for their time in investigating an incident?

Another more pressing question is who should take responsibility for managing animal welfare situations that get out of control. Take the case of the independent smallholder who is socially isolated and has an alcohol or drug-related problem. He can no longer look after his animals adequately and they are deteriorating. The way this is managed varies between countries. In some countries a government body steps in, or it may be a regional animal charity, or a local farmers group. These types of cases can be costly and sometimes there is reluctance to take on responsibility for rescuing or managing the situation. Evading responsibility leads nowhere.

Sheffield City College Library

Student Assignments

1. You have been asked to design an animal welfare auditing system for a broiler company's farms and processing plant. List the features that you would include in the audit. Distinguish those features that would involve recording objective measurements. (For this purpose an objective measurement includes the prevalence of a condition that might be assessed as present or absent by inspection.)

2. A city businessman with no previous experience of farming is considering setting up a beef-finishing enterprise. He plans on buying in store cattle in the spring and fattening them on lucern plus cocksfoot (orchard grass) pasture for sale during the late autumn and winter. The leased farm is in a coastal region, has 400 mm rainfall a year, is properly fenced and has a light sandy soil. Identify the potential animal welfare hazards.

3. Answer both parts.

    A. There is a consistent problem at a market with loading store cattle on to trucks. The animals refuse to step on to the tailboard ramp and into the vehicle. This leads to bullying by the farmers and truck drivers when trying to load the cattle. Discuss the possible reasons for this problem. Describe potential solutions.

    B. There is a consistent problem at a market with loading bobby calves on to vehicles. When pushed up to the tailboard of the vehicles, they fail to clear the edge with their feet and fold at the knees. They end up kneeling on the tailboard, and the stock people often push them from behind to encourage them to stand up and move on. What should be done?

4. You have been asked to advise a retail meat company on the production standards that should be used for organically produced veal. The company wishes to encompass welfare-friendly features in this label as well organic methods. How would you set about drafting the standards? Include a list of features that should be considered when drafting a set of standards.

5. A herdsman in semi-arid West Africa is employed to look after a herd comprising one bull, ten 1- to 3-year-old heifers and steers, and five cows with calves at foot. The herd is kept near a township, which is 20 km from where the owner of the cattle lives. The calves are growing very badly. What are the potential causes of their poor growth? The owner knows that the herdsman is milking the cows and selling the milk instead of allowing the calves to suckle from the cows. List the alternative courses of action the owner could take.

6. A supermarket company has asked you to provide an independent report on the standards at three feedlots that supply their shops with beef. What features would you examine, and what would you be looking for?

7. An abattoir has reported that some of the cull dairy cows it processes arrive with identification tags attached to the back of the udder instead of the ear. The tags pierce the skin in the udder in a similar way to an ear tag. You have been asked to consider whether this change in practice causes any welfare problems. How would you go about assessing this?

8. Visit your local livestock sale yard on a market day. Describe the strengths and weaknesses of the animal-handling methods that you see.

©N. Gregory 2007. *Animal Welfare and Meat Production*
(N. Gregory)

9. A pig farmer buys in weaners on a regular basis for fattening. In the past year the pigs have experienced problems with respiratory disease. The weaners are all right when they arrive at the farm, but start coughing about 7 days after arrival. About 6% die from pleuropneumonia within 3 weeks. Describe the potential contributory causes and ways of managing this problem.

10. Give brief answers to each of the following questions.

A. Sheep consistently slip on the concrete floor in the unloading bay as they come off the trucks, especially where there is a turn into a corridor. How can this be corrected?

B. A heifer goes down in the stunning pen before she is stunned and refuses to get up. What should be done?

C. There is a power failure (outage) at a pig abattoir that operates two shifts per day. Pigs are being delivered for the next shift, and the lairage is now full. The fans in the lairage are not working, and the pigs are getting hot. List the things that should be done in their order of importance.

D. An abattoir is using electrical stunning for large cattle with a nose-to-neck current, followed by a nose-to-brisket current to induce cardiac arrest. A batch of unusually large cattle is put through. After each animal is tipped out of the stunning pen it 'rises up'. They have to be shot because of concern that they are recovering. What could be causing this problem?

E. You visit a goose-processing plant that is using a water-bath stunner followed by manual neck cutting by knife. How would you recognize whether the geese are being properly electrically stunned and slaughtered?

References

Aalhus, J.L., Gariepy, C., Murray, A.C., Jones, S.D.M. and Tong, A.K.W. (1991) Stunning and shackling influences on quality of porcine longissimus dorsi and semimembranosus muscles. *Meat Science* 29, 323–334.

Aboagye, G.S., Twah, C.L. and Rege, J.E.O. (1994) Shorthorn cattle of West and Central Africa. 3. Physical, adaptive and special genetic characteristics. *World Animal Review* 78, 22–32.

Ahman, B., Nilsson, A., Eloranta, E. and Olsson, K. (2002) Wet belly in reindeer (*Rangifer tarandus tarandus*) in relation to body condition, body temperature and blood constituents. *Acta Veterinaria Scandinavica* 43, 85–97.

Ali, A.S.A., Jensen, J.F., Lawson, M.A. and Chwalibog, A. (2005) Variability in post-mortem pH values of broiler breast muscles due to electrical stunning voltages. *Archiv für Geflügelkunde* 69, 226–230.

Alvarado, C.Z. and Sams, A.R. (2000) Rigor mortis development in turkey breast muscle and the effect of electrical stunning. *Poultry Science* 79, 1694–1698.

Andersen, I.L. and Bøe, K.E. (1999) Straw bedding or concrete floor for loose-housed pregnant sows: consequences for aggression, production and physical health. *Acta Agriculturae Scandinavica* A 49, 190–195.

Anderson, J.F., Bates, D.W. and Jordan, K.A. (1978) Medical and engineering factors relating to calf health as influenced by the environment. *Transactions of the American Society of Agricultural Engineers* 21, 1169–1174.

Anderson, J.L. (1975) Embryonic and placental development during prolonged inanition in the pig. *American Journal of Physiology* 229, 1687–1694.

Andersson, H.K., Olsson, V., Hullberg, A. and Lundström, K. (2003) Effects of sex, feed and pre-slaughter routines on technological meat quality in carriers and non-carriers of the *RN⁻* allele. *Acta Agriculturae Scandinavica* A 53, 147–154.

Anil, L., Anil, S.S. and Deen, J. (2002) Relationship between postural behaviour and gestation stall dimensions in relation to sow size. *Applied Animal Behaviour Science* 77, 173–181.

Anil, M.H. (1991) Studies on the return of physical reflexes in pigs following electrical stunning. *Meat Science* 30, 13–21.

Anil, M.H. and McKinstry, J.L. (1998) Variations in electrical stunning tong placements and relative consequences in slaughter pigs. *Veterinary Journal* 155, 85–90.

Anil, M.H., McKinstry, J.L., Wotton, S.B. and Gregory, N.G. (1995) Welfare of calves – 1. Investigations into some aspects of calf slaughter. *Meat Science* 41, 101–112.

Anil, M.H., Whittington, P.E. and McKinstry, J.L. (2000) The effect of the sticking method on the welfare of slaughter pigs. *Meat Science* 55, 315–319.

Anil, M.H., Yeşildere, T., Aksu, H., Matur, E., McKinstry, J.L., Erdogan, O., Hughes, S. and Mason, C. (2004) Comparison of religious slaughter of sheep with methods that include pre-slaughter stunning, and the lack of difference in exsanguinations, packed cell volume and meat quality parameters. *Animal Welfare* 13, 387–392.

Anon. (1910) How to make meat tender. *Queensland Agricultural Journal* 24, 275.

Anon. (1992) *Nigerian Livestock Resources* Survey. Volume 2: *National Synthesis*. Federal Department of Livestock and Pest Control Services and Resource Inventory and Management Ltd, Abuja, Nigeria, and St Helier, Jersey, 440 pp.

Anon. (2004) Information bulletin on condemnations in Canada. Canadian Food Inspection Agency, Agriculture and Agri-Food Canada. www.agr.gc.ca/misb/aisd/poultry/condmn

Antia, R.E. and Alonge, D.O. (1982) Survey of abattoir data in southern Nigeria. *Tropical Animal Health and Production* 14, 119–120.

Applegate, A.L., Curtis, S.E., Groppel, J.L., McFarlane, J.M. and Widowski, T.M. (1988) Footing and gait of pigs on different concrete surfaces. *Journal of Animal Science* 66, 334–341.

Argüello, A., Castro, N., Capote, J. and Solomon, M. (2005) Effects of diet and live weight at slaughter on kid meat quality. *Meat Science* 70, 173–179.

Ash, A., Gross, J. and Stafford Smith, M. (2004) Scale, heterogeneity and secondary production in tropical rangelands. *African Journal of Range and Forage Science* 21, 137–145.

Assad, F., Bayoumi, M.T. and Khamis, H.S. (1997) Impact of long-term administration of saline water and protein shortage on the hemograms of camels and sheep. *Journal of Arid Environments* 37, 71–81.

Audigé, L., Wilson, P.R. and Morris, R.S. (2001) Disease and mortality on red deer farms in New Zealand. *Veterinary Record* 148, 334–340.

Auer, N., Covington, A.D., Evans, C.S., Natt, M. and Tozan, M. (1999) Enzymatic removal of dung from hides. *Journal of the Society of Leather Technologists and Chemists* 83, 215–219.

Avery, S.M., Small, A., Reid, C.A. and Buncic, S. (2002) Pulsed-field electrophoresis characterization of shiga toxin-producing *Escherichia coli* O157 from hide of cattle at slaughter. *Journal of Food Protection* 65, 1172–1176.

Ayantunde, A.A., Fernández-Rivera, S., Hiernaux, P.H.Y., van Keulen, H., Udo, H.M.J. and Chanono, M. (2001) Effect of timing and duration of grazing cattle in the West African Sahel on diet selection, faecal output, eating time, forage intake and live weight changes. *Animal Science* 72, 117–128.

Baba, S.S., Ambali, A.G., Zaria, L.T. and Kalra, S. (1994) Abattoir records of slaughtered camels (*Camelus dromedaries*) in Nigeria. *Bulletin of Animal Health and Production in Africa* 42, 253–257.

Babji, A.S., Froning, G.W. and Ngoka, D.A. (1982) The effect of preslaughter environmental temperature in the presence of electrolyte treatment on turkey meat quality. *Poultry Science* 61, 2385–2389.

Babot, D., Chavez, E.R. and Noguera, J.L. (2003) The effect of age at the first mating and herd size on lifetime productivity of sows. *Animal Research* 52, 49–64.

Bager, F., Devine, C.E. and Gilbert, K.V. (1988) Jugular blood flow in calves after head-only electrical stunning and throat-cutting. *Meat Science* 22, 237–243.

Bager, F., Braggins, T.J., Devine, C.E., Graafhuis, A.E., Mellor, D.J., Tavener, A. and Upsdell, M.P. (1992) Onset of insensibility at slaughter in calves: effects of electroplectic seizure and exsanguinations on spontaneous electrocortical activity and indices of cerebral metabolism. *Research in Veterinary Science* 52, 162–173.

Bager, F., Aarestrup, F.M. and Wegener, H.C. (2000) Dealing with antimicrobial resistance – the Danish experience. *Canadian Journal of Animal Science* 80, 223–228.

Bañon, S., Costa, E., Gil, M.D. and Garrido, M.D. (2003a) A comparative study of boar taint in cooked and dry-cured meat. *Meat Science* 63, 381–388.

Bañon, S., Gil, M.D. and Garrido, M.D. (2003b) The effect of castration on the eating quality of dry-cured hams. *Meat Science* 65, 1031–1037.

Bao, Y.M., Ru, Y.J., Glatz, P.C. and Miao, Z.H. (2004) The influence of weaning time on deer performance. *Asian–Australasian Journal of Animal Science* 17, 569–581.

Barber, C.L., Prescott, N.B., Wathes, C.M., Le Sueur, C. and Perry, G.C. (2004) Preference of growing ducklings and turkey poults for illuminance. *Animal Welfare* 13, 211–224.

Bareille, N., Beaudeau, F., Billon, S., Robert, A. and Faverdin, P. (2003) Effects of health disorders on feed intake and milk production in dairy cows. *Livestock Production Science* 83, 53–62.

Barham, A.R., Barham, B.L., Johnson, A.K., Allen, D.M., Blanton, J.R. and Miller, M.F. (2002) Effects of the transportation of beef cattle from the feedyard to the packing plant on prevalence levels of *Escherichia coli* O157 and *Salmonella* species. *Journal of Food Protection* 65, 280–283.

Barkocy-Gallagher, G.A., Arthur, T.M., Siragusa, G.R., Keen, J.E., Elder, R.O., Laegreid, W.W. and Koohmaraie, M. (2001) Genotypic analyses of *Escherichia coli* O157:H7 and O157 nonmotile isolates recovered from beef cattle and carcasses at processing plants in the Midwestern states of the United States. *Applied Environmental Microbiology* 67, 3810–3818.

Barnett, J.L., Cronin, G.M., McCallum, T.H. and Newman, E.A. (1993) Effects of pen size/shape and design on aggression when grouping unfamiliar adult pigs. *Applied Animal Behaviour Science* 36, 11–21.

Barton Gade, P. and Christensen, L. (1998) Effect of different stocking densities during transport on welfare and meat quality in Danish slaughter pigs. *Meat Science* 48, 237–247.

Batal, A.B. and Parson, C.M. (2002) Effect of fasting versus feeding Oasis after hatching on nutrient utilization in chicks. *Poultry Science* 81, 853–859.

Baumann, P., Oester, H. and Stauffacher, M. (2005) Effects of temporary nest box removal on maternal behaviour and pup survival in caged rabbits (*Oryctolagus cuniculus*). *Applied Animal Behaviour Science* 91, 167–178.

Beardmore, J.A., Mair, G.C. and Lewis, R.I. (2001) Monosex male production in finfish as exemplified by tilapia: applications, problems, and prospects. *Aquaculture* 197, 283–301.

Beattie, V.E., O'Connell, N.E., Kilpatrick, D.J. and Moss, B.W. (2000) Influence of environmental enrichment on welfare-related behavioural and physiological parameters in growing pigs. *Animal Science* 70, 443–450.

Beattie, V.R., Burrows, M.S., Moss, B.W. and Weatherup, R.N. (2002) The effect of food deprivation prior to slaughter on performance, behaviour and meat quality. *Meat Science* 62, 413–418.

Behnke, R. (1995) Natural resource management in pastoral Africa. In: Stiles, D. (ed.) *Social Aspects of Sustainable Dryland Management*. John Wiley & Sons, Chichester, UK, pp. 145–152.

Behnke, R. and Kerven, C. (1995) Redesigning for risks: tracking and buffering environmental variability in Africa's rangelands. In: Wilson, R.T., Ehui, S. and Mack, S. (eds) *Livestock Development Strategies for Low Income Countries*. FAO, Rome, Italy, pp. 47–69.

Benibo, B.S. and Farr, A.J. (1985) The effects of feed and water withdrawal and holding shed treatments on broiler yield parameters. *Poultry Science* 64, 920–924.

Benjamin, M.E., Gonyou, H.W., Ivers, D.L., Richardson, L.F., Jones, D.J., Wagner, J.R., Seneriz, R. and Anderson, D.B. (2001) Effect of handling method on the incidence of stress response in market swine in a model system. *Journal of Animal Science* 79 (Suppl. 1), 279 (Abstract).

Bennett, C.D., Classen, H.L. and Riddell, C. (2002) Feeding broiler chickens wheat and barley diets containing whole, ground and pelleted grain. *Poultry Science* 81, 995–1003.

Berends, B.R., Urlings, H.A.P., Snijders, J.M.A. and van Knapen, F. (1996) Identification and quantification of risk factors in animal management and transport regarding *Salmonella* spp in pigs. *International Journal of Food Microbiology* 30, 37–53.

Berg, T., Erikson, U. and Nordtvedt, T.S. (1997) Rigor mortis assessment of Atlantic salmon (*Salmo salar*) and effects of stress. *Journal of Food Science* 62, 439–446.

Berger, F., Dagorn, J., Ledenmat, M., Quillien, J.P., Vaudelet, J.C. and Signoret, J.P. (1998) The seasonal challenge. *Pig Progress* 14, 28–29.

Bernier, N.J., Brauner, C.J., Heath, J.W. and Randall, D.J. (2004) Oxygen and carbon dioxide transport during sustained exercise in diploid and triploid Chinook salmon (*Oncorhynchus tshawytscha*). *Canadian Journal of Fisheries and Aquatic Sciences* 61, 1797–1805.

Berri, C., Debut, M., Santé-Lhoutellier, V., Arnould, C., Boutten, B., Sellier, N., Baéza, E., Jehl, N., Jégo, Y., Duclos, M.J. and Le Bihan-Duval, E. (2005) Variations in chicken breast meat quality: implications of struggle and muscle glycogen content at death. *British Poultry Science* 46, 572–579.

Beyssen, C., Babile, R. and Fernandez, X. (2004) Electrocorticogram spectral analysis and somatosensory evoked potentials as tools to assess electrical stunning efficiency in ducks. *British Poultry Science* 45, 409–415.

Bharara, I.P., Mathur, Y.N. and Khan, W. (1999) Nomadism and pastoralism: socio-economic profile, grazing systems and feeding management in Rajasthan Desert. In: Hooja, R. and Joshi, R. (eds) *Desert, Drought and Development. Studies in Resource Management and Sustainability*. Rawat Publications, Jaipur, India, pp. 349–367.

Binns, S.H., Cox, I.J., Rizvi, S. and Green, L.E. (2002) Risk factors for lamb mortality on UK sheep farms. *Preventive Veterinary Medicine* 52, 287–303.

Biss, M.E. and Hathaway, S.C. (1992) Microbiological and visible contamination of lamb carcasses according to pre-slaughter presentation status: implications for HACCP. *Journal of Food Protection* 58, 776–783.

Bizeray, D., Estevez, I., Leterrier, C. and Faure, J.M. (2002a) Influence of increased environmental complexity on leg condition, performance, and level of fearfulness in broilers. *Poultry Science* 81, 767–773.

Bizeray, D., Leterrier, C., Constantin, P., Picard, M. and Faure, J.M. (2002b) Sequential feeding can increase activity and improve gait score in meat-type chickens. *Poultry Science* 81, 1798–1806.

Bjørnevik, M., Espe, M., Beattie, C., Nortvedt, R. and Kiessling, A. (2004) Temporal variation in muscle fibre area, gaping, texture, colour and collagen in triploid and diploid Atlantic salmon (*Salmo salar* L.). *Journal of the Science of Food and Agriculture* 84, 530–540.

Blackmore, D.K. (1984a) Differences in behaviour between sheep and cattle during slaughter. *Research in Veterinary Science* 37, 223–226.

Blackmore, D.K. (1984b) Insensibility during slaughter of pigs in comparison to other domestic stock. *New Zealand Veterinary Journal* 29, 219–220.

Blaxter, K.L. and Hamilton, W.J. (1980) Reproduction in farmed red deer. 2. Calf growth and mortality. *Journal of Agricultural Science* 95, 275–284.

Blaxter, K.L., Graham, N.M., Wainman, F.W. and Armstrong, D.G. (1959) Environmental temperature, energy metabolism and heat regulation in sheep. *Journal of Agricultural Science* 52, 25–40.

Blaxter, K.L., Fowler, V.R. and Gill, J.C. (1982) A study of the growth of sheep to maturity. *Journal of Agricultural Science* 98, 405–429.

Blench, R. and Marriage, Z. (1999) *Drought and Livestock in Semi-arid Africa and Southwest Asia*. ODI Working Paper 117, Overseas Development Institute, London, 138 pp.

Block, G. (2003) The moral reasoning of believers in animal rights. *Society and Animals* 11(2), 1–10.

Boissy, A., Terlow, C. and LeNeindre, P. (1999) Presence of phenomones from stressed conspecifics increases reactivity to aversive events in cattle, evidence for the existence of alarm substances in urine. *Physiology and Behavior* 4, 489–495.

Boleman, S.L., Boleman, S.J., Savell, J.W., Hale, D.S., Griffin, D.B., Smith, G.C., Dolezal, H.G., Morgan, J.B., Northcutt, S.L. and Gardner, B.A. (1995) Results of slaughter-floor audits from the NBQA. *Beef Cattle Research in Texas* 1995, 163–165.

Bond, J.J., Can, L.A. and Warner, R.D. (2004) The effect of exercise stress, adrenaline injection and electrical stimulation on changes in quality attributes and proteins in semimembranosus muscle of lamb. *Meat Science* 68, 469–477.

Bonde, M., Rousing, T., Badsberg, J.H. and Sørensen, J.T. (2004) Associations between lying-down behaviour problems and body condition, limb disorders and skin lesions of lactating sows housed in farrowing crates in commercial sow herds. *Livestock Production Science* 87, 179–187.

Boulianne, M., Hunter, D.B., Julian, R.J., O'Grady, M.R. and Physick-Sheard, P.W. (1992) Cardiac muscle mass distribution in the domestic turkey and relationship to electrocardiogram. *Avian Disease* 36, 582–589.

Braam, C.R., Ketelaars, J.J.M.H. and Smits, M.C.J. (1997) Effects of floor design and floor cleaning on ammonia emission from cubicle houses for dairy cattle. *Netherlands Journal of Agricultural Science* 45, 49–64.

Bradshaw, R.H., Parrott, R.F., Forsling, M.L., Goode, J.A., Lloyd, D.M., Rodway, R.G. and Broom, D.M. (1996) Stress and travel sickness in pigs: effects of road transport on plasma concentrations of cortisol, beta-endorphin and lysine vasopressin. *Animal Science* 63, 507–516.

Breukink, H.J., Wensing, T., van Weeren-Keverling Buisman, van Bruinessen-Kapsenberg, E.G. and de Visser, N.A.P.C. (1988) Consequence of failure of the reticular groove reflex in veal calves fed milk replacer. *Veterinary Quarterly* 10, 126–135.

Breuer, K., Hemsworth, P.H. and Coleman, G.J. (2003) The effect of positive or negative handling on the behavioural and physiological responses of nonlactating heifers. *Applied Animal Behaviour Science* 84, 3–22.

Brouns, F. and Edwards, S.A. (1994) Social rank and feeding behaviour of group-housed sows fed competitively or *ad libitum*. *Applied Animal Behaviour Science* 39, 225–235.

Bryant, M.J. and Ewbank, R. (1974) Effects of stocking rate upon the performance, general activity and ingestive behaviour of groups of growing pigs. *British Veterinary Journal* 130, 139–149.

Buchanan, S., Robertson, G.W. and Hocking, P.M. (1999) The relationship between vaginal collagen, plasma oestradiol and uterine prolapse in turkeys. *Research in Veterinary Science* 67, 153–157.

Buchanan, S., Robertson, G.W. and Hocking, P.M. (2000) Development of the reproductive system in turkeys with a high or low susceptibility to prolapse of the oviduct. *Poultry Science* 79, 1491–1498.

Buchwalder, T. and Huber-Eicher, B. (2003) A brief report on aggressive interactions within and between groups of domestic turkeys (*Melagris gallopavo*). *Applied Animal Behaviour Science* 84, 75–80.

Budgell, K.L. and Silversides, F.G. (2004) Incidence of bone breakage in three strains of end-of-lay hens. *Poultry Science* 83 (Suppl. 1), 83.

Buhr, R.J., Cason, J.A., Dickens, J.A., Hinton, A. and Ingram, K.D. (2000) Influence of flooring type during transport and holding on bacteria recovery from broiler carcass rinses before and after defeathering. *Poultry Science* 79, 436–441.

Buhr, R.J., Berrang, M.E., Cason, J.A. and Bourassa, D.V. (2005) Recovery of bacteria from broiler carcass respiratory tracts before and after immersion scalding. *Poultry Science* 84, 1769–1773.

Bureau, F., Uystepruyst, C., Coghe, J., van de Weerdt, M.-L. and Lekeux, P. (1999) Spirometric variables recorded after lobeline administration in healthy Friesian and Belgian White and Blue calves: normal values and effects of somatic growth. *Veterinary Journal* 157, 302–308.

Burley, J.R.W., Creeper, D.A. and Moulds, G.A. (1983) Damage to livestock caused by domestic dogs in Adelaide's urban fringe. *Department of Agriculture South Australia Technical Report* 24, 1–19.

Byrd, J.A., Corrier, D.E., Hume, M.E., Bailey, R.H., Stanker, L.H. and Hargis, B.M. (1998) Effect of feed withdrawal on *Campylobacter* in crops of market-age broiler chickens. *Avian Diseases* 42, 802–806.

Byrne, D.V., Bredie, W.L.P., Bak, L.S., Bertelsen, G., Martens, H. and Martens, M. (2001) Sensory and chemical analysis of cooked porcine meat patties in relation to warmed-over flavour and pre-slaughter stress. *Meat Science* 59, 229–249.

Byrne, D.V., O'Sullivan, M.G., Bredie, W.L.P., Andersen, H.J. and Martens, M. (2003) Descriptive sensory profiling and physical/chemical analyses of warmed-over flavour in pork patties from carriers and non-carriers of the *RN⁻* allele. *Meat Science* 63, 211–224.

Camacho, M.A., Suárez, M.E., Herrera, J.G., Cuca, J.M. and García-Bojalil, C.M. (2004) Effect of age of feed restriction and microelement supplementation to control ascites on production and carcass characteristics of broilers. *Poultry Science* 83, 526–532.

Campo, J.L. and Davila, S.G. (2002) Estimation of heritability for heterophil lymphocyte ratio in chickens by restricted maximum likelihood. Effects of age, sex, and crossing. *Poultry Science* 81, 1448–1453.

Cardinal, M., Cornet, J. and Vallet, J.L. (2002) Sensory characteristics of caviar from wild and farmed sturgeon. *International Review of Hydrobiology* 87, 651–659.

Carreras, I., Castellari, M., Valero, A., Regueiro, J.A.G. and Sárraga, C. (2005) Influence of enrofloxacin administration on the proteolytic and antioxidant enzyme activities of raw and cooked turkey products. *Journal of the Science of Food and Agriculture* 85, 2407–2412.

Carson, A.F., Dawson, L.E.R., Irwin, D. and Kilpatrick, D.J. (2004) The effect of management system at lambing and flock genetics on lamb output and labour requirements on lowland sheep farms. *Animal Science* 78, 439–450.

Carver, D.K., Fetrow, J., Gerig, T., Krueger, K.K. and Barnes, H.J. (2002) Hatchery and transportation factors associated with early poult mortality in commercial turkey flocks. *Poultry Science* 81, 1818–1825.

Castellini, C., Mugnai, C. and Dal Bosco, A. (2002) Effect of organic production system on broiler carcass and meat quality. *Meat Science* 60, 219–225.

Cava, R., Tárrega, R., Ramirez, M.R., Mingoarranz, F.J. and Carrasco, A. (2005) Effect of irradiation on colour and lipid oxidation of dry-cured hams from free-range reared and intensively reared pigs. *Innovative Food Science and Emerging Technologies* 6, 135–141.

Cave, J.G., Callinan, A.P.L. and Woonton, W.K. (2005) Mortalities in bobby calves associated with long distance transport. *Australian Veterinary Journal* 83, 82–84.

Chadfield, M.S., Christensen, J.P., Juhl-Hansen, J., Christensen, H. and Bisgaard, M. (2005) Characterization of *Enterococcus hirae* outbreaks in broiler flocks demonstrating increased mortality because of septicaemia and endocarditis and/or altered production parameters. *Avian Diseases* 49, 16–23.

Chambers, C., Powell, L., Wilson, E. and Green, L.E. (1995) A postal survey of tail biting in pigs in south west England. *Veterinary Record* 136, 147–148.

Chandra, B.S. and Das, N. (2001) The handling and short-haul road transportation of spent buffaloes in relation to bruising and animal welfare. *Tropical Animal Health and Production* 33, 155–163.

Channon, H.A., Pengelly, A.M. and Warner, R.D. (1997a) Effect of stunning method on pigmeat quality. In: *Manipulating Pig Production VI. Proceedings of the 6th Biennial Conference of the ASPA.* ASPA, Victoria, Australia, p. 124.

Channon, H.A., Trout, G.R., Pengelly, A.M. and Warner, R.D. (1997b) Stunning of pigs: effect of method, current level and duration on carcass and meat quality. In: *Manipulating Pig Production VI. Proceedings of the 6th Biennial Conference of the ASPA.* ASPA, Victoria, Australia, p. 125.

Channon, H.A., Payne, A.M. and Warner, R.D. (2000) Halothane genotype, pre-slaughter handling and stunning method all influence pork quality. *Meat Science* 56, 291–299.

Channon, H.A., Payne, A.M. and Warner, R.D. (2002) Comparison of CO₂ stunning with manual electrical stunning (50 Hz) of pigs on carcass and meat quality. *Meat Science* 60, 63–68.

Channon, H.A., Kerr, M.G. and Walker, P.J. (2004) Effect of Duroc content, sex and aging period on meat and eating quality attributes of pork loin. *Meat Science* 66, 881–888.

Chapman, R.E., Bennett, J.W. and Carter, N.B. (1984) Erythemal response of biologically denuded sheep to sunlight and the effects on skin structure and wool growth. *Australian Journal of Biological Sciences* 37, 217–235.

Chimonyo, M., Kusina, N., Hamudikuwanda, H., Nyoni, O. and Ncube, I. (2000) Effects of dietary supplementation and work stress on ovarian activity in non-lactating Mashona cows in a small-holder farming area of Zimbabwe. *Animal Science* 70, 317–323.

Chou, C.C., Jiang, D.D. and Hung, Y.P. (2004) Risk factors for cumulative mortality in broiler chicken flocks in the first week of life in Taiwan. *British Poultry Science* 45, 573–577.

Claeys, E., de Smet, S., Demeyer, D., Geers, R. and Buys, N. (2001) Effect of rate of pH decline on muscle enzyme activities in two pig lines. *Meat Science* 57, 257–263.

Clark, S., Hansen, G., McLean, P., Bond, P., Wakeman, W., Meadows, R. and Buda, S. (2002) Pododermatitis in turkeys. *Avian Diseases* 46, 1038–1044.

Coghe, J., Uystepruyst, C., Bureau, F., Detilleux, J., Art, T. and Lekeux, P. (2000) Validation and prognostic value of plasma lactate measurement in bovine respiratory disease. *Veterinary Journal* 160, 139–146.

Collins, M.N., Friend, T.H., Jousan, F.D. and Chen, S.C. (2000) Effects of density on displacement, falls, injuries, and orientation during horse transportation. *Applied Animal Behaviour Science* 67, 169–179.

Conington, J., Bishop, S.C., Grundy, B., Waterhouse, A. and Simm, G. (2001) Multi-trait selection indexes for sustainable UK hill sheep production. *Animal Science* 73, 413–423.

Contreras, C.C. and Beraquet, N.J. (2001) Electrical stunning, hot boning, and quality of chicken breast meat. *Poultry Science* 80, 501–507.

Corke, M.I. and Broom, D.M. (1999) The behaviour of sheep with sheep scab, *Psoroptes ovis* infestation. *Veterinary Parasitology* 83, 291–300.

Corr, S.A., Gentle, M.J., McCorquodale, C.C. and Bennett, D.D. (2003) The effect of morphology on walking ability in the modern broiler: a gait analysis study. *Animal Welfare* 12, 159–171.

Corrier, D.E., Purdy, C.W. and DeLoach, J.R. (1990) Effects of marketing stress on fecal excretion of *Salmonella* species in feeder calves. *American Journal of Veterinary Research* 51, 866–869.

Corrier, D.E., Byrd, J.A., Hargis, B.M., Hume, M.E., Bailey, R.H. and Stanker, L.H. (1999) Presence of *Salmonella* in the crop and ceca of broiler chickens before and after preslaughter feed withdrawal. *Poultry Science* 78, 45–49.

Costa, L.N., Fiego, D.P.L., Dall'Olio, S., Davoli, R. and Russo, V. (1999) Influence of loading method and stocking density during transport on meat and dry-cured ham quality in pigs with different halothane genotypes. *Meat Science* 51, 391–399.

Costa, L.N., Fiego, D.P.L., Dall'Olio, S., Davoli, R. and Russo, V. (2002) Combined effects of pre-slaughter treatments and lairage time on carcass and meat quality in pigs of different halothane genotype. *Meat Science* 61, 41–47.

Cox, L.N. and Cooper, J.J. (2001) Observations on the pre- and post-weaning behaviour of piglets reared in commercial indoor and outdoor environments. *Animal Science* 72, 75–86.

Crespo, R., Stover, S.M., Droual, R., Chin, R.P. and Shivaprasad, H.L. (1999) Femoral fractures in a young male turkey breeder flock. *Avian Diseases* 43, 150–154.

Crespo, R., Ghazikhanian, G.Y. and Hall, C.I. (2002a) Avulsion of the common retinaculum in meat turkeys. *Avian Diseases* 46, 245–248.

Crespo, R., Stover, S.M., Shivaprasaud, H.L. and Chin, R.P. (2002b) Microstructure and mineral content of femora in male turkeys with and without fractures. *Poultry Science* 81, 1184–1190.

Cronin, G.M. and Smith, J.A. (1992) Effects of accommodation type and straw bedding around parturition and during lactation on the behaviour of primiparous sows and survival and growth of piglets to weaning. *Applied Animal Behaviour Science* 33, 191–208.

Cruz, A.M. and Naylor, J.M. (1993) A retrospective study of frostbite injuries in calves (1982–1991). *Canadian Veterinary Journal* 34, 306–308.

Cummings, T.S. (2006) Stakeholder position paper: poultry. *Preventive Veterinary Medicine* 73, 209–212.

Dabiri, N., Holmes, C.W., McCutcheon, S.N., Parker, W.J. and Morris, S.T. (1995) Resistance to cold stress in sheep shorn by cover comb or standard comb. *Animal Science* 60, 451–456.

Dahl, K., Sandøe, P., Johnsen, P.F., Lassen, J. and Kornerup Hansen, A. (2003) Outline of a risk assessment: the welfare of future xeno-donor pigs. *Animal Welfare* 12, 219–237.

Dalton, D.C., Knight, T.W. and Johnson, D.L. (1980) Lamb survival in sheep breeds on New Zealand hill country. *New Zealand Journal of Agricultural Research* 23, 167–173.

Daly, C.C. (1987) Concussion stunning in red-meat species. In: Carter, H.E. and Carter, V.R. (eds) *Pre-slaughter Stunning of Food Animals. A Seminar Organised by the European Conference Group on the Protection of Farm Animals.* RSPCA, Horsham, UK, pp. 94–100.

Daly, C.C. and Whittington, P.E. (1986) Concussive stunning methods of pre-slaughter stunning in sheep: effects of captive bolt stunning in the poll position on brain function. *Research in Veterinary Science* 41, 353–355.

Daly, C.C. and Whittington, P.E. (1989) *A Survey of Commercial Practices Used in the Stunning of Cattle. Report to the RSPCA and MAFF, UK*, Bristol University, Langford, UK, 26 pp.

Danbury, T.C., Weeks, C.A., Chambers, J.P., Waterman-Pearson, A.E. and Kestin, S.C. (2000) Self-selection of the analgesic drug carprofen by lame broiler chickens. *Veterinary Record* 146, 307–311.

Davis, N.J., Prescott, N.B., Savory, C.J. and Wathes, C.M. (1999) Preference of growing fowls for different light intensities in relation to age, strain and behaviour. *Animal Welfare* 8, 193–203.

Dean, W.R.J. and Macdonald, I.A.W. (1994) Historical changes in stocking rates of domestic livestock as a measure of semi-arid and arid rangeland degradation in the Cape Province, South Africa. *Journal of Arid Environments* 26, 281–298.

Debut, M., Berri, C., Baéza, E., Sellier, N., Arnould, C., Guémené, D., Jehl, N., Boutten, B., Jego, Y., Beaumont, C. and Le Bihan-Duval, E. (2003) Variation of chicken technological meat quality in relation to genotype and preslaughter stress conditions. *Poultry Science* 82, 1829–1838.

de Decker, J.M., Ellis, M., Wolter, B.F., Corrigan, B.P., Curtis, S.E. and Hollis, G.R. (2005) Effect of stocking rate on pig performance in a wean-to-finish production system. *Canadian Journal of Animal Science* 85, 1–5.

Deeb, N., Shlosberg, A. and Cahaner, A. (2002) Genotype-by-environment interaction with broiler genotypes differing in growth rate. *Poultry Science* 81, 1454–1462.

de Graaf, G. and Janssen, H. (1996) *Artificial Reproduction and Pond Rearing of the African Catfish* Clarias gariepinus *in Sub-Saharan Africa*. Fisheries Technical Paper 362, FAO, Rome, 73 pp.

de Haan, C. (1995) Development support and livestock services. In: Wilson, R.T., Ehui, S. and Mack, S. (eds) *Livestock Development Strategies for Low Income Countries*. FAO, Rome, Italy, pp. 23–27.

de Jong, I.C., Prelle, I.T., van de Burgwal, J.A., Lambooij, E., Korte, S.M., Blokhuis, H.J. and Koolhaas, J.M. (2000) Effects of rearing conditions on behavioural and physiological responses of pigs to preslaughter handling and mixing at transport. *Canadian Journal of Animal Science* 80, 451–458.

de Jong, I.C., Enting, H., van Voorst, A. and Blokhuis, H.J. (2005) Do low-density diets improve broiler breeder welfare during rearing and laying? *Poultry Science* 84, 194–203.

de Kock, H.L., van Heerden, S.M., Heinze, P.H., Dijksterhuis, G.B. and Minaar, A. (2001) Reaction to boar odour by different South African consumer groups. *Meat Science* 59, 353–362.

de Leeuw, P.N. and Rey, B. (1995) Analysis of current trends in the distribution patterns of ruminant livestock in tropical Africa. *World Animal Review* 83, 47–59.

den Hartog, L.A., Backus, G.B.C. and Vermeer, H.M. (1993) Evaluation of housing systems for sows. *Journal of Animal Science* 67, 543–547.

de Smet, S., Bloemen, H., van de Voorde, G., Spincemaille, G. and Berckmans, D. (1998) Meat and carcass quality in two pig lines of different stress-susceptibility genotype and their crosses. *Animal Science* 66, 441–447.

Desta, Z.H. and Oba, G. (2004) Feed scarcity and livestock mortality in enset farming systems in the Bale highlands of southern Ethiopia. *Outlook on Agriculture* 33, 277–280.

Destefanis, G., Brugiapaglia, A., Barge, M.T. and Lazzaroni, C. (2003) Effect of castration on meat quality in Piedmontese cattle. *Meat Science* 64, 215–218.

Devine, C.E., Gilbert, K.V. and Ellery, S. (1983) Electrical stunning of lambs: the effect of stunning parameters and drugs affecting blood flow and behaviour on petechial haemorrhage incidence. *Meat Science* 9, 247–256.

Devlin, R.H., Yesaki, T.Y., Donaldson, E.M., Du, S.J. and Hew, C.-L. (1995) Production of germline transgenic Pacific salmonids with dramatically increased growth performance. *Canadian Journal of Fisheries and Aquatic Sciences* 52, 1376–1384.

Dionigi, C.P., Johnsen, P.B. and Vinyard, B.T. (2000) The recovery of flavour quality by channel catfish. *North American Journal of Aquaculture* 62, 189–194.

Dobson, K.N. and Glisson, J.R. (1992) Economic impact of a documented case of reovirus infection in broiler breeders. *Avian Diseases* 36, 788–791.

Dock, W. (1940) Vasoconstriction in renal hypertension abolished by pithing. *American Journal of Physiology* 130, 1–8.

Dodt, R.M., Anderson, B. and Horder, J.C. (1979) Bruising in cattle fasted prior to transport for slaughter. *Australian Veterinary Journal* 55, 528–530.

Donaldson, T.J. (1986) Pastoralism and drought. A case study of the Borana of southern Ethiopia. MPhil thesis, University of Reading, UK.

Donaldson, T., Newberry, R.C., Špinka, M. and Cloutier, S. (2002) Effects of early play experience on play behaviour of piglets after weaning. *Applied Animal Behaviour Science* 79, 221–231.

Donaldson, W.E., Clark, J. and Christensen, V.L. (1994) Protein, lipid and glycogen stores in newly-hatched turkey (*Meleagris gallopavo*) poults as affected by post-hatch stressors and holding time. *Comparative Biochemistry and Physiology* 107A, 559–562.

Doney, J.M., Gunn, R.G. and Griffiths, J.G. (1973) The effect of premating stress on the onset of oestrus and on ovulation rate in Scottish Blackface ewes. *Journal of Reproduction and Fertility* 35, 381–384.

Doney, J.M., Smith, W.F. and Gunn, R.G. (1976) Effects of post-mating environmental stress or administration of ACTH on early embryonic loss in sheep. *Journal of Agricultural Science* 87, 133–136.

Doreau, M., Michalet-Doreau, B. and Béchet, G. (2004) Effect of underfeeding on digestion in cows. Interaction with rumen degradable N supply. *Livestock Production Science* 88, 33–41.

Dougherty, C.T., Knapp, F.W., Burrus, P.B., Willis, D.C., Burg, J.G., Cornelius, P.L. and Bradley, N.W. (1993) Stable flies (*Stomoxys calcitrans* L.) and the behaviour of grazing beef cattle. *Applied Animal Behaviour Science* 35, 215–233.

Dozier, W.A., Thaxton, J.P., Branton, S.L., Morgan, G.W., Miles, D.M., Roush, W.B., Lott, B.D. and Vizzier-Thraxton, Y. (2005) Stocking density effects on growth performance and processing yields of heavy broilers. *Poultry Science* 84, 1332–1338.

D'Souza, D.N., Dunshea, F.R., Warner, R.D. and Leury, B.J. (1998) The effect of handling pre-slaughter and carcass processing rate post-slaughter on pork quality. *Meat Science* 50, 429–437.

D'Souza, D.N., Dunshea, F.R., Levry, B.J. and Warner, R. (1999) Effect of mixing boars during lairage and preslaughter on meat quality. *Australian Journal of Agricultural Research* 50, 109–113.

Dunlap, R.E. (1997) International opinion at the century's end: public attitudes toward environmental issues. In: Caldwell, L.K. and Bartlett, R.V. (eds) *Environmental Policy. Transnational Issues and National Trends*. Quorum Books, Connecticut, USA, pp. 201–224.

Dunlap, R.E. and Mertig, A.G. (1995) Global concern for the environment: is affluence a prerequisite? *Journal of Social Issues* 51 (4), 121–137.

Dunn, C.S. (1990) Stress reactions of cattle undergoing ritual slaughter using two methods of restraint. *Veterinary Record* 126, 522–525.

Dunne, P.G., O'Mara, F.P., Monahan, F.J., French, P. and Moloney, A.P. (2005a) Colour of muscle from 18-month-old steers given long-term daily exercise. *Meat Science* 71, 219–229.

Dunne, P.G., Monohan, F.J., O'Mara, F.P. and Moloney, A.P. (2005b) Colour stability, under simulated retail display conditions, of m. longissimus dorsi and m. semimembranosus from steers given long-term daily exercise and supplemented with vitamin E. *Meat Science* 71, 480–489.

Durrell, J.L., Beattie, V.E., Sneddon, I.A. and Kilpatrick, D. (2003) Pre-mixing as a technique for facilitating subgroup formation and reducing sow aggression in large dynamic groups. *Applied Animal Behaviour Science* 84, 89–99.

Dwyer, C.M., Lawrence, A.B. and Bishop, S.C. (2001) The effects of selection for lean tissue content on maternal and neonatal lamb behaviours in Scottish Blackface sheep. *Animal Science* 72, 555–571.

Edwards, A.J. and Landert, S.B. (1984) Economic evaluation of the use of abortifacients. *Bovine Practitioner* 19, 148–150.

Edwards, S.A., Armsby, A.W. and Large, J.W. (1988) Effects of feed station design on the behaviour of group-housed sows using an electronic individual feeding system. *Livestock Production Science* 19, 511–522.

Einen, O. and Thomassen, M.S. (1998) Starvation prior to slaughter in Atlantic salmon (*Salmo salar*). Ii. White muscle composition and evaluation of freshness, texture and colour characteristics in raw and cooked fillets. *Aquaculture* 169, 37–53.

Ekkel, E.D., Spoolder, H.A.M., Hulsegge, I. and Hopster, H. (2003) Lying characteristics as determinants for space requirements in pigs. *Applied Animal Behaviour Science* 80, 19–30.

Emmel, M.W. (1930) Insolation in ducks. *Journal of the American Veterinary Medical Association* 76, 842–843.

Erichsen Jones, J.R. (1952) The reactions of fish to water of low oxygen concentration. *Journal of Experimental Biology* 29, 403–415.

Erikson, U. (2001) Potential effects of preslaughter fasting, handling and transport. In: Kestin, S.C. and Warriss, P.D. (eds) *Farmed Fish Quality*. Fishing News Books, Oxford, UK, pp. 202–219.

Ersdal, C., Midtlyng, P.J. and Jarp, J. (2001) An epidemiological study of cataracts in sweater farmed Atlantic salmon *Salmo salar*. *Diseases of Aquatic Organisms* 45, 229–236.

Essén-Gustavsson, B., Jensen-Waern, M., Jonasson, R. and Andersson, L. (2005) Effect of exercise on proglycogen and macroglycogen content in skeletal muscles of pigs with the Rendement Napole mutation. *American Journal of Veterinary Research* 66, 1197–1202.

Estévez, M., Morcuende, D. and Cava, R.C. (2003) Oxidative and colour changes in meat from three lines of free-range reared Iberian pigs slaughtered at 90 kg live weight and from industrial pigs during refrigerated storage. *Meat Science* 65, 1139–1146.

Estévez, M., Morcuende, D., Ramírez, R., Ventanas, J. and Cava, R. (2004) Extensively reared Iberian pigs versus intensively reared white pigs for the manufacture of liver pâté. *Meat Science* 67, 453–461.

Ezanno, P., Ickowicz, A. and Bocquier, F. (2003) Factors affecting the body condition score of N'Dama cows under extensive range management in Southern Senegal. *Animal Research* 52, 37–48.

Ezanno, P., Ickowicz, A. and Lancelot, R. (2005) Relationships between N'Dama cow body condition score and production performance under an extensive range management system in Southern Senegal: calf weight gain, milk production, probability of pregnancy, and juvenile mortality. *Livestock Production Science* 92, 291–306.

Fa, J.E., Justa, J., Perez del Val, J. and Castroviejo, J. (1995) Impact of market hunting on mammal species in Equatorial Guinea. *Conservation Biology* 9, 1107–1115.

Færevik, G., Andersen, I.L. and Bøe, K.E. (2005) Preference of sheep for different types of pen flooring. *Applied Animal Behaviour Science* 90, 265–276.

Fahmy, M.H. and Dufour, J.J. (1976) Effects of post-weaning stress and feeding management on return to oestrus and reproductive traits during early pregnancy in swine. *Animal Production* 23, 103–110.

Fanguy, R.C., Misra, L.K., Vo, K.V., Blohowiak, C.C. and Krueger, W.F. (1980) Effect of delayed placement on mortality and growth performance on commercial broilers. *Poultry Science* 59, 1215–1220.

FAO (2001) *Pastoralism in the New Millennium.* Paper 150, FAO, Rome, Italy, 93 pp.

Faucitano, L. (1998) Preslaughter stressor effects on pork: a review. *Journal of Muscle Foods* 9, 293–303.

Faucitano, L. (2001) Causes of skin damage to pig carcasses. *Canadian Journal of Animal Science* 81, 39–45.

Faucitano, L., Marquardt, L., Oliveira, M.S., Coelho, H.S. and Terra, N.N. (1998) The effect of two handling and slaughter systems on skin damage, meat acidification and colour in pigs. *Meat Science* 50, 13–19.

Faure, J.-M., Guémené, D. and Guy, G. (2001) Is there avoidance of the force feeding procedure in ducks and geese? *Animal Research* 50, 157–164.

Feddes, J.J.R., Emmanuel, E.J. and Zuidhof, M.J. (2002) Broiler performance, body weight variance, feed and water intake, and carcass quality at different stocking densities. *Poultry Science* 81, 774–779.

Fegan, N., Vanderlinde, P., Higgs, G. and Desmarchelier, P. (2004) Quantification and prevalence of *Salmonella* in beef cattle presenting at slaughter. *Journal of Applied Microbiology* 97, 892–898.

Fehrenberg, C., von Mickwitz, G. and Reuter, G. (1991) Elektrische Betäubung von Schlachtschweinen. *Fleischwirtschaft* 71, 909–913.

Fell, L.R., Colditz, I.G., Walker, K.H. and Watson, D.L. (1999) Associations between temperament, performance and immune function in cattle entering a commercial feedlot. *Australian Journal of Experimental Agriculture* 39, 795–802.

Fenwick, D.C. (1969) Parturient paresis (milk fever) of cows. 1. The response to treatment and the effect of the duration of symptoms. *Australian Veterinary Journal* 45, 111–113.

Ferguson, N.S. and Gous, R.M. (1997) The influence of heat production on voluntary food intake in growing pigs given protein-deficient diets. *Animal Science* 64, 365–378.

Fernandes, T.H., Smith, W.C. and Armstrong, D.G. (1979) The administration of sugar solutions to pigs immediately prior to slaughter. *Animal Production* 29, 213–221.

Fernandez, X., Monin, G., Talmant, A., Mourot, J. and Lebret, B. (1999a) Influence of intramuscular fat content on the quality of pig meat – 1. Composition of the lipid fraction and sensory characteristics of m. longissimus lumborum. *Meat Science* 53, 59–65.

Fernandez, X., Monin, G., Talmant, A., Mourot, J. and Lebret, B. (1999b) Influence of intramuscular fat content on the quality of pig meat – 2. Consumer acceptability of m. longissimus lumborum. *Meat Science* 53, 67–72.

Fernandez, X., Mourot, J., Lebret, B., Gilbert, S. and Monin, G. (2000) Influence of intramuscular fat content on lipid composition, sensory qualities and consumer acceptability of cured cooked ham. *Journal of the Science of Food and Agriculture* 80, 705–710.

Fernandez, X., Santé, V., Baeza, E., Le Bihan-Duval, E., Berri, C., Rémignon, H., Babilé, R., Le Pottier, G., Millet, N., Berge, P. and Astruc, T. (2001) Post mortem muscle metabolism and meat quality in three genetic types of turkey. *British Poultry Science* 42, 462–469.

Fernandez, X., Gilbert, S. and Vendeuvre, J.-L. (2002a) Effects of halothane genotype and pre-slaughter treatment on pig meat quality. Part 2. Physico-chemical traits of cured-cooked ham and sensory traits of cured-cooked and dry-cured hams. *Meat Science* 62, 439–446.

Fernandez, X., Santé, V., Baeza, E., Le Bihan-Duval, E., Berri, C., Rémignon, H., Babilé, R., Le Pottier, G. and Astruc, T. (2002b) Effects of the rate of muscle post mortem pH fall on the technological quality of turkey meat. *British Poultry Science* 43, 245–252.

Field, R., McCormick, R., Balasubramanian, V., Sanson, D., Wise, J., Hixon, D., Riley, M. and Russell, W. (1996) Growth, carcass, and tenderness characteristics of virgin, spayed, and single-calf heifers. *Journal of Animal Science* 74, 2178–2186.

Fiems, L.O., de Campaneere, S., van Caelenbergh, W., de Boever, J.L. and Vanacker, J.M. (2003) Carcass and meat quality in double-muscled Belgian Blue bulls and cows. *Meat Science* 63, 345–352.

Fisher, A.D., Crowe, M.A., Prenderville, D.J. and Enright, W.J. (1997) Indoor space allowance effects on growth behaviour, adrenal and immune responses of finishing beef heifers. *Animal Science* 64, 53–62.

Fogarty, N.M., Safari, E., Taylor, P.J. and Murray, W. (2003) Genetic parameters for meat quality and carcass traits and their correlation with wool traits in Australian Merino sheep. *Australian Journal of Agricultural Research* 54, 715–722.

Forbes, A.B., Huckle, C.A. and Gibb, M.J. (2004) Impact of eprinomectin on grazing behaviour and performance in dairy cattle with sub-clinical gastrointestinal nematodes infections under continuous stocking management. *Veterinary Parasitology* 125, 353–364.

Franck, M., Svensson, M., von Seth, G., Jossell, Å., Figurer, P., Poirel, M.T. and Monin, G. (2003) Effect of stunning conditions on occurrence of PSE defects in hams of *rn+/RN*– pigs. *Meat Science* 64, 351–355.

Fraser, D. (1983) Testing the serving capacity of beef bulls. *New Zealand Veterinary Journal* 31, 20.

French, N.P. and Morgan, K.L. (1996) Role of neonatal and maternal risk factors in the faecal soiling of lambs. *Veterinary Record* 139, 460–465.

Frewer, L.J., Howard, C. and Shepherd, R. (1996) The influence of realistic product exposure on attitudes towards genetic engineering of food. *Food Quality and Preference* 7, 61–67.

Friend, T.H., Lay, D.C., Bushong, D.M. and Pierce, D.W. (1994) Wisconsin's 'stale calf' issue and a study designed to resolve some of the animal welfare concerns. *Journal of Animal Science* 72, 2260–2263.

Frisch, J.E. (1981) Changes in cattle as a consequence of selection for growth rate in a stressful environment. *Journal of Agricultural Science* 96, 23–38.

Gaili, E.S.E. (1978) A comparison of the development of body components in Sudan desert sheep and goats. *Tropical Animal Health and Production* 10, 103–108.

Galvin, S.L. and Herzog, H.A. (1992) Ethical ideology, animal rights activism, and attitudes toward the treatment of animals. *Ethics and Behavior* 2, 141–149.

Gardiner, E.E., Hunt, J.R., Newberry, R.C. and Hall, J.W. (1988) Relationships between age, body weight, and season of the year and the incidence of sudden death syndrome in male broiler chickens. *Poultry Science* 67, 1243–1249.

Garner, J.P., Falcone, C., Wakenell, P., Martin, M. and Mench, J.A. (2002) Reliability and validity of a modified gait scoring system and its use in assessing tibial dyschondroplasia in broilers. *British Poultry Science* 43, 355–363.

Garrett, R.P., Britain, K.G., Savell, J.W., Edwards, J.W. and Smith, S.B. (1999) Body composition of lambs receiving 30 or 60 days exercise training and (or) fenoterol treatment. *Meat Science* 52, 235–246.

Gatellier, P., Mercier, Y., Juin, H. and Renerre, M. (2005) Effect of finishing mode (pasture or mixed-diet) on lipid composition, colour stability and lipid oxidation in meat from Charolais cattle. *Meat Science* 69, 175–186.

Gaughan, J.B., Cameron, R.D.A., Dryden, G.McL. and Josey, M.J. (1995) Effect of selection for leanness on overall reproductive performance in Large White sows. *Animal Science* 61, 561–564.

Gauly, M., Mathjak, H., Hoffman, K., Kraus, M. and Erhardt, G. (2001) Estimating genetic variability in tempermental traits in German Angus and Simmental cattle. *Applied Animal Behaviour Science* 74, 109–119.

Geenty, K.G. (1986) Effect of early v. late lambing dates on ewe performance, lamb growth, and carcass composition in Canterbury. *New Zealand Journal of Experimental Agriculture* 14, 473–476.

Gerritzen, M.A., Lambooij, E., Hillebrand, S.J.W., Lankhaar, J.A.C. and Pieterse, C. (2000) Behavioural responses of broilers to different gaseous atmospheres. *Poultry Science* 79, 928–933.

Gerritzen, M.A., Lambooij, E., Reimert, H., Stegeman, A. and Spruijt, B. (2004) On-farm euthanasia of broiler chickens: effects of different gas mixtures on behaviour and brain activity. *Poultry Science* 83, 1294–1301.

Geverink, N.A., Schouten, W.G.P., Gort, G. and Wiegant, V.M. (2003) Individual differences in behaviour, physiology and pathology in breeding gilts, housed in groups or stalls. *Applied Animal Behaviour Science* 81, 29–41.

Geytenbeek, P.E. (1962) A survey of post-shearing losses. *Proceedings of the Australian Society of Animal Production* 4, 185–186.

Ghirotti, M. (1999) Making better use of animal resources in a rapidly urbanizing world: a professional challenge. *World Animal Review* 92, 2–14.

Gilbert, K.V. and Devine, C.E. (1982) Effect of electrical stunning methods on petechial haemorrhages and on the blood pressure of lambs. *Meat Science* 7, 197–207.

Gilbert, K.V., Devine, C.E., Hand, R. and Ellery, S. (1984) Electrical stunning and stillness of lambs. *Meat Science* 11, 45–58.

Gilbert, R.A., Tomkins, N., Padmanabha, J., Gough, J.M., Krause, D.O. and McSweeney, C.S. (2005) Effect of finishing diets of *Escherichia coli* populations and prevalence of enterohaemorrhagic *E. coli* virulence genes in cattle faeces. *Journal of Applied Microbiology* 99, 885–894.

Gispert, M., Faucitano, L., Oliver, M.A., Guàrdia, M.D., Coll, C., Siggens, K., Harvey, K. and Diestre, A. (2000) A survey of pre-slaughter conditions, halothane gene frequency, and carcass and meat quality in five Spanish pig commercial abattoirs. *Meat Science* 55, 97–106.

Gjedrem, T. (1997) Flesh quality improvement in fish through breeding. *Aquaculture International* 5, 197–206.

Gjedrem, T. (2000) Genetic improvement of cold-water fish species. *Aquaculture Research* 31, 25–33.

Gjestang, K.E. (1979) Comparison of slipperiness on rubber mats and concrete floors for dairy cows using behavioural method. *Applied Animal Ethology* 5, 294.

Godinez, D.D., Paschal, J.C., Hanselka, C.W. and Johnson, J. (1993) Stocker cattle production and management in South Texas. *Beef Cattle Research in Texas* 1993 (PR-5208), 184.

Gökdal, Ö., Aygün, T., Bingöl, M. and Karakuş, F. (2003) The effects of docking on performance and carcass characteristics of male Karakaş lambs. *South African Journal of Animal Science* 33, 185–192.

Goliomytis, M., Panopoulou, E. and Rogdakis, E. (2003) Growth curves for body weight and major component parts, feed consumption, and mortality of male broiler chickens raised to maturity. *Poultry Science* 82, 1061–1068.

Gomiero, T., Giampietro, M., Bukkens, S.G.F. and Paoletti, M. (1999) Environmental and socioeconomic constraints to the development of freshwater fish aquaculture in China. *Critical Reviews in Plant Science* 18, 359–371.

Goodson, K.J., Miller, R.K. and Savell, J.W. (2001) Carcass traits, muscle characteristics, and palatability attributes of lambs expressing the callipyge phenotype. *Meat Science* 58, 381–387.

Gottardo, F., Mattiello, S., Cozzi, G., Canali, E., Scanziani, E., Ravarotto, L., Ferrante, V., Verga, M. and Andrighetto, I. (2002) The provision of drinking water to veal calves for welfare purposes. *Journal of Animal Science* 80, 2362–2372.

Graham, W.C. and Price, M.A. (1982) Feedlot performance and carcass composition of cull cows of different ages. *Canadian Journal of Animal Science* 62, 845–854.

Grandin, T. (1981) Bruises on southwestern feedlot cattle. *Journal of Animal Science* 53 (Suppl. 1), 213 (Abstract).

Grandin, T. (1982) Pig behaviour studies applied to slaughter plant design. *Applied Animal Ethology* 6, 10–31.

Grandin, T. (1988) Double rail restrainer conveyor for livestock handling. *Journal of Agricultural Engineering Research* 41, 327–338.

Grandin, T. (1992) Observations of cattle restraint devices for stunning and slaughter. *Animal Welfare* 1, 85–91.

Grandin, T. (1994) Euthanasia and slaughter of livestock. *Journal of the American Veterinary Medical Association* 204, 1354–1360.

Grandin, T. (1996) Factors that impede animal movement in slaughter plants. *Journal of American Veterinary Medical Association* 209, 757–759.

Grandin, T. (1997) *Survey of Stunning and Handling in Federally Inspected Beef, Veal, Pork, and Sheep Slaughter Plants*. Project 3602-32000-002-08G, USDA/Agricultural Research Service, Beltsville, Maryland.

Grandin, T. (1998a) Objective scoring of animal handling and stunning practices at slaughter plants. *Journal of American Veterinary Medical Association* 212, 6–39.

Grandin, T. (1998b) The feasibility of vocalization scoring as an indicator of poor welfare during slaughter. *Applied Animal Behavior Science* 56, 121–128.

Grandin, T. (2000) Effect of animal welfare audits of slaughter plants by a major fast food company on cattle handling and stunning practices. *Journal of the American Veterinary Medical Association* 216, 848–851.

Grandin, T. (2001a) Solving return to sensibility problems after electrical stunning in commercial pork slaughter plants. *Journal of the American Veterinary Medical Association* 219, 608–611.

Grandin, T. (2001b) Cattle vocalizations are associated with handling and equipment problems at beef slaughter plants. *Applied Animal Behaviour Science* 71, 191–201.

Grandin, T. (2002a) Return to sensibility problems after penetrating captive bolt stunning of cattle in commercial slaughter plants. *Journal of the American Veterinary Medical Association* 221, 1258–1261.

Grandin, T. (2002b) Restaurant audits of stunning and handling in federally-inspected beef and pork slaughter plants. www.grandin.com

Grandin, T. (2003) Transferring results from behavioral research to industry to improve animal welfare on the farm, ranch, and slaughter plant. *Applied Animal Behavior Science* 81, 215–228.

Grandin, T. (2005a) Maintenance of good animal welfare standards in beef slaughter plants by use of auditory programs. *Journal of the American Veterinary Medical Association* 226, 370–373.

Grandin, T. (2005b) Recommended Animal Handling Guidelines of Audit Guide, 2005 Edition, American Meat Institute Foundation, Washington, DC. www.animalhandling.org, www.grandin.com

Grandin, T. (2007a) Introduction: effect of customer requirements, international standards and marketing system structure on the handling and transport of livestock and poultry. In: Grandin, T. (ed.) *Livestock Handling and Transport*, 3rd edn. CAB International, Wallingford, UK (in press).

Grandin, T. (ed.) (2007b) *Livestock Handling and Transport*, 3rd edn. CAB International, Wallingford, UK.

Grandin, T., Curtis, S.E. and Widowski, T.M. (1986) Electro-immobilization versus mechanical restraint in an avoid–avoid choice test. *Journal of Animal Science* 62, 1469–1480.

Grandin, T., McGee, K. and Lanier, J.L. (1999) Prevalence of severe welfare problems in horses that arrive at slaughter plants. *Journal of American Veterinary Medical Association* 214, 1531–1533.

Grandinson, K., Rydhmer, L., Strandberg, E. and Thodberg, K. (2003) Genetic analysis of on-farm tests of maternal behaviour in sows. *Livestock Production Science* 83, 141–152.

Grandinson, K., Rydhmer, L., Strandberg, E. and Solanes, F.X. (2005) Genetic analysis of body condition in the sow during lactation, and its relation to piglet survival and growth. *Animal Science* 80, 33–40.

Greaves, K. and Tuene, S. (2001) The form and context of aggressive behaviour in farmed Atlantic halibut (*Hippoglossus hippoglossus*). *Aquaculture* 193, 139–147.

Greer, A.W., Stankiewicz, M., Jay, N.P., McAnulty, R.W. and Sykes, A.R. (2005) The effect of concurrent corticosteroid induced immuno-suppression and infection with the intestinal parasite *Trichostrongylus colubriformis* on food intake and utilization in both immunologically naïve and competent sheep. *Animal Science* 80, 89–99.

Gregory, N.G. (1995) The role of shelterbelts in protecting livestock: a review. *New Zealand Journal of Agricultural Research* 38, 423–450.

Gregory, N.G. (1998) *Animal Welfare and Meat Science*. CAB International, Wallingford, Oxfordshire, UK, 298 pp.

Gregory, N.G. (2002) Meat safety issues. *Proceedings of the Australian Association of Cattle Veterinarians* 2002, 57–64.

Gregory, N.G. (2004) *Physiology and Behaviour of Animal Suffering*. Blackwell Publishing, Oxford, UK, 268 pp.

Gregory, N.G. (2005a) Recent concerns about stunning and slaughter. *Meat Science* 70, 481–491.

Gregory, N.G. (2005b) Bowhunting deer. *Animal Welfare* 14, 111–116.

Gregory, N.G. and Devine, C.D. (1999) Body condition in end-of-lay hens: some implications. *Veterinary Record* 145, 49.

Gregory, N.G. and Wilkins, L.J. (1992) Skeletal damage and bone defects during catching and processing. In: Whitehead, C.C. (ed.) *Bone Biology and Skeletal Disorders. 23rd Poultry Science Symposium of the WPSA*. Carfax Publishing Company, Abingdon, UK, pp. 313–328.

Gregory, N.G. and Wilkins, L.J. (1996) Effect of age on bone strength and the prevalence of broken bones in perchery laying hens. *New Zealand Veterinary Journal* 44, 31–32.

Gregory, N.G. and Wotton, S.B. (1984) Sheep slaughtering procedures. 2. Time to loss of brain responsiveness after exsanguinations or cardiac arrest. *British Veterinary Journal* 140, 354–366.

Gregory, N.G. and Wotton, S.B. (1990a) Effect of stunning on spontaneous physical activity and evoked activity in the brain. *British Poultry Science* 31, 215–220.

Gregory, N.G. and Wotton, S.B. (1990b) Comparison of neck dislocation and percussion of the head on visual evoked responses in the chicken's brain. *Veterinary Record* 126, 570–572.

Gregory, N.G. and Wotton, S.B. (1994) Effect of electrical stunning current on the duration of insensibility in hens. *British Poultry Science* 35, 463–465.

Gregory, N.G., Wilkins, L.J. and Wotton, S.B. (1991) Effect of electrical stunning frequency on ventricular fibrillation, downgrading and broken bones in broilers, hens and quails. *British Veterinary Journal* 147, 71–77.

Gregory, N.G., Anil, M.H., McKinstry, J.L. and Daly, C.C. (1996) Prevalence and duration of insensibility following electrical stunning in calves. *New Zealand Veterinary Journal* 44, 1–3.

Gregory, N.G., Haslett, S.J. and Pedley, J.C. (1999) Studies on lamb hypothermia using a model lamb. *New Zealand Journal of Agricultural Research* 42, 179–185.

Gregory, N.G., Jacobson, L.J., Nagle, T.A., Muirhead, R.W. and LeRoux, G.J. (2000) Effect of preslaughter feeding system on weight loss, gut bacteria, and the physico-chemical properties of digesta in cattle. *New Zealand Journal of Agricultural Research* 43, 351–361.

Grigorakis, K. and Alexis, M.N. (2005) Effects of fasting on the meat quality and fat deposition of commercial-sized farmed gilthead sea bream (*Sparus aurata* L.) fed different dietary regimes. *Aquaculture Nutrition* 11, 341–344.

Grimaud, P., Richard, D., Kanwé, A., Durier, C. and Doreau, M. (1998) Effect of undernutrition and refeeding on digestion in *Bos taurus* and *Bos indicus* in a tropical environment. *Animal Science* 67, 49–58.

Grizzle, J.M. and Lovshin, L.L. (1996) Injuries and serum enzyme activities of fingerling channel catfish (*Ictalurus punctatus*) harvested with a turbine pump. *Aquacultural Engineering* 15, 349–357.

Grommers, F.J., Elving, L. and van Eldick, P. (1985) Parturition difficulties in sheep. *Animal Reproduction Science* 9, 365–374.

Grunert, K.G., Bredahl, L. and Brunsø, K. (2004) Consumer perception of meat quality and implications for product development in the meat sector. *Meat Science* 66, 259–272.

Guàrdia, M.D., Estany, J., Balasch, S., Oliver, M.A., Gispert, M. and Diestre, A. (2004) Risk assessment of PSE condition due to pre-slaughter conditions and RYR1 gene in pigs. *Meat Science* 67, 471–478.

Guàrdia, M.D., Estany, J., Balasch, S., Oliver, M.A., Gispert, M. and Diestre, A. (2005) Risk assessment of DFD meat due to pre-slaughter conditions in pigs. *Meat Science* 70, 709–716.

Gue, M., Juniein, J.L. and Bueno, L. (1991) Conditioned emotional response in rats enhances colonic motility through the central release of corticotrophin-releasing factor. *Gastroenterology* 100, 964–970.

Guest, G.B. and Paige, J.C. (1991) The magnitude of the tissue residue problem with regard to consumer needs. *Journal of the American Veterinary Medical Association* 198, 805–808.

Gunn, R.G. (1970) A note on the effect of broken mouth on the performance of Scottish Blackface hill ewes. *Animal Production* 12, 517–520.

Gunter, G. and Furnham, A. (1992) *Consumer Profiles – an Introduction to Psychographics*. Routledge, London, p. 57.

Guo, S.-F., Gianola, D., Rekaya, R. and Short, T. (2001) Bayesian analysis of lifetime performance and prolificacy in Landrace sows using a linear mixed model with censoring. *Livestock Production Science* 72, 243–252.

Gustafsson, B. (1983) Effects of sow housing systems in practical pig production. *Transactions of the American Society of Agricultural Engineers* 26, 1181–1185.

Hall, D.G., Fogarty, N.M. and Gilmour, A.R. (1995) Performance of cross bred progeny of Trangie Fertility Merino and Booroola Merino rams and Poll Dorset ewes. 1. Lamb birth weight, survival and growth. *Australian Journal of Experimental Agriculture* 35, 1069–1074.

Hall, S.J.G., Kirkpatrick, S.M., Lloyd, D.M. and Broom, D.M. (1998) Noise and vehicular motion as potential stressors during the transport of sheep. *Animal Science* 67, 467–473.

Hambrecht, E., Eissen, J.J., Newman, D.J., Verstegen, M.W. and Hartog, L.A. (2005) Preslaughter handling affects pork quality and glycolytic potential of two muscles differing in fiber type organization. *Journal of Animal Science* 83, 900–907.

Hamilton, D.N., Ellis, M., Wolter, B.F., McKeith, F.K. and Wilson, E.R. (2003) Carcass and meat quality characteristics of the progeny of two swine sire lines reared under differing environmental conditions. *Meat Science* 63, 257–263.

Hamilton, G.F., Turner, A.S., Ferguson, J.G. and Pharr, J.W. (1978) Slipped capital femoral epiphysis in calves. *Journal of the American Veterinary Medical Association* 172, 1318–1322.

Hanlon, A.J., Rhind, S.M., Reid, H.W., Burrells, C. and Lawrence, A.B. (1995) Effects of changes in group composition on immune response, behaviour, adrenal activity and liveweight gain in farmed red deer yearlings. *Applied Animal Behaviour Science* 44, 57–64.

Hansen, L.L., Larsen, A.E., Jensen, B.B. and Hansen-Møller, J. (1997) Short time effect of zinc bacitracin and heavy fouling with faeces plus urine on boar taint. *Animal Science* 64, 351–363.

Harrington, G.N. and Pratchett, D. (1974) Stocking rate trials in Ankole, Uganda. 1. Weight gain of Ankole steers at intermediate and heavy stocking rates under different managements. *Journal of Agricultural Science* 82, 497–506.

Haslam, S.M., Brown, S.N., Wilkins, L.J., Kestin, S.C., Warriss, P.D. and Nicol, C.J. (2006) Preliminary studies to examine the utility of using foot burn or hock burn to assess aspects of housing conditions for broiler chicken. *British Poultry Science* 47, 13–18.

Hatziathanasiou, A., Paspatis, M., Houbart, M., Kestemont, P., Stefankis, S. and Kentouri, M. (2002) Survival, growth and feeding in early life stages of European sea bass (*Dicentrarchus labrax*) intensively cultured under different stocking densities. *Aquaculture* 205, 89–102.

Hazari, M.M., Misra, S.C., Panda, M.R., Bisoi, P. and Rao, A.T. (1991) Host's response to heavy *Boophilus microplus* infestation. *Indian Veterinary Journal* 68, 409–412.

Helfrich, L.A., Liston, C.R., Mefford, B. and Bark, R. (2001) Survival and injury of splittail and Chinook salmon passed through a large Hidrostal pump. *North American Journal of Fisheries Management* 21, 616–623.

Helfrich, L.A., Bark, R., Liston, C.R., Weigmann, D.L. and Mefford, B. (2004) Live transport of striped bass and rainbow trout using a Hidrostal pump. *Journal of World Aquaculture Society* 35, 268–273.

Hemsworth, P.H., Barnett, J.L., Hansen, C. and Winfield, C.G. (1986) Effects of social environment on welfare status and sexual behaviour of female pigs. *Applied Animal Behaviour Science* 16, 259–267.

Henckel, P., Karlsson, A., Jensen, M.T., Oksbjerg, N. and Petersen, J.S. (2002) Metabolic conditions in porcine longissimus muscle immediately pre-slaughter and its influence on peri- and post mortem energy metabolism. *Meat Science* 62, 145–155.

Herskin, M.S., Jensen, K.H. and Studnitz, M. (1998) Influence of timidity and social environment during lactation on maternal reactivity of outdoor sows. *Acta Agriculturae Scandinavica* A 48, 230–236.

Herzog, H.A. (1996) A test of the declining interest hypothesis. *American Psychologist* 51, 1184.

Hess, E. and Klinger, I. (1967) Quantitative Bestimmung des Blutgehaltes in der Muskulatur von Schlachtrindern nach Entbluten im Liegen und im Hängen. *Zentralblaat Veterinarmedizin* B 14, 685–697.

Hillebrand, S.J.W., Lambooy, E. and Veerkamp, C.H. (1996) The effects of alternative electrical and mechanical stunning methods on haemorrhaging and meat quality of broiler breast and thigh muscles. *Poultry Science* 75, 664–671.

Hindhede, J., Mogensen, L. and Sørensen, J.T. (1999) Effect of group composition and feeding system on behaviour, production and health of dairy heifers in deep bedding systems. *Acta Agriculturae Scandinavica* A 49, 211–220.

Hocking, P.M. (2006) High-fibre pelleted rations decrease water intake but do not improve physiological indexes of welfare in food-restricted female broiler breeders. *British Poultry Science* 47, 19–23.

Hocking, P.M. and Robertson, G.W. (2005) Limited effect of intense genetic selection for broiler traits on ovarian function and follicular sensitivity in broiler breeders at the onset of lay. *British Poultry Science* 46, 354–360.

Hocking, P.M., Maxwell, M.H. and Mitchell, M.A. (1996) Relationships between the degree of food restriction and welfare indices in broiler breeder females. *British Poultry Science* 37, 263–278.

Hocking, P.M., Bernard, R. and Wess, T.J. (1998) Comparative development of antitrochanteric disease in male and female turkeys of a traditional line and a contemporary sire-line fed *ad libitum* or with restricted quantities of food. *Research in Veterinary Science* 65, 29–32.

Hocking, P.M., Bernard, R. and Maxwell, M.H. (1999) Assessment of pain during locomotion and the welfare of adult male turkeys with degenerative cartilage loss of the hip joint. *British Poultry Science* 40, 30–34.

Hoffman, D.E., Spire, M.F., Schwenke, J.R. and Unruh, J.A. (1998) Effect of source of cattle and distance transported to a commercial slaughter facility on carcass bruising in mature beef cows. *Journal of the American Veterinary Medical Association* 212, 668–672.

Hoffman, L.C., Styger, E., Muller, M. and Brand, T.S. (2003) The growth and carcass and meat characteristics of pigs raised in a free-range or conventional housing system. *South African Journal of Animal Science* 33, 166–175.

Holechek, J., Galt, D., Joseph, J., Navarro, J., Kumalo, G., Molinar, F. and Thomas, M. (2003) Moderate and light cattle grazing effects on Chihuahuan desert rangelands. *Journal of Range Management* 56, 133–139.

Holst, P.J., Hegarty, R.S., Fogarty, N.M. and Hopkins, D.L. (1997) Fibre metrology and physical characteristics of lambskins from large Merino and crossbred lambs. *Australian Journal of Experimental Agriculture* 37, 509–514.

Homer, P.M. and Kahle, L.R. (1988) A structural equation test of the value–attitude–behavior hierarchy. *Journal of Personality and Social Psychology* 54, 638–646.

Homewood, K., Rodgers, W.A. and Arhem, K. (1987) Ecology of pastoralism in Ngorongoro Conservation area, Tanzania. *Journal of Agricultural Science* 108, 47–72.

Honkavaara, M., Rintasalo, E., Ylönen, J. and Pudas, T. (2003) Meat quality and transport stress of cattle. *Deutsche Tierärztliche Wochenschrift* 110, 125–128.

Hopkins, D.L. and Nicholson, A. (1999) Meat quality of wether lambs grazed on either saltbush (*Atriplex nummalaria*) plus supplements of lucerne (*Medicago sativa*). *Meat Science* 51, 91–95.

Horn, B.E., Hart, C.R. and Paisley, S.I. (2003) Management of rangeland livestock under drought. *Annals of Arid Zone* 41, 1–23.

Hossain, F. and Onyango, B. (2004) Product attributes and consumer acceptance of nutritionally enhanced genetically modified foods. *International Journal of Consumer Studies* 28, 255–267.

Hseu, J.-R. (2002) Effects of size difference and stocking density on cannibalism rate of juvenile grouper *Epinephelus coioides. Fisheries Science* 68, 1384–1386.

Huising, M.O., Guichelaar, T., Hoek, C., Verburg-van Kemenade, B.M.L., Flik, G., Savelkoul, H.F.J. and Rombout, J.H.W. (2003) Increased efficacy of immersion vaccination in fish with hyperosmotic pre-treatment. *Vaccine* 21, 4178–4193.

Hurd, H.S., Gailey, J.K., McKean, J.D. and Rostagno, M.H. (2001) Rapid infection in market-weight swine following exposure to a *Salmonella typhimurium*-contaminated environment. *American Journal of Veterinary Research* 62, 1194–1197.

Hutchinson, K.J. and McRae, B.H. (1969) Some factors associated with the behaviour and survival of newly shorn sheep. *Australian Journal of Agricultural Research* 20, 513–521.

Hutson, G.D. and Haskell, M.J. (1990) The behaviour of farrowing sows with free and operant access to an earth floor. *Applied Animal Behaviour Science* 26, 363–372.

Illius, A.W., Albon, S.D., Pemberton, J.M., Gordon, I.J. and Clutton-Brock, T.H. (1995) Selection for foraging efficiency during a population crash in Soay sheep. *Journal of Animal Ecology* 64, 481–492.

Imaeda, N. (2000) Influence of the stocking density and rearing season on incidence of sudden death syndrome in broiler chickens. *Poultry Science* 79, 201–204.

Immonen, K., Ruusunen, M., Hissa, K. and Puolanne, E. (2000a) Bovine muscle glycogen concentration in relation to finishing diet, slaughter and ultimate pH. *Meat Science* 55, 25–31.

Immonen, K., Ruusunen, M. and Puolanne, E. (2000b) Some effects of normal pH beef. *Meat Science* 55, 33–38.

Iordanidis, P., Lekkas, S., Georgopoulou, J. and Artopios, E. (1998) Case report. Non-infective epiphysioloysis associated with nutritional muscular dystrophy in broiler chickens. *Israel Journal of Veterinary Medicine* 53, 94–98.

Jahan, K., Paterson, A. and Spickett, C.M. (2004) Fatty acid composition, antioxidants and lipid oxidation in chicken breasts from different production regimes. *International Journal of Food Science and Technology* 39, 443–453.

Jana, B.B. and Jana, S. (2003) The potential and sustainability of aquaculture in India. *Journal of Applied Aquaculture* 13, 283–316.

Janczak, A.M., Pedersen, L.J., Rydhmer, L. and Bakken, M. (2003) Relation between early fear- and anxiety-related behaviour and maternal ability in sows. *Applied Animal Behaviour Science* 82, 121–135.

Jarvis, S., McLean, K.A., Chirnside, J., Deans, L.A., Calvert, S.K., Molony, V. and Lawrence, A.B. (1997) Opioid-mediated changes in nociceptive threshold during pregnancy and parturition in the sow. *Pain* 72, 153–159.

Jeppsson, K.-H. (1999) Volatilization of ammonia in deep-litter systems with different bedding materials for young cattle. *Journal of Agricultural Engineering Research* 73, 49–57.

Jittinandana, S., Kenney, P.B., Slider, S.D., Mazik, P., Bebak-Williams, J. and Hankins, J.A. (2003) Effect of fish attributes and handling stress on quality of smoked arctic char fillets. *Journal of Food Science* 68, 57–63.

Johnson, W.L. and Djajanegara, A. (1989) A pragmatic approach to improving small ruminant diets in the Indonesian humid tropics. *Journal of Animal Science* 67, 3068–3079.

Johnston, D.E. and Sloss, V. (1967) The causes and treatment of dystocia in beef cattle in Western Victoria. 1. Obstetrical procedures. *Australian Veterinary Journal* 43, 7–12.

Johnston, D.J., Reverter, A., Ferguson, D.M., Thompson, J.M. and Burrow, H.M. (2003) Genetic and phenotypic characterisation of animal, carcass and meat quality traits from temperate and tropically adapted beef breeds. 3. Meat quality traits. *Australian Journal of Agricultural Research* 54, 135–147.

Johnstone, R. (1996) Experience with salmonid sex reversal and triploidisation technologies in the United Kingdom. *Bulletin of the Aquaculture Association of Canada* 96 (2), 9–13.

Jones, J.B., Burgess, L.R., Webster, A.J.F. and Wathes, C.M. (1996) Behavioural responses of pigs to atmospheric ammonia in a chronic choice test. *Animal Science* 63, 437–445.

Jones, J.B., Wathes, C.M., Persand, K.C., White, R.P. and Jones, R.B. (2001) Acute and chronic exposure to ammonia and olfactory acuity for *n*-butanol in the pig. *Applied Animal Behaviour Science* 71, 13–28.

Jones, T.A., Donnelly, C.A. and Dawkins, M.S. (2005) Environmental and management factors affecting the welfare of chickens on commercial farms in the United Kingdom and Denmark stocked at five densities. *Poultry Science* 84, 1155–1165.

Josell, Å., Martinsson, L. and Tornberg, E. (2003) Possible mechanism for the effect of the *RN⁻* allele on pork tenderness. *Meat Science* 64, 341–350.

Joyce, J.P., Clarke, J.N., MacLean, K.S., Lynch, R.J. and Cox, E.H. (1976) The effect of level of nutrition on the productivity of sheep of different genetic origin. *Proceedings of the New Zealand Society of Animal Production* 36, 170–178.

Juell, J.-E. (1995) The behaviour of Atlantic salmon in relation to efficient cage-rearing. *Reviews in Fish Biology* 5, 320–335.

Juncher, D., Rønn, B., Mortensen, E.T., Henckel, P., Karlsson, A., Skibsted, L.H. and Bertelsen, G. (2001) Effect of pre-slaughter physiological conditions on the oxidative stability of colour and lipid during chill storage of pork. *Meat Science* 58, 347–357.

Juncher, D., Rønn, B., Hansen, T.B., Henckel, P., Karlsson, A., Skibsted, L.H. and Bertelsen, G. (2003) Effect of pre-slaughter physiological conditions on the oxidative stability of colour and lipid during chill storage of sliced, retail packed roast ham. *Meat Science* 63, 151–159.

Kabir, J., Umoh, J.U. and Umoh, V.J. (2002) Characterisation and screening for antimicrobial substances of slaughtered cattle in Zaria, Nigeria. *Meat Science* 61, 435–439.

Kalita, D. (1997) Surgical correction of gid in goats. *Indian Veterinary Journal* 74, 682–684.

Karim, S.A. and Rawat, P.S. (1996) Responses of sheep and goats during summer and winter to graded feed restriction and efficiency of regain to graded feed restriction and efficiency of regain on alimentation. *World Review of Animal Production* 31, 93–103.

Kempston, J.W., Lewitus, A.J., Deeds, J.R., Law, J.McH. and Place, A.R. (2002) Toxicity of *Karlodinium micrum* (*Dionophyceae*) associated with a fish kill in a South Carolina brackish retention pond. *Harmful Algae* 1, 233–241.

Kestin, S.C., Wotton, S.B. and Gregory, N.G. (1991) Effect of slaughter by removal from water on visual evoked activity in the brain and reflex movement of rainbow trout (*Oncorhynchus mykiss*). *Veterinary Record* 128, 443–446.

Kestin, S.C., Knowles, T.G., Tinch, A.E. and Gregory, N.G. (1992) Prevalence of leg weakness in broiler chickens and its relationship with genotype. *Veterinary Record* 131, 190–194.

Khan, A.G. (1996) Indigenous poultry and strategies for improving sustainable rural poultry production in India. *Indian Farming* 46 (6), 19–24.

Kieffer, J.D. (2000) Limits to exhaustive exercise in fish. *Comparative Biochemistry and Physiology* A 126, 161–179.

Kingston, D.J. (1979) Some hatchery factors involved in early chick mortality. *Australian Veterinary Journal* 55, 418–421.

Kirton, A.H. and Frazerhurst, L.F. (1983) Effects of normal, light/normal or double stunning on the incidence and severity of blood splash in lambs. *Meat Science* 8, 1–6.

Kirton, A.H., Frazerhurst, L.F., Woods, E.G. and Chrystall, B.B. (1980) Effect of electrical stunning method and cardiac arrest on bleeding efficiency, residual blood and blood splash in lambs. *Meat Science* 5, 347–353.

Klont, R.E., Barnier, V.M.H., Smulders, F.J.M., van Dijk, A., Hoving-Bolink, A.H. and Eikelenboom, G. (1999) Post-mortem variation in pH, temperature and colour profiles of veal carcasses in relation to breed, blood haemoglobin content, and carcass characteristics. *Meat Science* 53, 195–202.

Klug, B. (1989) Ritual murmur: the undercurrent of protest against religious slaughter of animals in Britain in the 1980s. *Patterns of Prejudice* 23, 15–28.

Knierim, U. and Gocke, A. (2003) Effect of catching broilers by hand or machine on rates of injuries and dead-on-arrivals. *Animal Welfare* 12, 63–73.

Knight, T.W., Dalton, D.C. and Hight, G.K. (1980) Changes in the median lambing dates and lambing pattern with variation in time of joining and breed of teasers. *New Zealand Journal of Agricultural Research* 23, 281–285.

Knight, T.W., Hall, D.R.H., Lynch, P.R. and Hockey, H.-U.P. (1988) Effects of pre-joining shearing, stress, pasture allowance, and haemoglobin type on reproductive performance of Romney and Marshall Romney ewes. *New Zealand Journal of Agricultural Research* 31, 249–258.

Knowles, T.G., Brown, S.N., Warriss, P.D., Butterworth, A. and Hewitt, I. (2004) Welfare aspects of chick handling in broiler and laying hen hatcheries. *Animal Welfare* 13, 409–418.

Kongsted, A.G. (2004) Stress and fear as possible mediators of reproduction problems in group housed sows: a review. *Acta Agriculturae Scandinavica* A 54, 58–66.

Koorts, R. (1991) The development of a restraining system to accommodate the Jewish method of slaughter (shechita). M.Dip.Tech., Technikon Witwatersrand Johannesburg, pp. 72–81.

Kotula, A.W. and Helbacka, N.V. (1966) Blood volume of live chickens and influence of slaughter technique on blood loss. *Poultry Science* 45, 684–688.

Kranen, R.W., Lambooy, E., Veerkamp, C.H., van Kuppevelt, T.H. and Veerkamp, J.H. (2000a) Histological characterization of hemorrhages in muscles of broiler chickens. *Poultry Science* 79, 110–116.

Kranen, R.W., Lambooij, E., Veerkamp, C.H., van Kuppevelt, T.H. and Veerkamp, J.H. (2000b) Haemorrhages in muscles of broiler chickens. *World's Poultry Science Journal* 56, 93–126.

Kreuter, U.P. and Workman, J.P. (1994) Costs of overstocking on cattle and wildlife ranches in Zimbabwe. *Ecological Economics* 11, 237–248.

Kuber, P.S., Duckett S.K., Busboom, J.R., Snowder, G.D., Dodson, M.V., Virck, J.L. and Bailey, J.F. (2003) Measuring the effects of phenotype and mechanical restraint on proteolytic degradation and rigor shortening in callipyge lamb longissimus dorsi muscle during extended aging. *Meat Science* 63, 325–331.

Kudi, A.C., Umoh, J.U., Eduvie, L.O. and Gefu, J.O. (1998) Causes of illness and death in cattle managed traditionally in Northern Nigeria. *Bulletin of Animal Health and Production in Africa* 46, 235–240.

Kuit, H.G., Traore, A. and Wilson, R.T. (1986) Livestock production in central Mali: ownership, management and productivity of poultry in the traditional sector. *Tropical Animal Health and Production* 18, 222–231.

Kumar, G.S. and Parihar, N.S. (1998) Respiratory ailments in Indian buffaloes. *Buffalo Journal* 14, 109–114.

Kumar, S. and Pant, K.P. (2003) Development perspective of goat rearing in India: status, issues and strategies. *Indian Journal of Agricultural Economics* 58, 752–767.

Kuo, C.C. and Chu, C.Y. (2003) Quality characteristics of Chinese sausages made from PSE pork. *Meat Science* 64, 441–449.

Kuzmanvić, Ž. and Elabjer, I. (2000) Influence of preslaughter resting time of bulls on meat temperature, colour and pH value. *Czech Journal of Animal Science* 45, 511–518.

Kwak, W.S., Kim, K.H., Kim, C.W., Jeon, B.T. and Lee, S.M. (1994) Deer farming in Korea: on-farm survey in Kjung-Kee and Chung-Buk Provinces. *Asian-Australasian Journal of Animal Sciences* 7, 347–355.

Lambe, N.R., Connington, J., Bishop, S.C., Waterhouse, A. and Simm, G. (2001) A genetic analysis of maternal behaviour score in Scottish Blackface sheep. *Animal Science* 72, 415–425.

Lambooij, E., Pieterse, C., Potgieter, C.M., Snymand, J.D. and Nortjé, G.L. (1999) Some neural and behavioural aspects of electrical and mechanical stunning in ostriches. *Meat Science* 52, 339–345.

Lambooij, E., van de Vis, J.W., Kloosterboer, R.J. and Pieterse, C. (2002a) Evaluation of head-only and head-to-tail electrical stunning of farmed eels (*Anguilla anguilla* L.) for the development of a humane slaughter method. *Aquaculture Research* 33, 323–331.

Lambooij, E., van de Vis, J.W., Kuhlmann, H., Münker, W., Oehlenschläger, J., Kloosterboer, R.J. and Pieterse, C. (2002b) A feasible method for humane slaughter of eels (*Anguilla anguilla* L.): electrical stunning in fresh water prior to gutting. *Aquaculture Research* 33, 643–652.

Lambooij, E., Hulsegge, B., Klont, R.E., Winkelman-Goedhart, H.A., Reimert, H.G.M. and Kranen, R.W. (2004) Effects of housing conditions of slaughter pigs on some post mortem muscle metabolites and pork quality characteristics. *Meat Science* 66, 855–862.

Lambooy, E. (1982) Electrical stunning of sheep. *Meat Science* 6, 123–135.

Lambooy, E. (1985) *Automatische elektrische bedwelming van vleeskalveren in een V-vormige restrainer.* Rapport B-273, Instituut voor Veeteeltkundig Onderzoek 'Schoonoord', Zeist, the Netherlands, 45 pp.

Lambooy, E. and Van Voorst, N. (1985) Electroanaesthesia of calves and sheep. In: Eikenboom, G. (ed.) *Stunning Animals for Slaughter.* Martinus Nijhoff, Dordrecht, the Netherlands, pp. 117–122.

Landau, S., Perevolotsky, A., Bonfil, D., Barki, D. and Silanikove, N. (2000) Utilization of low quality resources by small ruminants in Mediterranean agro-pastoral systems: the case of browse and aftermath cereal stubble. *Livestock Production Science* 64, 39–49.

Laros, F.J.M. and Steenkamp, J.-B.E.M. (2004) Importance of fear in the case of genetically modified food. *Psychology and Marketing* 21, 889–908.

Larsen, S.T., Hurd, H.S., McKean, J.D., Griffith, R.W. and Wesley, I.V. (2004) Effect of short-term lairage on the prevalence of *Salmonella enterica* in cull sows. *Journal of Food Protection* 67, 1489–1493.

Le Bihan-Duval, E., Millet, N. and Remignon, H. (1999) Broiler meat quality: effect of selection for increased carcass quality and estimates of genetic parameters. *Poultry Science* 78, 822–826.

Le Bihan-Duval, E., Berri, C., Baéza, E., Santé, V., Astruc, T., Rémignon, H., Le Pottier, G., Beaumont, C. and Fernandez, X. (2003) Genetic parameters of meat technological quality traits in a grand-parental commercial line of turkeys. *Genetics Selection Evolution* 35, 623–635.

Lebret, B. and Guillard, A.-S. (2005) Outdoor rearing of cull sows: effects on carcass, tissue composition and meat quality. *Meat Science* 70, 247–257.

Lebret, B., Massabie, P., Granier, R., Juin, H., Mourot, J. and Chevillon, P. (2002) Influence of outdoor rearing and indoor temperature on growth performance, carcass, adipose tissue and muscle traits in pigs, and on the technological and eating quality of dry-cured hams. *Meat Science* 62, 447–455.

Le Cozler, Y., Ringmar-Cederberg, E., Rydhmer, L., Lundeheim, N., Dourmad, J.Y. and Neil, M. (1999) Effect of feeding level during rearing and mating strategy on performance of Swedish Yorkshire sows. 2. Reproductive performance, food intake, backfat changes and culling rate during the first two parities. *Animal Science* 68, 365–377.

Leet, N.G., Devine, C.E. and Gavey, A.B. (1977) The histology of blood splash in lamb. *Meat Science* 1, 229–234.

Lefaucheur, L., Le Roy, P., Ecolan, P., Clochefert, N., Gogué, J., Sanchez, M.-P. and Fernandez, X. (2000) Divergent selection on contractile properties of longissimus muscle fibers in the pig. In: *Proceedings of the 46th International Congress of Meat Science and Technology*, ICoMST, Buenos Aires, pp. 94–95.

Lensink, B.J., Fernandez, X., Boivin, P., Pradel, P., Le Neindre, P. and Veissier, I. (2000) The impact of gentle contacts on ease of handling, welfare, and growth of calves and on quality of veal meat. *Journal of Animal Science* 78, 1219–1226.

Lensink, B.J., Fernandez, X., Cozzi, G., Florand, L. and Veissier, I. (2001a) The influence of farmers' behaviour on calves' reactions to transport and quality of veal meat. *Journal of Animal Science* 79, 642–652.

Lensink, B.J., Veissier, I. and Florand, L. (2001b) The farmers' influence on calves' behaviour, health and production of a veal unit. *Animal Science* 72, 105–116.

Leonard, F.C., O'Connell, J.M. and O'Farrell, K.J. (1996) Effect of overcrowding on claw health in first-calved Friesian heifers. *British Veterinary Journal* 152, 459–472.

Lessard, M., Taylor, A.A., Braithwaite, L. and Weary, D.M. (2002) Humoral and cellular immune responses of piglets after castration at different ages. *Canadian Journal of Animal Science* 82, 519–526.

Lewis, P.D., Perry, G.C. and Sherwin, C.M. (1998) Effect of intermittent light regimens on the performance of intact male turkeys. *Animal Science* 67, 627–636.

Ley, S.J., Waterman, A.E. and Livingston, A. (1995) A field study of the effect of lameness on mechanical nociceptive thresholds in sheep. *Veterinary Record* 137, 85–87.

Lin, M., Cavinato, A.G., Mayes, D.M., Smiley, S., Huang, Y., Al-Holy, M. and Rasco, B.A. (2003) Bruise detection in pacific salmon (*Oncorhynchus gorbuscha*) by visible and short-wavelength near-infrared (SW-NIR) spectroscopy (600-1100 nm). *Journal of Agricultural and Food Chemistry* 51, 6404–6408.

Lippens, M., Room, G., de Groote, G. and Decuypere, E. (2000) Early and temporary quantitative food restriction of broiler chickens. 1. Effects on performance characteristics, mortality and meat quality. *British Poultry Science* 41, 343–354.

Listrat, A., Rakadjiyski, N., Jurie, C., Picard, B., Touraille, C. and Geay, Y. (1999) Effect of the type of diet on muscle characteristics and meat palatability of growing Salers bulls. *Meat Science* 53, 115–124.

Lorenzen, C.L., Hale, D.S., Griffin, D.B., Savell, J.W., Belk, K.E., Frederick, T.L., Miller, M.F., Montgomery, T.H. and Smith, G.C. (1993) National beef quality audit: survey of producer-related defects and carcass quality and quantity attributes. *Journal of Animal Science* 71, 1495–1502.

Lorenzen, C.L., Griffin, D.B., Morgan, W.W., Harris, J.J. and Savell, J.W. (1994) Cull cow bruising in Texas: a survey. *Beef Cattle Research in Texas* 1994 (PR 5256), 180–184.

Losinger, W.C., Bush, E.J., Smith, H.A. and Corso, B.A. (1998) Mortality in the grower/finisher phase on farrowing-to-finish swine operations in the United States. *Annales de Zootechnie* 47, 99–105.

Lott, B.D., May, J.D., Simmons, J.D. and Branton, S.L. (2001) The effect of nipple height on broiler performance. *Poultry Science* 80, 408–410.

Lott, D.F. and Hart, B.L. (1977) Aggressive domination of cattle by Fulani herdsmen and its relation to aggression in Fulani culture and personality. *Ethos* 5 (2), 174–186.

Love, R.J. (1978) Definition of a seasonal infertility problem in pigs. *Veterinary Record* 103, 443–446.

Lowe, T.E., Gregory, N.G., Fisher, A.D. and Payne, S.R. (2002) The effects of temperature elevation and water deprivation on lamb physiology, welfare, and meat quality. *Australian Journal of Agricultural Research* 53, 707–714.

Lund, A., Wallgren, P., Rundgren, M., Artursson, K., Thomke, S. and Forsum, C. (1998) Performance, behaviour and immune capacity of domestic pigs reared for slaughter as siblings or transported and reared in mixed groups. *Acta Agriculturae Scandinavica* A 48, 103–112.

McArthur, I.D. (1980) Pre-lambing supplementation of gadic ewes in western Afghanistan. *Journal of Agricultural Science* 95, 39–45.

McCarthy, M., de Boer, M., O'Reilly, S. and Cotter, L. (2003) Factors influencing intention to purchase beef in the Irish market. *Meat Science* 65, 1071–1083.

McCutcheon, S.N., Holmes, C.W. and McDonald, M.F. (1981) The starvation-exposure syndrome and neonatal lamb mortality: a review. *Proceedings of the New Zealand Society of Animal Production* 41, 209–217.

Macer, D.R.J. (1992) *Attitudes to Genetic Engineering. Japanese and International Comparisons.* Eubios Ethics Institute, Christchurch, New Zealand, 165 pp.

Mack, S. (1996) Pasture – the overstocking issue. *World Animal Review* 87, 1, 85.

MacKenzie, A.J., Thwaites, C.J. and Edey, T.N. (1975) Oestrous, ovarian and adrenal response of the ewe to fasting and cold stress. *Australian Journal of Agricultural Research* 26, 545–551.

McKinstry, J.L. and Anil, M.H. (2004) The effect of repeat application of electrical stunning on the welfare of pigs. *Meat Science* 67, 121–128.

Mackintosh, C.G. (1998) Deer health and disease. *Acta Veterinaria Hungarica* 46, 381–394.

McLean, J.A., Savory, C.J. and Sparks, N.A.C. (2002) Welfare of male and female broiler chickens in relation to stocking density, as indicated by performance, health and behaviour. *Animal Welfare* 11, 55–73.

McLeese, J.M., Patience, J.F., Wolynetz, M.S. and Christison, G.I. (1991) Evaluation of the quality of ground water supplies used on Saskatchewan swine farms. *Canadian Journal of Animal Science* 71, 191–203.

McMeekan, C.M., Stafford, K.J., Mellor, D.J., Bruce, R.A., Ward, R.N. and Gregory, N.G. (1998) Effects of regional analgesia and/or a non-steroidal anti/inflammatory analgesic on the acute cortisol response to dehorning in calves. *Research in Veterinary Science* 64, 147–150.

McNamee, P.T., McCullagh, J.J., Thorp, B.H., Ball, H.J., Graham, D., McCullough, S.J., McConaghy, D. and Smyth, J.A. (1998) Study of leg weakness in two commercial broiler flocks. *Veterinary Record* 143, 131–135.

McNeal, W.D. and Fletcher, D.L. (2003) Effects of high frequency electrical stunning and decapitation on early rigor development and meat quality of broiler breast meat. *Poultry Science* 82, 1352–1355.

McNeal, W.D., Fletcher, D.L. and Buhr, R.J. (2003) Effects of stunning and decapitation on broiler activity during bleeding, blood loss, carcass, and breast meat quality. *Poultry Science* 82, 163–168.

McNeill, D.M., Kelly, R.W. and Williams, I.H. (1999) Maternal fatness influences fetal size in ewes underfed in late pregnancy. *Australian Journal of Agricultural Research* 50, 1171–1177.

McQueen, C. (1972) Hyperthermia in cattle in Northern Australia. *Australian Veterinary Journal* 48, 128.

Mader, T.L., Holt, S.M., Hahn, G.L., Davis, M.S. and Spiers, D.E. (2002) Feeding strategies for managing heat load in feedlot cattle. *Journal of Animal Science* 80, 2373–2382.

Madruga, M.S., Arruda, S.G.B., Narain, N. and Souza, J.G. (2000) Castration and slaughter age effects on panel assessment and aroma compounds of the 'mestiço' goat meat. *Meat Science* 56, 117–125.

Malan, D.D., Scheele, C.W., Buyse, J., Kwakernaak, C., Siebrits, F.K., van der Klis, J.D. and Decuypere, E. (2003) Metabolic rate and its relationship with ascites in chicken genotypes. *British Poultry Science* 44, 309–315.

Mallia, J.G., Barbut, S., Vaillancourt, J.-P., Martin, S.W. and McEwen, S.A. (2000a) A dark, firm dry-like condition in turkeys condemned for cyanosis. *Poultry Science* 79, 281–285.

Mallia, J.G., Vaillancourt, J.-P., Martin, S.W. and McEwen, S.A. (2000b) Risk factors for abattoir condemnation of turkey carcasses due to cyanosis in southern Ontario. *Poultry Science* 79, 831–837.

Malone, F., McParland, P.J. and O'Hagan, J. (1985) Causes of mortality in an intensive lamb fattening unit. *Irish Veterinary Journal* 39, 86–90.

Manefield, G.W. and Tinson, A.H. (1996) *Camels – a Compendium.* University of Sydney Post Graduate Foundation in Veterinary Science, Sydney, Australia, 372 pp.

Manjeli, Y., Tchoumboue, J., Teguia, A. and Zango, P. (1996) Productivity of West African Dwarf goats under traditional management in the western highlands of Cameroon. *World Review of Animal Production* 31, 87–92.

Manjeli, Y., Tchoumboue, J.T., Njwe, R.M. and Tenguin, A. (1998) Guinea-pig productivity under traditional management. *Tropical Animal Health and Production* 30, 115–122.

Marchant, J.N. and Broom, D.M. (1996) Factors affecting posture changing in loose-housed and confined gestating sows. *Animal Science* 63, 477–485.

Marchant, J.N., Broom, D.M. and Corning, S. (2001) The influence of sow behaviour on piglet mortality due to crushing in an open farrowing system. *Animal Science* 72, 19–28.

Maree, C. (1986) The influence of high level feeding on the duration of parturition and the incidence of dystocia in dairy cows. *Journal of the South African Veterinary Association* 57, 151–153.

María, G.A., Villarroel, M., Sañudo, C., Olleta, J.L. and Gebresenbet, G. (2003) Effect of transport time and ageing on aspects of beef quality. *Meat Science* 65, 1335–1340.

María, G.A. Villareol, M., Chacon, G. and Gebresenbet, G. (2004) Scoring system for evaluating stress of cattle during commercial loading and unloading. *Veterinary Record* 154, 818–821.

Marteniuk, J., Alwynelle, M.S., Ahl, S. and Bartlett, P.C. (1988) Compliance with recommended drug withdrawal requirements for dairy cows sent to market in Michigan. *Journal of the American Veterinary Medical Association* 193, 404–407.

Martoft, L., Lomholt, L., Kothoff, C., Rodriguez, B.E., Jensen, E.W., Jørgensen, P.F., Pedersen, H.D. and Forslid, A. (2002) Effects of CO_2 anaesthesia on central nervous system activity in swine. *Laboratory Animals* 36, 115–126.

Martrenchar, A. (1999) Animal welfare and intensive production of turkey broilers. *World's Poultry Science Journal* 55, 143–152.

Martrenchar, A., Huonnic, D., Cotte, J.P., Boilletot, E. and Morisse, J.P. (1999) Influence of stocking density on behavioural, health and productivity traits of turkeys in large flocks. *British Poultry Science* 40, 323–331.

Martrenchar, A., Huonnic, D. and Cotte, J.P. (2001) Influence of environmental enrichment on injurious pecking and perching behaviour in young turkeys. *British Poultry Science* 42, 161–170.

Martrenchar, A., Boilletot, E., Huonnic, D. and Pol, F. (2002) Risk factors for foot-pad dermatitis in chicken and turkey broilers in France. *Preventive Veterinary Medicine* 52, 213–226.

Mast, M.G. and MacNeil, J.H. (1983) Effect of kosher vs conventional processing on yield quality, and acceptability of broiler chickens. *Journal of Food Science* 48, 1013–1015, 1024.

May, G.J., van Tassell, L.W., Waggoner, J.W. and Smith, M.A. (1999) Relative costs and feeding strategies associated with winter/spring calving. *Journal of Range Management* 52, 560–568.

Mekonnen, G.A. (2004) Meat production potential of Issa type camels under traditional management system in Eastern Ethiopia. *Bulletin of Animal Health and Production in Africa* 52, 215–220.

Mellado, M., Cantú, L. and Suárez, J.E. (1996) Effects of body condition, length of breeding period, buck : doe ratio, and month of breeding on kidding rates in goats under extensive conditions in arid zones of Mexico. *Small Ruminant Research* 23, 29–35.

Mellado, M., Valdez, R., Lara, L.M. and Lopez, R. (2003) Stocking rate effects on goats: a research observation. *Journal of Range Management* 56, 167–173.

Mellado, M., Rodríguez, A., Olvera, A., Villareal, J.A. and Lopez, R. (2004) Age and body condition score and diets of grazing goats. *Journal of Range Management* 57, 517–523.

Mellor, D. and Stafford, K. (1999) Assessing and minimising the distress caused by painful husbandry procedures in ruminants. *In Practice* 21, 436–446.

Mersie, A. and Bekele, M. (1994) Causes of hide damage in eastern Ethiopia. *World Animal Review* 79, 55–57.

Mesías, F.J., Escribano, M., Rodríguez de Ledesma, A. and Pulido, F. (2005) Consumers' preferences for beef in the Spanish region of Extremadura: a study using conjoint analysis. *Journal of the Science of Food and Agriculture* 85, 2487–2494.

Meyer, A., Cloette, S.W.P., Brown, C.R. and van Schalkwyk, S.J. (2002) Declawing ostrich (*Struthio camelus domesticus*) chicks to minimize skin damage during rearing. *South African Journal of Animal Science* 32, 192–200.

Meyer, A., Cloette, S.W.P., Brown, C.R. and van Schalkwyk, S.J. (2003) The persistence to slaughter age of scars resulting from damage inflicted to ostrich skins during the grow-out phase. *South African Journal of Animal Science* 33, 32–37.

Meyer, H.H. and Clarke, J.N. (1978) Genetic and environmental effects on incidence and causes of lamb mortality. *Proceedings of the New Zealand Society of Animal Production* 38, 181–184.

Miles, D.M., Branton, S.J. and Lott, B.D. (2004) Atmospheric ammonia is detrimental to the performance of modern commercial broilers. *Poultry Science* 83, 1650–1654.

Millman, S.T., Duncan, I.J.H. and Widowski, T.M. (2000) Male broiler breeder fowl display high levels of aggression toward females. *Poultry Science* 79, 1233–1241.

Minihan, D., O'Mahoney, M., Whyte, P. and Collins, J.D. (2003) An investigation on the effect of transport and lairage on the faecal shedding prevalence of *Escherichia coli* O157 in cattle. *Journal of Veterinary Medicine* B 50, 378–382.

Mitchell, K.D., Stookey, J.M., Laturnas, D.K., Watts, J.M., Haley, D.B. and Huyde, T. (2004) The effects of blindfolding on behavior and heartrate in beef cattle during restraint. *Applied Animal Behaviour Science* 85, 233.

Miyashita, S., Sawada, Y., Hattori, N., Nakatsukasa, H., Okada, T., Murata, O. and Kumai, H. (2000) Mortality of Northern Bluefin tuna *Thunnus thynnus* due to trauma caused by collision during growout culture. *Journal of the World Aquaculture Society* 31, 632–639.

Mobolaji-Bukola, P.U., Allison-Oguru, E.A., Berepubo, N.A. and Oruwari, B.M. (2002) Bio-economic evaluation of the performance of rabbits raised under two different housing systems. *Tropicultura* 20, 176–180.

Moelich, E.I., Hoffman, L.C. and Conradie, P.J. (2003) Sensory and functional meat quality characteristics of pork derived from three halothane genotypes. *Meat Science* 63, 333–338.

Mohan Raj, A.B. and Gregory, N.G. (1990) Investigation into the batch stunning/killing of chickens using carbon dioxide or argon-induced hypoxia. *Research in Veterinary Science* 49, 364–366.

Mohan Raj, A.B., Gregory, N.G. and Wotton, S.B. (1991) Changes in the somatosensory evoked potentials and spontaneous electroencephalogram of hens during stunning in argon-induced anoxia. *British Veterinary Journal* 147, 322–330.

Mohan Raj, A.B., Wotton, S.B. and Gregory, N.G. (1992) Changes in the somatosensory evoked potentials and spontaneous electroencephalogram of hens during stunning with a carbon dioxide and argon mixture. *British Veterinary Journal* 148, 147–156.

Moinard, C., Lewis, P.D., Perry, G.C. and Sherwin, C.M. (2001) The effects of light intensity and light source on injuries due to pecking of male domestic turkeys (*Meleagris gallopavo*). *Animal Welfare* 10, 131–139.

Mojto, J., Chrenek, J., Palanská, O., Kmet, J., Zaujec, K. and Pavlič, M. (1998) Nutritive and physical and technological quality of meat and fat in slaughter bulls of Holstein breed and in crosses with Belgian White-Blue breed. *Czech Journal of Animal Science* 43, 483–487.

Mokantla, E., McCrindle, C.M.E., Sebei, J.P. and Owen, R. (2004) An investigation into the causes of low calving percentage in communally grazed cattle in Jericho, North West Province. *Journal of the South African Veterinary Association* 75, 30–36.

Monsón, F., Sañudo, C. and Sierra, I. (2004) Influence of cattle breed and ageing time on textural meat quality. *Meat Science* 68, 595–602.

Moore, G.H., Cowie, G.M. and Bray, A.R. (1985) Herd management of farmed red deer. In: Fennessy, P.F. and Drew, K.R. (eds) *Biology of Deer Production. Royal Society of New Zealand Bulletin* 22, 343–355.

Morgan, P.D., Arnold, G.W. and Boundy, C.A.P. (1972) The behaviour of ewes and their lambs in a hot climate. *Proceedings of the Australian Society of Animal Production* 9, 371–375.

Morris, S.T. and McCutcheon, S.N. (1997) Selective enhancement of growth in twin foetuses by shearing ewes in early gestation. *Animal Science* 65, 105–110.

Morzel, M. and van de Vis, H. (2003) Effect of the slaughter method on the quality of raw and smoked eels (*Anguilla anguilla* L.). *Aquaculture Research* 34, 1–11.

Mossman, D.H. (1983) Serving capacity test. *New Zealand Veterinary Journal* 31, 123–124.

Mott, I. and Ivarie, R. (2002) Expression of myostatin is not altered in lines of poultry exhibiting myofiber hyper- and hypoplasia. *Poultry Science* 81, 799–804.

Mouchonière, M., Le Pottier, G. and Fernandez, X. (1999) The effect of current frequency during waterbath stunning on the physical recovery and rate and extent of bleed out in turkeys. *Poultry Science* 78, 485–489.

Mouthon, G., Longin, C. and Magat, A. (1976) Effet du transport sur le pH, la pCO_2 at la pO_2 dans le sang de jeunes bovines de boucherie. *Bulletin de la Societé des Sciences Vétérinaire et Médecine Comparée* 78, 333–336.

Mouttotou, N., Hatchell, F.M. and Green, L.E. (1999) Foot lesions in finishing pigs and their associations with the type of floor. *Veterinary Record* 144, 629–632.

Mpiti-Shakhane, L.M., Kanyama-Phiri, G.Y., Odenya, W.O. and Banda, J.W. (2002) Merino sheep responses to different stocking rates and seasonal variations under rangeland conditions of Lesotho. *African Journal of Range and Forage Science* 19, 71–75.

Muggli-Cockett, N.E., Cundiff, L.V. and Gregory, K.E. (1992) Genetic analysis of bovine respiratory disease in beef calves during the first year of life. *Journal of Animal Science* 70, 2013–2019.

Mukasa-Mugerwa, E., Anindo, D., Lahlou-Kasi, A., Umunna, N.N. and Tegegne, A. (1997) Effect of body condition and energy utilization on the length of post-partum anoestrus in PRID-treated and untreated post-partum *Bos indicus* (zebu) cattle. *Animal Science* 67, 17–24.

Murray, A.C. and Johnson, C.P. (1998) Impact of the halothane gene on muscle quality and pre-slaughter deaths in Western Canadian pigs. *Canadian Journal of Animal Science* 78, 543–548.

Murray, R.D., Downham, D.Y., Clarkson, M.J., Faull, W.B., Hughes, J.W., Manson, F.J., Merritt, J.B., Russell, W.B., Sutherst, J.E. and Ward, W.R. (1996) Epidemiology of lameness in dairy cattle: description and analysis of foot lesions. *Veterinary Record* 138, 586–591.

Murray, R.D., Cartwright, T.A., Downham, D.Y., Murray, M.A. and de Kruif, A. (2002) Comparison of external and internal pelvic measurements of Belgian Blue cattle from sample herds in Belgium and the United Kingdom. *Reproduction in Domestic Animals* 37, 1–7.

Mustafa, A., Rankaduwa, W. and Campbell, P. (2001) Estimating the cost of sea lice to salmon aquaculture in eastern Canada. *Canadian Veterinary Journal* 42, 54–56.

Nagle, T.A., Gregory, N.G. and Lowe, T.E. (2000) Effect of preslaughter heat stress, exercise or adrenaline injection on post-mortem heat-stable pinkness, texture and water holding capacity in chicken. *Proceedings of the 46th International Congress of Meat Science and Technology* 2.II, ICoMST, Buenos Aires, pp. 146–147.

Nanni Costa, L., Lo Fiego, D.P., Cassanelli, M.G., Tassone, F. and Russo, V. (2003) Effect of journey time and environmental condition on bull behaviour and beef quality during road transport in Northern Italy. *Deutsche Tierärztliche Wochenschrift* 110, 107–110.

Naudé, R.T., van Rensburg, A.J.J., Smit, M.C., Stiemie, S., Dreyer, J.H., Rossouw, E.J. and de Jager, A.L. (1979) *Report on Certain Characteristics of Ostrich Muscle and Meat.* Meat Research Laboratory Report, ADRI, Irene, 12 pp.

Newberry, R.C. (1992) Influence of increasing photoperiod and toe clipping on breast buttons of turkeys. *Poultry Science* 71, 1471–1479.

Newell, D.G. and Fearnley, C. (2003) Sources of *Campylobacter* colonization in broiler chickens. *Applied and Environmental Microbiology* 69, 4343–4351.

Nfi, A.N. and Ndamukong, K. (1997) Health problems in small ruminant farms of North West Province, Cameroon. *World Animal Review* 88, 56–58.

Nibert, D. (2002) *Animal Rights/Human Rights.* Rowman and Littlefield Publishers, Lanham, USA, 269 pp.

Nicholson, M.J. (1987) Effects of night enclosure and extensive walking on the productivity of zebu cattle. *Journal of Agricultural Science* 109, 445–452.

Nielsen, B.L. (2004) Breast blisters in groups of slow-growing broilers in relation to strain and availability and use of perches. *British Poultry Science* 45, 306–315.

Nielsen, E.K. and Ingvartsen, K.L. (2000) Effects of cereal disintegration method, feeding method and straw as bedding on stomach characteristics including ulcers and performance in growing pigs. *Acta Agriculturae Scandinavica* A 50, 30–38.

Nijdam, E., Delezie, E., Lambooij, E., Nabuurs, M.J.A., Decuypere, E. and Stegeman, J.A. (2005) Comparison of bruises and mortality, stress parameters, and meat quality in manually and mechanically caught broilers. *Poultry Science* 84, 467–474.

Northcutt, J.K., Buhr, R.J. and Young, L.L. (1998) Influence of preslaughter stunning on turkey breast muscle quality. *Poultry Science* 77, 487–492.

Northcutt, J.K., Berrang, M.E., Dickens, J.A., Fletcher, D.L. and Cox, N.A. (2003) Effect of broiler age, feed withdrawal, and transportation on levels of coliforms, *Campylobacter*, *Escherichia coli*, and *Salmonella* on carcasses before and after immersion chilling. *Poultry Science* 82, 169–173.

Noss, A.J. (1997) The economic importance of communal net hunting among the BaAka of the Central African Republic. *Human Ecology* 25, 71–89.

Ogut, H. and Reno, P. (2004) Prevalence of furunculosis in Chinook salmon depends on density of the host exposed by cohabitation. *North American Journal of Aquaculture* 66, 191–197.

Oksbjerg, N., Henckel, P., Andersen, S., Pedersen, B. and Nielsen, B. (2004) Genetic variation of *in vivo* muscle glycerol, glycogen, and pigment in Danish purebred pigs. *Acta Agriculturae Scandinavica* A 54, 187–192.

Oliván, M., Martínez, A., Osoro, K., Sañudo, C., Begoña, P., Oletta, J.L., Campo, M.M., Oliver, M.À., Serra, X., Gil, M. and Piedrafita, J. (2004) Effect of muscular hypertrophy on physico-chemical, biochemical and texture traits of meat from yearling bulls. *Meat Science* 68, 567–575.

Olkowski, A.A. and Classen, H.L. (1998) Progressive bradycardia, a possible factor in the pathogenesis of ascites in fast growing broiler chickens raised at low altitude. *British Poultry Science* 39, 139–146.

Olsen, A.W., Vestergaard, E.-M. and Dybkjær, L. (2000) Roughage as additional rooting substrates for pigs. *Animal Science* 70, 451–456.

Olsson, V., Andersson, K., Hansson, I. and Lundström, K. (2003) Differences in meat quality between organically and conventionally produced pigs. *Meat Science* 64, 287–297.

Omotayo, A.M., Adu, I.F. and Aina, A.B. (1999) The evolving sedentary lifestyle among nomadic pastoralists in the humid zone of Nigeria: implications for land-use policy. *International Journal of Sustainable Development and World Ecology* 6, 220–228.

O'Neill, D.J., Lynch, P.B., Troy, D.J., Buckley, D.J. and Kerry, J.P. (2003a) Influence of the time of year on the incidence of PSE and DFD in Irish pigmeat. *Meat Science* 64, 105–111.

O'Neill, D.J., Lynch, P.B., Troy, D.J., Buckley, D.J. and Kerry, J.P. (2003b) Effects of PSE on the quality of cooked hams. *Meat Science* 64, 113–118.

Önenç, A. and Kaya, A. (2004) The effects of electrical stunning and percussive captive bolt stunning on meat quality of cattle processed by Turkish slaughter procedures. *Meat Science* 66, 809–815.

Ouattara, N.I., Teugels, G.G., N'Douba, V. and Philppart, J-C. (2003) Aquaculture potential of the black-chinned tilapia, *Sarotherodon melanotheron* (Cichlidae). Comparative study of the effect of stocking density on growth performance of landlocked and natural populations under cage culture in Lake Ayame (Côte d'Ivoire). *Aquaculture Research* 34, 1223–1229.

Owen, B.L., Montgomery, J.L., Ramsey, C.B. and Miller, M.F. (2000) Preslaughter resting and hot-fat trimming effects on the incidence of pale, soft and exudative (PSE) pork and ham processing characteristics. *Meat Science* 54, 221–229.

Owens, C.M. and Sams, A.R. (2000) The influence of transportation on turkey meat quality. *Poultry Science* 79, 1204–1207.

Owens, C.M., McKee, S.R., Matthews, N.S. and Sams, A.R. (2000a) The development of pale, exudative meat in two genetic lines of turkeys subjected to heat stress and its prediction by halothane screening. *Poultry Science* 79, 430–435.

Owens, C.M., Hirschler, E.M., McKee, S.R., Martinez-Dawson, R. and Sams, A.R. (2000b) The characterization and incidence of pale, soft, exudative turkey meat in a commercial plant. *Poultry Science* 79, 553–558.

Pajor, E.A., Weary, D.M., Caceres, C., Fraser, D. and Kramer, D.L. (2002) Alternative housing for sows and litters. Part 3. Effects of piglet diet quality and sow-controlled housing on performance and behaviour. *Applied Animal Behaviour Science* 76, 267–277.

Parrott, R.F., Hall, S.G., Lloyd, D.M., Goode, J.A. and Broom, D.M. (1998) Effects of a maximum permissible journey time (31 h) on physiological responses of fleeced and shorn sheep to transport, with observations on behaviour during short (1 h) rest-stop. *Animal Science* 66, 197–207.

Parry, M.A. (1986) The effect of confinement on behaviour and reproductive performance in the sow. *Agricultural Progress* 61, 46–54.

Pascoe, P.J. (1986) Humaneness of electro-immobilization unit for cattle. *American Journal of Veterinary Research* 10, 2252–2256.

Pattamarakha, K., Tanapanyarachwong, J. and Saithanoo, S. (1997) The use of recommended goat husbandry practices by farmers in southern Thailand. *Asian–Australasian Journal of Animal Sciences* 10, 587–592.

Paulick, C., Stolle, F.A. and von Mickwitz, G. (1989) The influence of different stunning methods on the quality of sheep meat. *Fleischwirtschaft* 69, 227–230.

Pearse, A.J. and Drew, K.R. (1998) Ecologically sound management: aspects of modern sustainable deer farming systems. *Acta Veterinaria Hungarica* 46, 315–328.

Peek, C.W., Bell, N.I. and Dunham, C.C. (1996) Gender, gender ideology and animal rights advocacy. *Gender and Society* 10, 464–478.

Pegram, R.G., Tatchell, R.J., de Castro, J.J., Chizyuka, H.G.B., Creek, M.J., McCosker, P.J., Moran, M.C. and Nigarura, G. (1993) Tick control: new concepts. *World Animal Review* 74–75, 2–11.

Pérez, M.P., Palacio, J., Santolaria, M.P., Aceña, M.C., Chacón, G., Gascón, M., Calvo, J.H., Zaragoza, P., Beltran, J.A. and Garćia-Belenguer, S. (2002) Effect of transport time on welfare and meat quality in pigs. *Meat Science* 61, 425–433.

Perry, D. and Thompson, J.M. (2005) The effect of growth rate during backgrounding and finishing on meat quality traits in beef cattle. *Meat Science* 69, 691–702.

Petersen, G.V. and Blackmore, D.K. (1982) The effect of different slaughter methods on the post mortem glycolysis of muscle in lambs. *New Zealand Veterinary Journal* 30, 195–198.

Petracci, M., Fletcher, D.L. and Northcutt, J.K. (2001) The effect of holding temperature on live shrink, processing yield, and breast meat quality of broiler chickens. *Poultry Science* 80, 670–675.

Petty, D.B., Hattingh, J., Ganhao, M.F. and Bezuidenhout, L. (1994) Factors which affect blood variables of slaughtered cattle. *Journal of the South African Veterinary Association* 65, 41–45.

Pfeiffer, W. (1962) The fright reaction of fish. *Biological Reviews* 37, 495–511.

Phillips, P.A., Fraser, D. and Thompson, B.K. (1991) Preference of sows for a partially enclosed farrowing crate. *Applied Animal Behaviour Science* 32, 35–43.

Phillips, P.A., Fraser, D. and Thompson, B.K. (1992) Sow preference for farrowing-crate width. *Canadian Journal of Animal Science* 72, 745–750.

Pirottin, D., Grobert, L., Adamantidis, A., Farnir, F., Herens, C., Schrøder, H.D. and Georges, M. (2005) Transgenic engineering of male-specific muscular hypertrophy. *Proceedings of the National Academy of Sciences* 102, 6413–6418.

Plous, S. (1991) An attitude survey of animal rights activists. *Psychological Science* 2, 194–196.

Plous, S. (1998) Signs of change within the animal rights movement: results from a follow-up survey of activists. *Journal of Comparative Psychology* 112, 48–54.

Pollard, J.C. and Littlejohn, R.P. (1998) Effects of winter housing, exercise, and dietary treatments on the behaviour and welfare of red deer (*Cervus elaphus*) hinds. *Animal Welfare* 7, 45–56.

Pollard, J.C., Littlejohn, R.P. and Suttie, J.M. (1992) Behaviour and weight change of red deer calves during different weaning procedures. *Applied Animal Behaviour Science* 35, 23–33.

Pollard, J.C., Littlejohn, R.P. and Suttie, J.M. (1993) Effects of isolation and mixing of social groups on heart rate and behaviour of red deer stags. *Applied Animal Behaviour Science* 38, 311–322.

Pollard, J.C., Stevenson-Barry, J.M. and Littlejohn, R.P. (1999) Factors affecting behaviour, bruising and pH$_u$ in a deer slaughter premises. *Proceedings of the New Zealand Society of Animal Production* 59, 148–151.

Poppe, T.T. and Breck, O. (1997) Pathology of Atlantic salmon *Salmo salar* intraperitoneally immunized with oil-adjuvanted vaccine. *Diseases of Aquatic Organisms* 29, 219–226.

Porcher, J., Cousson-Gélie, F. and Dantzer, R. (2004) Affective components of the human–animal relationship in animal husbandry: development and validation of a questionnaire. *Psychological Reports* 95, 275–290.

Pordomingo, A.J. (2002) Early weaning and length of supplementation effects on beef calves. *Journal of Range Management* 55, 327–335.

Pottinger, T.G. (2001) Effects of husbandry stress on flesh quality indicators in fish. In: Kestin, S.C. and Warriss, P.D. (eds) *Farmed Fish Quality*. Fishing News Books, Oxford, UK, pp. 145–160.

Potturi, P.V.L., Patterson, J.A. and Applegate, T.J. (2005) Effects of delayed placement on intestinal characteristics in turkey poults. *Poultry Science* 84, 816–824.

Prawl, Z.I. and Owens, F.N. (1998) Activity patterns of feedlot steers. *Animal Science Research Report. Oklahoma Agricultural Experiment Station* P965, 57–63.

Price, S., Sibly, R.M. and Davies, M.H. (1993) Effects of behaviour and handling on heart rate in farmed red deer. *Applied Animal Behaviour Science* 37, 111–123.

Priolo, A., Micol, D., Agabriel, J., Prache, S. and Dransfield, E. (2002) Effect of grass or concentrate feeding systems on lamb carcass and meat quality. *Meat Science* 62, 179–185.

Prunier, A., Mounier, A.M. and Hay, M. (2005) Effects of castration, tooth resection, or tail docking on plasma metabolites and stress hormones in young pigs. *Journal of Animal Science* 83, 216–222.

Pugliese, C., Bozzi, R., Campodoni, G., Acciaioli, A., Franci, O. and Gandinin, G. (2005) Performance of Cinta Senese pigs reared outdoors and indoors: 1. Meat and subcutaneous fat characteristics. *Meat Science* 69, 459–464.

Puit, M.A., Leroy, H.P. and Nsangou, I.N. (2004) Dynamique de la filière viande de brousse dans la partie continentale du Rio Muni en Guinée équatoriale. *Tropicultura* 22, 204–210.

Quigley, D.T.G. (2003) Early maturation in the Irish salmon farming industry. *Bulletin of the Aquaculture Association of Canada* 103, 27–35.

Quiniou, N., Renaudeau, D., Dubois, S. and Noblet, J. (2000) Influence of high ambient temperatures on food intake and feeding behaviour of multiparous lactating sows. *Animal Science* 70, 471–479.

Quinn, A.P., Kettlewell, P.J., Mitchell, M.A. and Knowles, T. (1998) Air movement and thermal microclimate observed in poultry lairages. *British Poultry Science* 39, 469–476.

Rahman, S.A. (2004) Animal welfare: a developing country perspective. In: *Global Conference on Animal Welfare: an OIE Initiative*. Office International des Épizooties, Paris, France, pp. 101–118.

Raj, A.B.M. (1994) An investigation into batch killing of turkeys in their transport containers using mixtures of gases. *Research in Veterinary Science* 56, 325–331.

Raj, A.B.M. and Gregory, N.G. (1995) Welfare implications of the gas stunning of pigs. 1. Determination of aversion to the initial inhalation of carbon dioxide or argon. *Animal Welfare* 4, 273–280.

Raj, A.B.M. and O'Callaghan M. (2004) Effect of amount and frequency of head-only stunning currents on the electroencephalogram and somatosensory evoked potentials in broilers. *Animal Welfare* 13, 159–170.

Raj, A.B.M., Wilkins, L.J., O'Callaghan, M. and Phillips, A.J. (2001) Effect of electrical stun/kill method, interval between killing and neck cutting and blood vessels cut on blood loss and meat quality in broilers. *British Poultry Science* 42, 51–56.

Raj, A.B.M., O'Callaghan, M. and Knowles, T.G. (2006a) The effects of amount and frequency of alternating current used in water bath stunning and of slaughter methods on electroencephalograms in broilers. *Animal Welfare* 15, 7–18.

Raj, A.B.M., O'Callaghan, M. and Hughes, S.I. (2006b) The effects of amount and frequency of pulsed direct current used in water bath stunning and of slaughter methods on spontaneous electroencephalograms in broilers. *Animal Welfare* 15, 19–24.

Raj, A.B.M., O'Callaghan, M. and Hughes, S.I. (2006c) The effects of pulse width of a direct current used in water bath stunning and of slaughter methods on spontaneous electroencephalograms in broilers. *Animal Welfare* 15, 25–30.

Ramírez, J.A., Oliver, M.A., Pla, M., Guerrero, L., Ariño, P., Blasco, A., Pascual, M. and Gil, M. (2004) Effect of selection for growth rate on biochemical, quality and texture characteristics of meat from rabbits. *Meat Science* 67, 617–624.

Randall, J.M. and Bradshaw, R.H. (1998) Vehicle motion and motion sickness in pigs. *Animal Science* 66, 239–245.

Rantzer, D. and Svendsen, J. (2001a) Slatted versus solid floors in the dung area of farrowing pens: effects on hygiene and pig performance, birth to weaning. *Acta Agriculturae Scandinavica* A 51, 167–174.

Rantzer, D. and Svendsen, J. (2001b) Slatted versus solid floors in the dung area: comparison of pig production system (moved versus not moved) and effects on hygiene and pig performance, weaning to four weeks after weaning. *Acta Agriculturae Scandinavica* A 51, 175–183.

Rathore, G.S. (1986) *Camels and their Management*. Indian Council of Agricultural Research, New Delhi, India, 228 pp.

Regan, T. (1983) *The Case for Animal Rights*. University of California Press, Berkeley, California.

Regenstein, J.M. and Grandin, T. (1992) Religious slaughter and animal welfare – an introduction for animal scientists. *Reciprocal Meat Conference Proceedings* 45, 155–159.

Regost, C., Arzel, J., Cardinal, M., Laroche, M. and Kaushik, S.J. (2001) Fat deposition and flesh quality in seawater reared, triploid brown trout (*Salmo trutta*) as affected by dietary fat levels and starvation. *Aquaculture* 193, 325–345.

Reiter, K. (2004) Einfluss der Entfernung zwischen Futtertrog und Trance auf die Bewegungsaktivität und Beinschäden bei Broilern. *Archiv für Geflügelkunde* 68, 98–105.

Renner, P.A., Nestor, K.E. and Havenstein, G.B. (1989) Effects on turkey mortality and body weight of type of beak trimming, age at rimming, and injection of poults with vitamin and electrolytes solution at hatching. *Poultry Science* 68, 369–373.

Rennie, T., Light, D., Rutherford, A., Miller, M., Fisher, I., Pratchett, D., Capper, B., Buck, N. and Trail, J. (1977) Beef cattle productivity under traditional and improved management in Botswana. *Tropical Animal Health and Production* 9, 1–6.

Reverte, D., Xiong, Y.L. and Moody, W.G. (2003) Properties of restructured beef steaks from forage- and grain-fed cattle as affected by antioxidant and flavouring agents. *Meat Science* 65, 539–546.

Reverter, A., Johnston, D.J., Ferguson, D.M., Perry, D., Goddard, M.E., Burrow, H.M., Oddy, V.H., Thompson, J.M. and Bindon, B.M. (2003) Genetic and phenotypic characterisation of animal, carcass, and meat quality traits from temperate and tropically adapted beef breeds. 4. Correlations among animal, carcass, and meat quality traits. *Australian Journal of Agricultural Research* 54, 149–158.

Rhee, K.S., Lupton, C.J., Ziprin, Y.A. and Rhee, K.C. (2003) Effects of sheep production systems on oxidative storage stability of lean lamb patties. *Meat Science* 65, 701–706.

Rhind, S.M., Reid, H.W., McMillen, S.R. and Palmarini, G. (1998) The role of cortisol and ß-endorphin in the response of the immune system to weaning in lambs. *Animal Science* 66, 397–402.

Ridell, J. and Korkeala, H. (1993) Special treatment during slaughtering in Finland of cattle carrying an excessive load of dung: meat hygienic aspects. *Meat Science* 35, 223–228.

Rinaldo, D. and Mourot, J. (2001) Effects of tropical climate and season on growth, chemical composition of muscle and adipose tissue and meat quality in pigs. *Animal Research* 50, 507–521.

Ritz, C.W., Webster, A.B. and Czarick, M. (2005) Evaluation of hot weather thermal environment and incidence of mortality associated with broiler live haul. *Journal of Applied Poultry Research* 14, 594–602.

Rivoal, K., Ragimbeau, C., Salvak, G., Colin, P. and Ermel, G. (2005) Genomic diversity of *Campylobacter coli* and *Campylobacter jejuni* isolates recovered from free-range broiler farms and comparison with isolates of various origins. *Applied and Environmental Microbiology* 71, 6216–6227.

Robb, D.H.F. and Kestin, S.C. (2002) Methods used to kill fish: field observations and literature reviewed. *Animal Welfare* 11, 269–282.

Robb, D.H.F., Wotton, S.B., McKinstry, J.L., Sørensen, N.K. and Kestin, S.C. (2000) Commercial slaughter methods used on Atlantic salmon: determination of the onset of brain failure by electroencephalography. *Veterinary Record* 147, 298–303.

Robb, D.H.F., Kestin, S.C., Warriss, P.D. and Nute, G.R. (2002a) Muscle lipid content determines the eating quality of smoked and cooked Atlantic salmon (*Salmo salar*). *Aquaculture* 205, 345–358.

Robb, D.H.F., O'Callaghan, M., Lines, J.A. and Kestin, S.C. (2002b) Electrical stunning of rainbow trout (*Oncorhynchus mykiss*): factors that affect stun duration. *Aquaculture* 205, 359–371.

Robb, D.H.F., Phillips, A.J. and Kestin, S.C. (2003) Evaluation of methods for determining the prevalence of blood spots in smoked Atlantic salmon and the effect of exsanguinations method on prevalence of blood spots. *Aquaculture* 217, 125–138.

Robinson, D.W. (1969) Preliminary observations on the heat tolerance of shorn and nutritionally depleted sheep in a tropical environment. *British Veterinary Journal* 125, 112–120.

Robinson, J.J. and Ørskov, E.R. (1975) An integral approach to improving the biological efficiency of sheep meat production. *World Review of Animal Production* 11 (3), 63–76.

Roderick, S. (1995) Pastoralist cattle productivity in a tsetse infested area of south west Kenya. Ph.D. thesis, University of Reading, UK, 274 pp.

Roderick, S., Stevenson, P. and Ndungu, J. (1998) The production parameters influencing the composition and structure of pastoral cattle herds in a semi-arid area of Kenya. *Animal Science* 66, 585–594.

Roderick, S., Stevenson, P. and Ndung'u, J. (1999) Factors influencing the production of milk from pastoral cattle herds in Kenya. *Animal Science* 68, 201–209.

Rose, J.D. (2002). The neurobehavioural nature of fishes and the question of awareness and pain. *Reviews in Fisheries Science* 10, 1–38.

Rosenvold, K. and Andersen, H.J. (2003) The significance of pre-slaughter stress and diet on colour and colour stability of pork. *Meat Science* 63, 199–209.

Rosenvold, K., Lærk, H.N., Jensen, S.K., Karlsson, A.H., Lundström, K. and Andersen, H.J. (2001) Strategic finishing feeding as a tool in the control of pork quality. *Meat Science* 59, 397–406.

Rostagno, M.H., Hurd, H.S., McKean, J.D., Zimer, C.J., Gailey, J.K. and Leite, R.C. (2003) Preslaughter holding environment in pork plants is highly contaminated with *Salmonella enterica*. *Applied and Environmental Microbiology* 69, 4489–4494.

Roth, B., Moeller, D., Veland, J.O., Imsland, A. and Slinde, E. (2002) The effect of stunning methods on rigor mortis and texture properties of Atlantic salmon (*Salmo salar*). *Journal of Food Science* 67, 1462–1466.

Roth, B., Moeller, D. and Slinde, E. (2004) Ability of electric field strength, frequency, and current duration to stun farmed Atlantic salmon and Pollack and relations to observed injuries using sinusoidal and square wave alternating current. *North American Journal of Aquaculture* 66, 208–216.

Roth, J.A. (1984) Immunosuppression and immunomodulation in bovine respiratory disease. In: Loan, R.W. (ed.) *Bovine Respiratory Disease – a Symposium.* Texas A&M University Press, College Station, Texas, pp. 20–49.

Rothfuß, U., Müller, E. and Grashorn, M. (1984) Einfluß unterschiedlicher Betäubungsspannungen auf Parameter des Stoffwechsels und der Fleischaffenheit beim Schwein. *Fleischwirtschaft* 64, 833–837.

Rowland, D., Bobbitt, J., Phillips, M., Haines, H., Whitehouse, J., Barlow, S., Isaac, J., Kondekas, N., Isgro, D. and Coates, K. (2000) Preparation of clean livestock for slaughter: effect of cleaning practices on the microbiological quality of beef carcasses. In: *Proceedings of the 46th International Congress of Meat Science and Technology*, ICoMST, Buenos Aires, pp. 678–679.

Ruff, N., Fitzgerald, R.D., Cross, T.F., Teurtrie, G. and Kerry, J.P. (2002) Slaughtering method and dietary α-tocopherol acetate supplementation affect rigor mortis and fillet shelf-life of turbot *Scophthalmus maximus* L. *Aquaculture Research* 33, 703–714.

Rushen, J. (1986) Aversion of sheep to electro-immobilization and physical restraint. *Applied Animal Behaviour Science* 15, 315.

Russell, S.M. (2003) The effect of airsacculitis on bird weights, uniformity, fecal contamination, processing errors, and populations of *Campylobacter* spp and *Escherichia coli*. *Poultry Science* 82, 1326–1331.

Sadler, J., Pankhurst, P.M. and King, H.R. (2001) High prevalence of skeletal deformity and reduced gill surface area in triploid Atlantic salmon. *Aquaculture* 198, 369–386.

Sambraus, H.H. (1998) Normal and disturbed behaviour of ostriches. In: *Proceedings of 8th World Conference on Animal Production* 1, WAAP, Seoul, pp. 270–271.

Sánchez, J.P., Baselga, M., Pieró, R. and Silvestre, M.A. (2004) Analysis of factors influencing longevity of rabbit does. *Livestock Production Science* 90, 227–234.

Sandercock, D.A., Hunter, R.R., Nute, G.C., Mitchell, M.A. and Hocking, P.M. (2001) Acute heat stress-induced alterations in blood acid–base status and skeletal muscle membrane integrity in broiler chickens at two ages: implications for meat quality. *Poultry Science* 80, 418–425.

Santé, V., Le Pottier, G., Astruc, T., Mouchonière, M. and Fernandez, X. (2000) Effect of stunning current frequency on carcass downgrading and meat quality of turkey. *Poultry Science* 79, 1208–1214.

Sañudo, C., Santolaria, M.P., María, G., Osorio, M. and Sierra, I. (1996) Influence of carcass weight on instrumental and sensory lamb meat quality in intensive production systems. *Meat Science* 42, 195–202.

Sather, A.P., Jeremiah, L.E. and Squires, E.J. (1999) Effects of castration on live performance, carcass yield, and meat quality of male pigs fed wheat or corn based diets. *Journal of Muscle Foods* 10, 245–259.

Satterlee, D.G., Parker, L.H., Castille, S.A., Cadd, G.G. and Jones, R.B. (2000) Struggling behaviour in shackled male and female broiler chickens. *Poultry Science* 79, 652–655.

Savenije, B., Schreurs, F.J.G., Winkelman-Goedhart, H.A., Gerritzen, M.A., Korf, J. and Lambooij, E. (2002) Effects of feed deprivation and electrical, gas, and captive needle stunning on early post-mortem muscle metabolism and subsequent meat quality. *Poultry Science* 81, 561–571.

Savolainen, R., Ruohonen, K. and Railo, E. (2004) Effect of stocking density on growth, survival and cheliped injuries of stage 2 juvenile signal crayfish *Pasifastacus leniusculus* Dana. *Aquaculture* 231, 237–248.

Savory, C.J. and Lariviere, J.-M. (2000) Effects of qualitative and quantitative food restriction treatments on feeding motivational state and general activity level of growing broiler breeders. *Applied Animal Behaviour Science* 69, 135–147.

Savory, C.J., Carlisle, A., Maxwell, M.H., Mitchell, M.A. and Robertson, G.W. (1993) Stress, arousal and opioid peptide-like immunoreactivity in restricted and ad lib fed broiler breeder fowls. *Comparative Biochemistry and Physiology* 106A, 587–594.

Schillhorn van Veen, T.W. (1997) Sense or nonsense? Traditional methods of animal parasitic disease control. *Veterinary Parasitology* 71, 177–194.

Schlink, K. (1995) Naturalistic inquiry of hot-iron jaw branding: what do industry leaders believe consumers think? *Beef Cattle Research in Texas* 1995, 30–35.

Schmidt, T.B., Unruh, L., Perino, L.J. and Montgomery, T.H. (2000) Hail bruising of fed cattle. *Bovine Practitioner* 34, 60–62.

Schouten, J.M., van de Giessen, A.W., Frankena, K., de Jong, M.C.M. and Graat, E.A.M. (2005) *Escherichia coli* O157 prevalence in Dutch poultry, pig finishing and veal herds and risk factors in Dutch veal herds. *Preventive Veterinary Medicine* 70, 1–15.

Schwartzkopf-Genswein, K.S., Stookey, J.M. and Welford, R. (1997) Behavior of cattle during hot-iron and freeze branding and the effects on subsequent handling ease. *Journal of Animal Science* 75, 2064–2072.

Scott, G.B. and Kelly, M. (1989) Cattle cleanliness in different housing systems. *Farm Building Progress* 95 (Jan.), 21–24.

Sebei, P.J., McCrindle, C.M.E. and Webb, E.C. (2004) An economic analysis of communal goat production. *Journal of the South African Veterinary Association* 75, 19–23.

Selwyn, P. and Hathaway, S. (1990). A study of the prevalence and economic significance of diseases and defects of slaughtered farmed deer. *New Zealand Veterinary Journal* 38, 94–97.

Sen, A.R., Santra, A. and Karim, S.A. (2004) Carcass yield, composition and meat quality attributes of sheep and goat under semiarid conditions. *Meat Science* 66, 757–763.

Serenius, T. and Stalder, K.J. (2004) Genetics of length of productive life and lifetime prolificacy in the Finnish Landrace and Large White pig populations. *Journal of Animal Science* 82, 3111–3117.

Serenius, T., Sevon-Aimónen, M.-L., Kause, A., Mäntysaari, E.A. and Mäki-Tanila, A. (2004) Genetic associations of prolificacy with performance, carcass, meat quality, and leg conformation traits in the Finnish Landrace and Large White pig populations. *Journal of Animal Science* 82, 2301–2306.

Shaw, F.D., Weidemann, J.F. and Baxter, R.I. (1971) Vaso-active drugs and the occurrence of intramuscular ecchymotic haemorrhages in the electrically stunned rat. *Research in Veterinary Science* 12, 480–483.

Shelton, M. (1964) Relation of environmental temperature during gestation to birth weight and mortality of lambs. *Journal of Animal Science* 23, 360–364.

Shima, T., Suzuki, N., Yamamoto, T. and Furuita, H. (2001) A comparative study of self-feeder and automatic feeder: effects on the growth performance of rainbow trout fry. *Aquaculture Research* 32 (Suppl. 1), 142–146.

Short, R.E., MacNeil, M.D., Grosz, M.D., Gerrard, D.E. and Grings, E.E. (2002) Pleitropic effects in Hereford, Limousin and Piedmontese F_2 crossbred calves of genes controlling muscularity including the Piedmontese myostatin allele. *Journal of Animal Science* 80, 1–11.

Shorthose, W.R. and Shaw, F.D. (1977) Plasma constituents of 'Downer' sheep slaughtered at an abattoir. *Australian Veterinary Journal* 53, 330–333.

Sillence, M.N., Matthews, M.L., Badran, T.W. and Pegg, G.G. (2000) Effects of clenbuterol on growth in underfed cattle. *Australian Journal of Agricultural Research* 51, 401–406.

Simmins, P.H. (1993) Reproductive performance of sows entering stable and dynamic groups after mating. *Animal Production* 57, 293–298.

Singer, P. (1991) *Animal Liberation*, 2nd edn. Thorsons, London, UK, 320 pp.

Singh, R. and Trehan, P.K. (1994) Genetic and phenotypic parameters of body and muscle weights and abdominal fat in meat-type chicken. *Indian Journal of Animal Sciences* 64, 388–392.

Skinner-Noble, D.O., Jones, R.B. and Teeter, R.G. (2003) Components of feed efficiency in broiler breeding stock: is improved feed conversion associated with increased docility and lethargy in broilers? *Poultry Science* 82, 532–537.

Skjervold, P.O., Fjæra, S.O. and Østby, P.B. (1999) Rigor in Atlantic salmon as affected by crowding stress prior to chilling before slaughter. *Aquaculture* 175, 93–101.

Šloss, V. and Dufty, J.H. (1980) In: *Handbook of Bovine Obstetrics*. Williams and Wilkins, Baltimore, USA, p. 99.

Smidt, D. and Huth, F.W. (1979) Survey of the incidence of calving problems, calf mortality, and their economic importance: dairy and dual-purpose cattle. In: Hoffman, B., Mason, I.L. and Schmidt, J. (eds) *Calving Problems and Early Viability of the Calf*. Martinus Nijhoff Publishers, The Hague, the Netherlands, pp. 3–29.

Smith, C.A. (1959) Studies on the Northern Rhodesia *Hyparrhenia* veld. Part 1. The grazing behaviour of indigenous cattle grazed at light and heavy stocking rates. *Journal of Agricultural Science* 52, 369–375.

Smith, D.P., Lyon, C.E. and Lyon, B.G. (2002) The effect of age, dietary carbohydrate source, and feed withdrawal on broiler breast fillet color. *Poultry Science* 81, 1584–1588.

Smith, W.C., Ellis, M., Clark, J.B.K. and Innes, N. (1983) A comparison of boars and castrates for bacon manufacture. *Animal Production* 37, 17–23.

Sneddon, L.U., Braithwaite, V.A. and Gentle, M.J. (2002) Do fishes have nociceptors? Evidence for the evolution of a vertebrate sensory system. *Proceedings of the Royal Society* B 270, 1115–1121.

Solaiman, M. and Azad, A.K. (1995) Poultry rearing in Comilla. *Journal of Rural Development* 25 (2), 17–36.

Sonesson, A.K., de Greef, K.H. and Meuwissen, T.H.E. (1998) Genetic parameters and trends of meat quality, carcass composition and performance traits in two selected lines of Large White pigs. *Livestock Production Science* 57, 23–32.

Southwell, C.J. and Pickles, G.S. (1999) Abundance, distribution and rate of increase of feral goats in western Australia. *Rangeland Journal* 15, 334–338.

Sowińska, J., Brzostowski, H., Tański, Z. and Czaja, K. (2001) The weaning stress response in lambs of different age. *Czech Journal of Animal Science* 46, 465–468.

Spinu, M., Benveneste, S. and Degen, A.A. (2003) Effect of density and season on stress and behaviour in broiler breeder hens. *British Poultry Science* 44, 170–174.

Splan, R.K., Cundiff, L.V. and van Vleck, L.D. (1998) Genetic parameters for sex-specific traits in beef cattle. *Journal of Animal Science* 76, 2272–2278.

Spoolder, H.A.M., Burbidge, J.A., Edwards, S.A., Simmins, P.H. and Lawrence, A.B. (1995) Provision of straw as a foraging substrate reduces the development of excessive chain chewing and bar manipulation in food restricted sows. *Applied Animal Behaviour Science* 43, 249–262.

Spoolder, H.A.M., Edwards, S.A. and Corning, S. (1999) Effects of group size and feeder space on welfare in finishing pigs. *Animal Science* 69, 481–489.

Squires, V.R. and Sidahmed, A. (1997) Livestock management in dryland pastoral systems: prospects and problems. *Annals of Arid Zone* 36, 79–96.

Stafford, K.J. and Mellor, D.J. (2005) Dehorning and disbudding distress and its alleviation in calves. *Veterinary Journal* 169, 337–349.

Stafford, K.J., Mellor, D.J., Todd, S.E., Bruce, R.A. and Ward, R.N. (2002) Effects of local anaesthesia or local anaesthesia plus a non-steroidal anti-inflammatory drug on the acute pain response of calves to five different methods of castration. *Research in Veterinary Science* 73, 61–70.

Stalder, K.J., Maya, J., Christian, L.L., Moeller, S.J. and Prusa, K.J. (1998) Effects of preslaughter management on the quality of carcasses from Porcine Stress Syndrome heterozygous market hogs. *Journal of Animal Science* 76, 2435–2443.

Steelman, C.D., White, T.W. and Schilling, P.E. (1972) Effects of mosquitoes on the average daily gain of feedlot steers in southern Louisiana. *Journal of Economic Entomology* 65, 462–466.

Stern, S., Heyer, A., Andersson, H.K., Rydhmer, L. and Lundstrom, K. (2003) Production results and technological meat quality for pigs in indoor and outdoor rearing systems. *Acta Agriculturae Scandinavica* A 53, 166–174.

Støier, S., Aaslyng, M.D., Olsen, E.V. and Henckel, P. (2001) The effect of stress during lairage and stunning on muscle metabolism and drip loss in Danish pork. *Meat Science* 59, 127–131.

Stokka, G.L., Lechtenberg, K., Edwards, T., MacGregor, S., Voss, K., Griffin, D., Grotelueschen, D.M., Smith, R.A. and Perino, L.J. (2001) Lameness in feedlot cattle. *Veterinary Clinics of North America: Food Animal Practice* 17, 189–202.

Strydom, P.E., Naude, R.T., Smith, M.F., Scholtz, M.M. and van Wyk, J.B. (2000) Characterisation of indigenous African cattle breeds in relation to meat quality traits. *Meat Science* 55, 79–88.

Suguna, C., Reddy, M.S. and Reddy, G.R. (1997) Effect of electrical stimulation on quality characteristics of spent hen carcasses. *Indian Journal of Poultry Science* 32, 277–281.

Sun, X., McElroy, A., Webb, K,E., Sefton, A.E. and Novak, C. (2005) Broiler performance and intestinal alterations when fed drug-free diets. *Poultry Science* 84, 1294–1302.

Sutherland, M.M. and Ames, J.M. (1995) The effect of castration on headspace aroma components of cooked lamb. *Journal of the Science of Food and Agriculture* 69, 403–413.

Sutton, G.D., Fourie, P.D. and Retief, J.S. (1967) The behaviour of cattle in transit by rail. *Journal of the South African Veterinary Medical Association* 38, 153–156.

Svendsen, J., Bengtsson, A.C. and Svendsen, L.S. (1986) Occurrence and causes of traumatic injuries in neonatal pigs. *Pig News and Information* 7, 159–170.

Swallow, B.M., Brokken, R.F., Motsamai, M., Sopeng, L. and Storey, G.G. (1987) *Livestock Development and Range Utilization in Lesotho*. Research Division Report RD-R-82, Ministry of Agriculture, Maseru, Lesotho, 83 pp.

Swan, J.E. and Boles, J.A. (2002) Processing characteristics of beef roasts made from high and normal pH bull inside rounds. *Meat Science* 62, 399–403.

Swanenburg, M., Urlings, H.A.P., Keuzenkamp, D.A. and Snijders, J.M.A. (2001) *Salmonella* in the lairage of pig slaughterhouses. *Journal of Food Protection* 64, 12–16.

Tanida, H., Miura, A., Tanaka, T. and Yoshimoto, T. (1996) Behavioral responses of piglets to darkness and shadows. *Applied Animal Behaviour Science* 49, 173–183.

Tanner, J.C., Holden, S.J., Owen, E., Winugroho, M. and Gill, M. (2001) Livestock sustaining intensive smallholder crop production through traditional feeding practices for generating high quality manure-compost in upland Java. *Agriculture, Ecosystems and Environment* 84, 21–30.

Tarrant, P.V. (1990) Transportation of cattle by road. *Applied Animal Behaviour Science* 28, 153–170.

Taylor, A.A. and Weary, D.M. (2000) Vocal responses of piglets to castration: identifying procedural sources of pain. *Applied Animal Behaviour Science* 70, 17–26.

Terosky, T.L., Wilson, L.L., Stull, C.L. and Stricklin, W.R. (1997) Effects of individual housing design and size on special-fed Holstein veal calf growth performance, hematology, and carcass characteristics. *Journal of Animal Science* 75, 1697–1703.

Thompson, M., Rideout, K., Trenholm, R. and Gillet, B. (2002) Evaluation of flesh quality in ranched cod. *Bulletin of the Aquaculture Society of Canada* 102 (1), 27–30.

Tigner, J.R. and Larson, G.E. (1977) Sheep losses on selected ranches in Southern Wyoming. *Journal of Range Management* 30, 244–252.

Tillyard, R.J. and Seddon, H.R. (1933) *The Sheep Blowfly Problem in Australia*. Pamphlet 37, CSIR, Canberra, Australia, 136 pp.

Ting, S.T.L., Earley, B., Veissier, I., Gupta, S. and Crowe, M.A. (2005) Effects of age of Holstein–Friesian calves on plasma cortisol, acute-phase proteins, immunological function, scrotal measurements and growth in responses to Burdizzo castration. *Animal Science* 80, 377–386.

Troeger, K. (1990) Schlachten: Tierschutz und Fleischqualität. *Fleischwirtschaft* 70, 266–272.

Troeger, K. and Woltersdorf, W. (1988) Optimierte Elektrobetäubungsverfahren für Schlachtschweine. *Mitteilungsblatt BAFF Kulmbach* 1.6 (100), 7909–7917.

Troeger, K. and Woltersdorf, W. (1991) Gas anaesthesia of slaughter pigs. 1. Stunning experiments under laboratory conditions with fat pigs of known halothane reaction type: meat quality, animal protection. *Fleischwirtschaft* 71, 1063–1068.

Uhlhorn, H., Dalin, G., Lundeheim, N. and Ekman, S. (1995) Osteochondrosis in wild boar–Swedish Yorkshire crossbred pigs (F2 generation). *Acta Veterinaria Scandinavica* 36, 41–53.

Urbinati, E.C. and Carneiro, P.C.F. (2001) Metabolic and hormonal responses of matrinxã (*Brycon cephalus*) to transport stress under influence of benzocaine. *Journal of Aquaculture in the Tropics* 16, 75–85.

Urdaneta-Rincon, M. and Leeson, S. (2002) Quantitative and qualitative feed restriction on growth characteristics of male broiler chickens. *Poultry Science* 81, 679–688.

USDA (2006) www.fsis.usda.gov/OPHS/adrscadx

van der Wal, P.G., Engel, B. and Reimert, H.G.M. (1999) The effect of stress, applied immediately before stunning, on pork quality. *Meat Science* 53, 101–106.

van de Water, G., Verjans, F. and Geers, R. (2003) The effect of short distance transport under commercial conditions on the physiology of slaughter calves; pH and colour profiles of veal. *Livestock Production Science* 82, 171–179.

Van Donkersgoed, J., Jericho, K.W.F., Grogan, H. and Thorlakson, B. (1997) Preslaughter hide status of cattle and the microbiology of the carcasses. *Journal of Food Protection* 60, 1502–1508.

Van Dresser, W.R. and Wilcke, J.R. (1989) Drug residues in food animals. *Journal of the American Veterinary Medical Association* 194, 1700–1710.

Van Putten, G. and Elshof, J. (1978) Observations of the effects of transport on the well being and lean quality of slaughter pigs. *Animal Regulation Studies* 1, 247–271.

Varley, M.A., Peaker, R.E. and Atkinson, T. (1984) Effect of lactation length of the sow on plasma progesterone, oestradiol 17-ß and embryonic survival. *Animal Production* 38, 113–119.

Veerkamp, C.H. (1978) The influence of fasting and transport on yields of broilers. *Poultry Science* 57, 634–638.

Velarde, A., Gispert, M., Faucitano, L., Manteca, X. and Diestre, A. (2000) The effect of stunning method on the incidence of PSE meat and haemorrhages in pork carcasses. *Meat Science* 55, 309–314.

Velarde, A., Gispert, M., Faucitano, L., Alonso, P., Manteca, X. and Diestre, A. (2001) Effects of the stunning procedure and the halothane genotype on meat quality and incidence of haemorrhages in pigs. *Meat Science* 58, 313–319.

Verbeke, W., van Oeckel, M.J., Warnants, N., Viaene, J. and Boucqué, C.V. (1999) Consumer perception, facts and possibilities to improve acceptability of health and sensory characteristics of pork. *Meat Science* 53, 77–99.

Vergara, H. and Gallego, L. (1999) Effect of type of suckling and length of lactation period on carcass and meat quality in intensive lamb production systems. *Meat Science* 53, 211–215.

Vergara, H. and Gallego, L. (2000) Effect of electrical stunning on meat quality of lamb. *Meat Science* 56, 345–349.

Vergara, H., Linares, M.B., Berrunga, M.I. and Gallego, L. (2005) Meat quality in suckling lambs: effect of pre-slaughter handling. *Meat Science* 69, 473–478.

Vestergaard, E.-M. and Danielsen, V. (1998) Dietary fibre for sows: effects of large amounts of soluble and insoluble fibres in the pregnancy period on the performance of sows during three reproductive cycles. *Animal Science* 68, 355–362.

Vestergaard, M., Therkildsen, M., Henckel, P., Jensen, L.R., Andersen, H.R. and Sejrsen, K. (2000) Influence of feeding intensity, grazing and finishing feeding on meat and eating quality of young bulls and the relationship between muscle fibre characteristics, fibre fragmentation and meat tenderness. *Meat Science* 54, 187–195.

Vieville-Thomas, R.K. and Signoret, J.P. (1992) Pheromonal transmission of aversive substances in domestic pigs. *Journal of Chemical Endocrinology* 18, 1551.

Villé, H., Bertels, S., Geers, R., Janssens, S., Goedseels, V., Parduyns, G., van Bael, J., Goossens, K., Bosschaerts, L., de Ley, J. and Heylen, L. (1993) Electrocardiogram parameters of piglets during housing, handling and transport. *Animal Production* 56, 211–216.

Vipond, J.E., Fitzsimons, J., Robb, J. and Horgan, G. (1992) Effects of fish-meal supplementation and shearing in store/finishing systems for lambs. *Animal Production* 55, 79–87.

von Zweigbergk, A.-J., Lunström, K. and Hansson, I. (1989) The incidence of high internal reflectance 45 minutes pm with different stunning methods. In: *Proceedings of 35th International Congress of Meat Science and Technology*, ICoMST, Copenhagen, pp. 1145–1148.

Waghorn, G.C., Gregory, N.G., Todd, S.E. and Wesselink, R. (1999) Dags in sheep: a look at faeces and reasons for dag formation. *Proceeding of New Zealand Grassland Association* 61, 43–49.

Wahra, G.N., Haider, Z. and Shahnaz, F. (1999) *Bangladesh Disaster Report 1998*. Disaster Forum, Dhaka, Bangladesh, 252 pp.

Wall, A.J. (2001) Ethical considerations in the handling and slaughter of farmed fish. In: Kestin, S.C. and Warriss, P.D. (eds) *Farmed Fish Quality*. Fishing News Books, Oxford, UK, pp. 108–115.

Wallace, D.S., Bairden, K., Duncan, J.L., Eckersall, P.D., Fishwick, G., Holmes, P.H., McKellar, Q.A., Mitchell, S., Murray, M., Parkins, J.J. and Stear, M.J. (1999) The influence of increased feeding on the susceptibility of sheep to infection with *Haemonchus contortus*. *Animal Science* 69, 457–463.

Wallace, J.S., Stanley, K.N. and Jones, K. (1998) The colonization of turkeys by thermophilic campylobacters. *Journal of Applied Microbiology* 85, 224–230.

Warnock, J.P., Caple, J.W., Halpin, C.G. and McQueen, C.S. (1978) Metabolic changes associated with the 'downer' condition in dairy cows at abattoirs. *Australian Veterinary Journal* 54, 566–569.

Warren, J.T., Mysterud, I. and Lynnebakken, T. (2001) Mortality of lambs in free-ranging domestic sheep (*Ovis aries*) in northern Norway. *Journal of Zoology* 254, 195–202.

Warriss, P.D. and Leach, T.M. (1978) The influence of slaughter method on the residual blood content of meat. *Journal of the Science of Food and Agriculture* 29, 608–610.

Warriss, P.D., Brown, S. and Adams, S.J.M. (1994) Relationship between subjective and objective assessment of stress at slaughter and meat quality in pigs. *Meat Science* 38, 329–340.

Warriss, P.D., Brown, S.N., Barton Gade, P., Santos, C., Costa, L.N., Lambooij, E. and Geers, R. (1998a) An analysis of data relating to pig carcass quality and indices of stress collected in the European Union. *Meat Science* 49, 137–144.

Warriss, P.D., Brown, S.N., Knowles, T.G., Edwards, J.E., Kettlewell, P.J. and Guise, H.J. (1998b) The effect of stocking density in transit on the carcass quality and welfare of slaughter pigs. 2. Results from the analysis of blood and meat samples. *Meat Science* 50, 447–456.

Warriss, P.D., Brown, S.N., Edwards, J.E. and Knowles, T.G. (1998c) Effect of lairage time on levels of stress and meat quality in pigs. *Animal Science* 66, 255–261.

Warriss, P.D., Wilkins, L.J., Brown, S.N., Phillips, A.J. and Allen, V. (2004) Defecation and weight of the gastro-intestinal tract contents after feed and water withdrawal in broilers. *British Poultry Science* 45, 61–66.

Warriss, P.D., Pagazaurtundua, A. and Brown, S.N. (2005) Relationship between maximum daily temperature and mortality of broiler chickens during transport and lairage. *British Poultry Science* 46, 647–651.

Wassmuth, R., Wallbaum, F. and Langholz, H.-J. (1999) Outdoor wintering of suckler cows in low mountain ranges. *Livestock Production Science* 61, 193–200.

Weary, D.M., Pajor, E.A., Fraser, D. and Honkanen, A.M. (1996) Sow body movements that crush piglets: a comparison between two types of farrowing accommodation. *Applied Animal Behaviour Science* 49, 149–158.

Webster, A.B. and Fletcher, D.L. (2001) Reactions of laying hens and broilers to different gases used for stunning poultry. *Poultry Science* 80, 1371–1377.

Webster, J.R., Corson, I.D. and Suttie, J.M. (1997) The effect of housing and food restriction during winter on growth of male red deer calves. *Animal Science* 64, 171–176.

Weeks, C.A., Danbury, T.D., Davies, H.C., Hunt, P. and Kestin, S.C. (2000) The behaviour of broiler chickens and its modification by lameness. *Applied Animal Behaviour Science* 67, 111–125.

Weeks, C.A., McNally, P.W. and Warriss, P.D. (2002) Influence of the design of facilities at auction markets and animal handling procedures on bruising in cattle. *Veterinary Record* 150, 743–748.

Wegener, H., Hald, T., Wong, D.L.F., Madsen, M., Korsgaard, H., Bager, F., Gerner-Smidt, P. and Mølbak, K. (2003) Salmonella control programs in Denmark. *Emerging Infectious Diseases* 9, 774–780.

Welchman, D.deB. and Baust, G.N. (1987) A survey of abomasal ulceration in veal calves. *Veterinary Record* 121, 586–590.

Wesley, I.V., Muraoka, W.T., Trampel, D.W. and Hurd, H.S. (2005) Effect of preslaughter events on prevalence of *Campylobacter jejuni* and *Campylobacter coli* in market-weight turkeys. *Applied and Environmental Microbiology* 71, 2824–2831.

Westervelt, R.G. and Stouffer, J.R. (1978) Relationship among spinal cord severing, electrical stimulation and post-mortem quality characteristics of the porcine carcass. *Journal of Animal Science* 46, 1206–1211.

White, R. (1993) *Livestock Development and Pastoral Production on Communal Rangeland in Botswana.* The Botswana Society, Gaborone, Botswana, 72 pp.

White, R.G., DeShazer, J.A., Tressler, C.J., Borcher, G.M., Davey, S., Waninge, A., Parkhurst, A.M., Milanuk, M.J. and Clems, E.T. (1995) Vocalizations and physiological response of pigs during castration with and without anesthetic. *Journal of Animal Science* 73, 381–386.

Whitmore, J.S. (2000) *Drought Management on Farm Land.* Kluwer Academic Publishers, Dordrecht, the Netherlands, 360 pp.

Whittaker, X., Spoolder, H.A.M., Edwards, S.A., Lawrence, A.B. and Corning, S. (1998) The influence of dietary fibre and the provision of straw on the development of stereotypic behaviour in food restricted pregnant sows. *Applied Animal Behaviour Science* 61, 89–102.

Whittington, F.M., Nute, G.R., Hughes, S.I., McGivan, J.D., Lean, I.J., Wood, J.D. and Doran, E. (2004) Relationship between skatole and androstenone accumulation and cytochrome P4502E1 expression in Meishan × Large White pigs. *Meat Science* 67, 569–576.

Whyte, P., Collins, J.D., McGill, K. and O'Mahony, H. (2001) The effect of transportation stress on excretion rates of campylobacters in market-age broilers. *Poultry Science* 80, 817–820.

Wiener, G., Lee, G.J. and Woolliams, J.A. (1992) Effects of rapid inbreeding and of crossing of inbred lines on the body weight growth of sheep. *Animal Production* 55, 89–99.

Wiepkema, P.R., van Hellemond, K.K., Roessingh, P. and Romberg, H. (1987) Behaviour and abomasal damage in individual veal calves. *Applied Animal Behaviour Science* 18, 257–268.

Wiklund, E., Malmfors, G. and Lundström, K. (1997) The effects of pre-slaughter selection of reindeer bulls (*Rangifer tarandus tarandus* L.) on technological and sensory meat quality, blood metabolites and abomasal lesions. *Rangifer* 17, 65–72.

Wiklund, E., Sampels, S., Manley, T.R., Pickova, J. and Littlejohn, R.P. (2006) Effects of feeding regimen and chilled storage on water-holding capacity, colour stability, pigment content and oxidation in red deer (*Cervus elaphus*) meat. *Journal of the Science of Food and Agriculture* 86, 98–106.

Wikner, I., Gebresenbet, G. and Tolo, E. (2003) Dynamic performances of cattle transporting vehicle on Scandinavian roads and behavioural response of animals. *Deutsche Tierärztliche Wochenschrift* 110, 114–120.

Wikse, S.E., Herd, D.B., Field, R.W., Holland, P.S., McGrann, J.M. and Thompson, J.A. (1994) *Beef Cattle Research in Texas* 1994 (PR 5256), 14–23.

Wilkie, D.S. and Carpenter, J.F. (1999) Bushmeat hunting in the Congo basin: an assessment of impacts and options for mitigation. *Biodiversity and Conservation* 8, 927–955.

Wilkins, L.J. and Wotton, S.B. (2002) Effect of frequency of the stunning current waveform on carcase and meat quality of turkeys processed in a commercial plant in the UK. *British Poultry Science* 43, 231–237.

Wilkins, L.J., Gregory, N.G., Wotton, S.B. and Parkman, I.D. (1998) Effectiveness of electrical stunning applied using a variety of waveform–frequency combinations and consequences for carcase quality in broiler chickens. *British Poultry Science* 39, 511–518.

Willis, W.L., Murray, C. and Raczkowski, C.W. (1996) The influence of feed and water withdrawal on *Campylobacter jejuni* detection and yield of broilers. *Journal of Applied Poultry Research* 5, 210–214.

Wilson, R.T. (1978) Studies on the livestock of southern Darfur, Sudan. 5. Notes on camels. *Tropical Animal Health and Production* 10, 19–25.

Wilson, R.T. (1983) Livestock production in central Mali. The Macina wool sheep of the Niger inundation zone. *Tropical Animal Health and Production* 15, 17–31.

Wilson, R.T., Traore, A., Kuit, H.G. and Slingerland, M. (1987) Livestock production in central Mali: reproduction, growth and mortality of domestic fowl under traditional management. *Tropical Animal Health and Production* 19, 229–236.

Woelfel, R.L., Owens, C.M., Hirschler, E.M., Martinez-Dawson, R. and Sams, A.R. (2002) The characterization and incidence of pale, soft, and exudative broiler meat in a commercial processing plant. *Poultry Science* 81, 579–584.

Woldemeskel, M. and Gumi, B. (2001) A study on the diseases of one-humped camels (*Camelus dromedaries*) in Borana, southern Ethiopia. *Bulletin of Animal Health and Production in Africa* 49, 179–182.

Wood, C.M. (2004) Dogmas and controversies in the handling of nitrogenous wastes. Is exogenous ammonia a growth stimulant in fish? *Journal of Experimental Biology* 207, 2043–2054.

Woolaston, R.R. (1993) Factors affecting the prevalence and severity of footrot in a Merino flock selected for resistance to *Haemonchus contortus*. *Australian Veterinary Journal* 70, 365–369.

Woolums, A.R., McAllister, T.A., Loneragan, G.H. and Gould, D.H. (2001) Etiology of acute interstitial pneumonia in feedlot cattle: non-infectious causes. *Compendium of Continuing Education* 23, S86–S93.

Worobec, E.K., Duncan, I.J.H. and Widowski, T.M. (1999) The effects of weaning at 7, 14 and 28 days on piglet behaviour. *Applied Animal Behaviour Science* 67, 111–125.

Wotton, S.B. and Wilkins, L.J. (1999) Effect of very low pulsed direct currents at high frequency on the return of neck tension in broilers. *Veterinary Record* 145, 393–396.

Wotton, S.B., Gregory, N.G., Whittington, P.E. and Parkman, I.D. (2000) Electrical stunning of cattle. *Veterinary Record* 147, 681–684.

Wulf, D.M., Tatum, J.D., Green, R.D., Morgan, J.B., Golden, B.L. and Smith, G.C. (1996) Genetic influences on beef longissimus palatability in Charolais- and Limousin-sired steers. *Journal of Animal Science* 74, 2394–2405.

Yahav, S., Straschnow, A., Vax, E., Razpakovski, V. and Shinder, D. (2001) Air velocity alters broiler performance under harsh environmental conditions. *Poultry Science* 80, 724–726.

Yakoob, A., Holmes, P.H. and Armour, J. (1983) Pathophysiology of gastrointestinal trichostrongyles in sheep: plasma loss and changes in plasma pepsinogen levels associated with parasite challenge of immune animals. *Research in Veterinary Science* 34, 305–309.

Yalcin, S., Özkan, S., Türkmut, L. and Siegel, P.B. (2001) Responses to heat stress in commercial and local broiler stocks. 1. Performance traits. *British Poultry Science* 42, 149–152.

Young, J.F., Stagsted, J., Jensen, S.K., Karlsson, A.H. and Henckel, P. (2003) Ascorbic acid, α-tocopherol, and oregano supplements reduce stress-induced deterioration of chicken meat quality. *Poultry Science* 82, 1343–1351.

Young, O.A., Thomson, R.D., Merhtens, V.G. and Loeffen, M.P.F. (2004) Industrial application to cattle of a method for the early determination of meat ultimate pH. *Meat Science* 67, 107–112.

Yuan, Y., Jansen, J., Charles, B. and Zanella, A.J. (2004) The influence of weaning age on post-mixing agonistic interactions in growing pigs. *Applied Animal Behaviour Science* 88, 39–41.

Zekarias, B., Songserm, T., Post, J., Kok, G.L., Pol, J.M.A., Engel, B. and ter Huurne, A.A.H.M. (2002) Development of organs and interstitial mucosa leucocytes in four broiler lines that differ in susceptibility to malabsorption syndrome. *Poultry Science* 81, 1283–1288.

Zhang, L. and Barbut, S. (2005a) Effects of regular and modified starches on cooked pale, soft and exudative; normal; and dry, firm, and dark breast meat batters. *Poultry Science* 84, 789–796.

Zhang, L. and Barbut, S. (2005b) Rheological characteristics of fresh and frozen PSE, normal and DFD chicken breast muscle. *British Poultry Science* 46, 687–693.

Zulandt Schneider, R.A. and Moore, P.A. (2000) Urine as a source of conspecific disturbance signals in the crayfish *Procambarus clarkii*. *Journal of Experimental Biology* 203, 765–771.

Zulkifli, I., Radee, A., Nor Syaadah, O. and Che Norma, M.T. (1998) Daylength effects and fear responses in broiler chickens. *Asian–Australasian Journal of Animal Science* 11, 751–754.

Zulkifli, I., Che Norma, M.T., Israf, D.A. and Omar, A.R. (2000) The effect of early age feed restriction on subsequent response to high environmental temperatures in female broiler chickens. *Poultry Science* 79, 1401–1407.

Zulkifli, I., Norbaiyah, B. and Siti Nor Azah, A. (2004) Growth performance, mortality and immune response of two commercial broiler strains subjected to early age feed restriction and heat conditioning under hot, humid tropical environment. *Archiv für Geflügelkunde* 68, 253–258.

Index